Measurement Theory

GIAN-CARLO ROTA, *Editor*

ENCYCLOPEDIA OF MATHEMATICS AND ITS APPLICATIONS

Volume 7

Section: Mathematics and the Social Sciences
Fred S. Roberts, *Section Editor*

Measurement Theory
with Applications to Decisionmaking, Utility, and the Social Sciences

Fred S. Roberts
Rutgers University
New Brunswick, New Jersey

CAMBRIDGE
UNIVERSITY PRESS

CAMBRIDGE UNIVERSITY PRESS
Cambridge, New York, Melbourne, Madrid, Cape Town, Singapore, São Paulo, Delhi

Cambridge University Press
The Edinburgh Building, Cambridge CB2 8RU, UK

Published in the United States of America by Cambridge University Press, New York

www.cambridge.org
Information on this title: www.cambridge.org/9780521102438

First published 1985
This digitally printed version 2009

A catalogue record for this publication is available from the British Library

ISBN 978-0-521-30227-2 hardback
ISBN 978-0-521-10243-8 paperback

In memory of my father
LOUIS ROBERTS

Contents

Editor's Statement

A large body of mathematics consists of facts that can be presented and described much like any other natural phenomenon. These facts, at times explicitly brought out as theorems, at other times concealed within a proof, make up most of the applications of mathematics, and are the most likely to survive changes of style and of interest.

This ENCYCLOPEDIA will attempt to present the factual body of all mathematics. Clarity of exposition, accessibility to the non-specialist, and a thorough bibliography are required of each author. Volumes will appear in no particular order, but will be organized into sections, each one comprising a recognizable branch of present-day mathematics. Numbers of volumes and sections will be reconsidered as times and needs change.

It is hoped that this enterprise will make mathematics more widely used where it is needed, and more accessible in fields in which it can be applied but where it has not yet penetrated because of insufficient information.

GIAN-CARLO ROTA

Foreword

Our society faces difficult problems, for instance, fairly allocating scarce resources, making good societal decisions about energy use and pollution, reasonably disseminating urban and social services, and understanding and eliminating social inequalities. These problems involve in an important way issues in the social sciences. Increasingly, mathematics is being used, at least in small ways, to tackle these problems, and also to develop the necessary underlying principles of such fields as sociology, psychology, and political science.

This volume is the first in the section, Mathematics and the Social Sciences. The section is expected to present mathematical treatments of social scientific problems as well as theories and mathematical tools, techniques, and questions motivated by problems of the social sciences. It is anticipated that this classification will concern itself with such fundamental topics as learning theory, perception, signal detection, scaling and measurement, social networks, social mobility theory, voting behavior, social choice, and utility theory. At the same time, the section will concern itself with applications to problems of society, and deal with such problems as environment, transportation, urban affairs, and energy, from a societal point of view.

The problems of the social sciences in general, and the problems facing society in particular, are extremely complex. We should not expect too much of mathematics when it comes to solving these problems. On the other hand, mathematics, as the language of science, has a very important role to play. As the most precise language ever invented by man, mathematics is a tool which can be used to carefully formulate social scientific problems, give us insight into the nature of these problems, and suggest potential approaches. At the same time, problems of the social sciences can

ENCYCLOPEDIA OF MATHEMATICS and Its Applications, Gian-Carlo Rota (ed.). Vol. 7: Fred S. Roberts, Measurement Theory

be and have been the stimulus for the development of new mathematics, which is often very interesting in its own right and is, as yet, not very well known even in the mathematical community. Because the social sciences are a stimulus to new mathematics, this series is concerned with mathematics *and* the social sciences, not just with mathematics *in* the social sciences.

If mathematics is the language of science, perhaps one of the crucial reasons is that it helps us to measure things. Since the ability to measure is critical to the development of science and to the solution of many problems, it is appropriate that the first volume in this section be concerned with the theory of measurement. This volume is concerned with foundations of measurement, in particular with putting measurement in the social sciences on a firm mathematical foundation. It takes a very broad point of view of the nature of measurement in particular and of mathematical treatment of the social sciences in general. The volume takes the attitude that treating a problem mathematically—even performing measurement—does not require the assignment of *numbers*. Rather, it involves the use of precisely defined mathematical objects and relations among them to reflect empirical objects and observed relations among these objects. This attitude will be reflected in other volumes in this classification as well.

This section is begun while the jury is still out on the role of mathematics vis-à-vis the social sciences. There can be no doubt that mathematics is already finding widespread use in the social sciences. It is hoped that this will lead to some really important breakthroughs in the understanding of social and societal problems. At the same time, these problems have already been a stimulus to the development of new mathematics, and should continue to be in the future.

FRED S. ROBERTS
General Editor, Section on
Mathematics and the Social Sciences

Preface

1 Measurement Theory

There is a large body of research work in a gray area which seems to have no disciplinary home; it can be called measurement theory. This work has been performed by philosophers of science, physicists, psychologists, economists, mathematicians, and others. In the past several decades, much of this work has been stimulated by the need to put measurement in the social sciences on a firm foundation. As well as being closely tied to applications, measurement theory has a very interesting and serious mathematical component, which, surprisingly, has escaped the attention of most of the mathematical community.

This book presents an introduction to measurement theory from a representational point of view. The emphasis is on putting measurement in the social and behavioral sciences on a firm foundation, and the applications will be chosen from a variety of problems in decision theory, economics, psychophysics, policy science, etc. The purpose of this book is to present an introduction to the theory of measurement in a form appropriate for the nonspecialist. I hope that both mathematics students and practicing mathematicians with no prior exposure to the subject will find this material interesting, both as mathematics in its own right and because of its applications. I hope, indeed, that a number of mathematicians will find this subject interesting enough to solve some of the open problems posed in the text. I also hope that nonmathematicians with sufficient mathematical background will find the work thought-provoking and useful. I believe that the results, especially those on scale type, meaningfulness, organization of data, etc., should be of interest to psychologists, economists, statisticians, philosophers, policymakers, and others. The results on decisionmaking and utility are potentially of interest to executives, policy advisors, managers, etc. The theory of measurement has, I believe, some hope of assisting some social scientists in the construction of theories and other social scientists in organizing data, reporting observations, and reasoning about the phenomena they study. The theory also has the potential to assist decisionmakers in making better decisions about such problems as energy, transportation, pollution, and public health. When systematically applied to such problems, techniques of measurement as discussed in this book can lead to an intuitive or qualitative understanding, which is often more important than the formal results obtained. (Keeney and Raiffa [1976] make exactly this point in their very interesting book.)

2 Other Books on the Subject

This material has benefited from earlier books on measurement theory with much the same philosophy, in particular the books by Pfanzagl [1968] and by Krantz *et al.* [1971]. The work does not intend to approach the scope and depth of the latter book, which is highly recommended to those who wish to get further into the field of measurement. Rather, this work is written at an introductory level, with an attempt to mention a variety of topics and discuss their applications. It is aimed at introducing the nonspecialist to the problems with which measurement theory is concerned, the mathematical concepts with which measurement theory deals, the questions that measurement theorists ask, and the applications that measurement theory has. The book is also concerned with applications of measurement-theoretic techniques to decisionmaking, and to problems of public policy and society. It is in the increased emphasis on these sorts of applications that the book differs from those mentioned earlier, and falls closer to the books by Raiffa [1968] and by Keeney and Raiffa [1976]. Finally, in its emphasis on utility, the book has been influenced by such works as Fishburn [1970].

3 Use as a Textbook

The material in this book has been used for several undergraduate and graduate courses in mathematics at Rutgers. The course for undergraduates was a course in mathematical models in the social sciences, taught at the junior level, and used measurement theory as one topic. The material covered was that in Chapters 1 and 2 and Sections 3.1 and 6.1, with most proofs omitted or simplified. The graduate course has been given several times, at a level appropriate for first-year graduate students in mathematics, and covers most of the book. The order of presentation of topics varies from time to time, as the material in later chapters is relatively independent and can be presented in different orders. For example, I often present Chapter 6 immediately after Chapter 3. (See Section 6 of the Introduction for a discussion of interdependencies of the chapters.) A course for nonmathematics students could be given with this material if many of the proofs were omitted or if some proofs were included and students with especially strong mathematical background were enrolled.

Specific prerequisites for any of these courses are few: elementary concepts of sets (union, intersection, etc.), logical notation and principles of inference (implication, quantification, contrapositive, etc.), and properties of the real number system (order, Archimedean, etc.). However, the presentation will be on much too high a level for most students without the sophistication of undergraduate mathematics beyond the calculus. Parts of the book use more difficult mathematics: modern algebra (elementary group theory), notions of density, continuity, and metrics from analysis, and point set topology. However, it is easy enough to skip these parts, and they are usually designated as candidates for skipping. As for nonmathematical prerequisites, it is hard to specify them. Certainly the reader is not expected to be an expert in any of the social sciences. Different sections will use terminology and concepts from physics, psychology, economics, environmental science, etc. It will probably be necessary for the reader to look up some background material from time to time. Indeed, this is to be encouraged.

The exercises form an integral part of this book. They review ideas presented in the text, present concrete examples and generalizations, and introduce new material. Some of the exercises are of a routine mathematical nature, while others emphasize applications or ask the reader to speculate about the applicability of some abstract idea or the reasonableness of some assumption. For these latter kinds of exercises, of course, there is not necessarily a "right" answer. Indeed, the same can be said about many of the potential applications of the theory of measurement discussed in this book. On many occasions, rather than provide a "right" answer, measurement theory will, I hope, provide the user with intuition, insight, and understanding about some phenomenon he is trying to study or some complex decision he is trying to make.

4 Acknowledgments

I first learned about the theory of measurement in a course given in the Fall of 1964 at Stanford University by Patrick Suppes. This course was based in part on the foundational paper by Suppes and Zinnes [1963], which, along with the important paper by Scott and Suppes [1958], laid the groundwork for much of the philosophy that underlies this book. My views about measurement theory have been greatly influenced by Dana Scott and Pat Suppes, during the years I spent at Stanford and afterward. My views have also been strongly influenced by Duncan Luce, and it was extremely helpful and rewarding for me to spend parts of two years working close to him, first at the University of Pennsylvania and then at the Institute for Advanced Study. I owe a large debt of gratitude to Professors Luce, Scott, and Suppes. I have also benefited from conversations and correspondence with Zoltan Domotor, Peter Fishburn, David

Krantz, Michael Levine, Amos Tversky, and others, and from the opportunity to sit in on a course in measurement at the University of California at Los Angeles, given by Eric Holman during the Fall of 1970.

This book was started during the year 1971–1972, while I was at the Institute for Advanced Study in Princeton. An early version of the material was used as notes for a series of lectures I gave at Cornell University during the summer of 1973. I wish to thank Bill Lucas for giving me this and other forums to talk about measurement theory (and other topics) over the years. The project was interrupted while I worked on a book, *Discrete Mathematical Models, with Applications to Social, Biological, and Environmental Problems*, which used a rather abridged version of the measurement material as one chapter.

This work has benefited from extensive comments on earlier drafts by Victor Klee and Duncan Luce, and on later drafts by Zoltan Domotor and some anonymous referees. James Dalton, Rochelle Leibowitz, and Robert Opsut made very helpful comments on the exercises and on the text, and Midge Cozzens, Chen-Shung Ko, and Robert Opsut helped with the proof reading. Many other people, too numerous to mention, have made contributions, large and small. However, I alone am responsible for any errors that may remain.

I would like to thank the Institute for Advanced Study for their support during my year there, and Prentice-Hall for permission to borrow freely from the measurement material of Chapter 8 of my *Discrete Mathematical Models* book.

I would like to thank Lily Marcus for her great help during the final stages of the preparation of this book.

I would also like to thank my parents for their love and understanding.

Finally, I would like to thank my wife Helen, for her continuing support, assistance, and inspiration, throughout all the years this book was always in the background or the foreground. I would also like to thank my daughter Sarah, whose development during the final stages of this work was so dramatic that it took no theory to measure it.

FRED S. ROBERTS

References

Fishburn, P. C., *Utility Theory for Decision Making*, Wiley, New York, 1970.

Keeney, R. L., and Raiffa, H., *Decisions with Multiple Objectives: Preferences and Value Tradeoffs*, Wiley, New York, 1976.

Krantz, D. H., Luce, R. D., Suppes, P., and Tversky, A., *Foundations of Measurement*, Vol. I, Academic Press, New York, 1971.

Pfanzagl, J., *Theory of Measurement*, Wiley, New York, 1968.

Raiffa, H., *Decision Analysis*, Addison-Wesley, Reading, Massachusetts, 1968.

Scott, D., and Suppes, P., "Foundational Aspects of Theories of Measurement," *J. Symbolic Logic*, **23** (1958), 113–128.

Suppes, P., and Zinnes, J. L., "Basic Measurement Theory," in R. D. Luce, R. R. Bush, and E. Galanter (eds.), *Handbook of Mathematical Psychology*, Vol. 1, Wiley, New York, 1963, pp. 1–76.

Measurement Theory

Introduction

1 Measurement

A major difference between a "well-developed" science such as physics and some of the less "well-developed" sciences such as psychology or sociology is the degree to which things are measured. In this volume, we develop a theory of measurement that can act as a foundation for measurement in the social and behavioral sciences. Starting with such classical measurement concepts of the physical sciences as temperature and mass, we extend a theory of measurement to the social sciences. We discuss the measurement of preference, loudness, brightness, intelligence, and so on. We also apply measurement to such societal problems as air and noise pollution, weather forecasting, and public health, and comment on the development of pollution indices and consumer price indices.

Throughout, we apply the results to decisionmaking. The decisionmaking applications deal with transportation, consumer behavior, environmental problems, energy use, and medicine, as well as with laboratory situations involving human and animal subjects.

In this introduction, we try to give the reader a quick preview of the contents and organization of the book, and of the problems we shall address. The reader might prefer to read the introduction rather quickly the first time, and to return to it later.

The following questions are some of those we shall ask. A few of these are stated here in very general terms, and of course we shall try to be more specific in what follows.

1. What does it mean to measure preference, likes and dislikes, etc.?
2. When does it make sense to say goal a is twice as important as goal b? Or twice as worthwhile?

ENCYCLOPEDIA OF MATHEMATICS and Its Applications, Gian-Carlo Rota (ed.). Vol. 7: Fred S. Roberts, Measurement Theory

3. Does it make sense to assert that the average IQ of one group of individuals is twice the average IQ of a second group?

4. Is it possible to measure air pollution with one index that takes account of many different pollutants? If so, does it make sense to assert that the pollution level today is 20% lower than it was yesterday?

5. Is it meaningful to assert that the consumer price index has increased by 20% in a given period?

6. How can we quantitatively relate subjective judgments of loudness of a sound to the physical intensity of the sound? And how can we use such quantitative relationships to develop indices of noise pollution?

7. How can we use expert judges to judge the relative importance, or significance, or merit of alternative candidates, and then combine the judgments of the experts into one measure of importance, significance, or merit?

8. How can we choose between two different treatments for a disease, given that we are not sure exactly what the outcomes of the treatments will be?

9. Can subjective judgments that one event is more likely to occur than another be quantified?

10. Is measurement still possible if judgments of preference, relative importance, loudness, etc., are inconsistent?

At an early stage of scientific development, measurement is usually performed at only the crudest level, that of classification. Some philosophers of science (e.g., Torgerson [1958]) do not even wish to call this measurement. What are the advantages of performing measurement that goes beyond simple classification? Hempel [1952] describes some of these. First, if we can measure things, we can begin to differentiate more than we can by simply classifying. For example, we can do more than simply distinguishing between warm objects and cold ones; we can assign degrees of warmth. Greater descriptive flexibility leads to greater flexibility in the formulation of general laws. (One should imagine trying to state the laws of physics using only classifications!) Measurement is usually performed by assigning numbers—though we shall argue below that this is not a prerequisite for measurement. Assignment of numbers makes possible the application of the concepts and theories of mathematics. General laws can now be stated in mathematical language—for example, as formal relations among quantities. Mathematical tools for analysis of numbers can help us to reason about objects and their properties, and to deduce general principles describing these properties. Thus, the existence of and experience with centuries of mathematical reasoning is a large part of the reason we find it useful to measure things.

We shall study two quite different types of measurement, fundamental measurement and derived measurement. (See Table 1.) Fundamental

Table 1.   Measurement

A. Types of Measurement
1. Fundamental
2. Derived
B. Problems in Fundamental Measurement
1. Representation
2. Uniqueness
C. Axioms in a Representation Theorem
1. Prescriptive or normative
2. Descriptive

measurement, as we shall describe it, takes place at an early stage of scientific development, when several fundamental concepts are measured for the first time. Mass, temperature, and volume are fundamental measures. Derived measurement takes place later, when some concepts have already been measured, and new measures are defined in terms of existing ones. Density can be thought of as a derived measure, defined as mass divided by volume. (Derived measurement is usually a relative matter. As we shall point out, the same scale—for example, density—can be introduced either as a derived scale or as a fundamental one.) Much of this work will be concerned with fundamental measurement. However, large parts of Chapters 2, 4, and 6 deal with derived measurement.

In the development of a scientific discipline, fundamental measurement is not usually performed in as formalistic a way as we describe in this volume. We usually do not attempt to describe an actual process that is undergone as techniques of measurement are developed. We are not interested in a measuring apparatus and in the interaction between the apparatus and the objects being measured. Rather, we attempt to describe how to put measurement on a firm, well-defined foundation. A number of the results, however, are of potential practical significance, and we shall attempt to mention practical techniques for actually computing measures or scales whenever possible. Once measurement has been put on a firm foundation, we develop a variety of tools for analyzing statements made in terms of scale values. These techniques have immediate application in a variety of practical problems.

Putting measurement on a firm foundation is not a terribly important activity in the modern-day physical sciences; many physicists would probably not consider it physics, but only "philosophy of physics." Practicing physicists usually take measurement for granted, and anyway measurements are usually based on powerful, well-established theories. In the social sciences, on the other hand, much present activity can be categorized as the search for appropriate scales of measurement to describe behavior, aid in decisions, etc. Putting measurement on a firm foundation can potentially play a very important role in the development of the social sciences.

Since much of this volume is devoted to putting measurement on a firm foundation, it is not surprising that much of the discussion is axiomatic in nature. We try to develop axioms or conditions under which measurement is possible. These are usually conditions about an individual's judgments, preferences, reactions, and so on. For example, we shall show that preferences can be measured in a consistent way provided that they satisfy two axioms:

(A) If you prefer a to b, you do not prefer b to a.
(B) If you do not prefer a to b, and do not prefer b to c, then you do not prefer a to c.

Such axioms can be looked at in two ways. The *prescriptive* or *normative* interpretation looks at the axioms as conditions of rationality. A truly rational man, given ideal conditions (unlimited computational ability, unlimited resources, etc.), should make judgments that satisfy the axioms, or else he is not acting rationally. Theories can be built based on the definition of a rational man, and procedures for a rational man to make decisions based on his judgments can be developed.* A second interpretation is that these axioms are *descriptive*. They give conditions on behavior which, if satisfied, allow measurement to take place. Whether or not measurement can take place then depends on whether or not an individual's judgments satisfy the axioms, and we hope that this is a testable question. Whenever possible in this volume, we try to discuss tests of the axioms presented. Although it is not fair to generalize, it is often true that the prescriptive axiomatic theories of measurement in the social sciences have been developed in the economic literature, and the descriptive theories in the psychological literature. The theories of physical measurement in general are simultaneously prescriptive and descriptive, and have largely philosophical significance. However, both the prescriptive and descriptive theories of measurement in the social sciences propose new theories and lead to new laws (the axioms), suggest experiments to distinguish between these laws, and give rise to practical techniques of measurement where none was possible before. (See Krantz [1968, Section 2.1] for a more detailed discussion of this point.)

The problem of finding axioms under which measurement is possible will be called in this volume the *representation problem*. Once measurement has been accomplished, we also consider the *uniqueness problem*: How

---

*Keeney and Raiffa [1976] distinguish between the normative and the prescriptive. The normative interpretation refers to an "idealized,…superrational being with an all-powering intellect," while the prescriptive refers to "normally intelligent people who want to think hard and systematically…" We shall not make this distinction.

unique is the resulting measure or scale? This problem will be very important in telling us what kinds of comparisons and what kinds of mathematical manipulations are possible with the measures obtained. For example, we shall ask whether it is meaningful to say that one group's average IQ is twice that of a second group's, or that the average air pollution level in one city is greater than that in a second city. We shall discuss the uniqueness problem in depth in Chapter 2, and return to it often.

2 The Measurement Literature

Our approach to measurement theory follows very closely that of Scott and Suppes [1958], Suppes and Zinnes [1963], Pfanzagl [1968], Krantz [1968], and Krantz *et al.* [1971]. There is a long-standing literature on the nature of measurement, and the reader might wish to consult some of this literature for other points of view. Much of this literature concerns itself with measurement in the physical sciences. Early influential books were written by Campbell [1920, 1928, 1938]. Campbell and such writers as Helmholtz [1887], Cohen and Nagel [1934], Guild [1938], Reese [1943], and Ellis [1966] did not accept measurement unless it involved some sort of concatenation or addition operation of the type we shall describe in Section 3.2. Although this approach to measurement was quite appropriate for physics, it was not really broad enough to encompass the measurement problems of the social sciences. However, as late as 1940, a committee of the British Association for the Advancement of Science questioned specifically whether psychologists such as S. S. Stevens, who were measuring human sensations such as loudness, were really performing measurement, since they used no concept of addition (Final Report of the British Association for the Advancement of Science [1940]). Our approach is broader than that of the classical writers, and considers a measurement theory that can act as a basic foundation for measurement in the social sciences. Some other important references on measurement theory, closer to our point of view, are Stevens [1946, 1951, 1959, 1968] and Adams [1965]. Stevens was the first to observe that uniqueness of a measurement assignment defined scale type, in the sense we describe in Section 2.3. Although chemistry and biology have used scales not much different from those of physics, the social sciences, in measurement of preference, intelligence, etc., have given rise to entirely new types of scales, with somewhat different character. We shall explore these in detail.

3 Decisionmaking

A major application of measurement theory is to problems of decisionmaking. Various authors (for example, Luce and Raiffa [1957]) have classified decisionmaking problems in several ways. (See Table 2.) The first

Table 2.   Classification of Decisionmaking Problems

A. Who Makes the Decision?

    1. Individual
    2. Group

B. How Much Information about Consequences of Actions Is Known?

    1. Certainty
    2. Risk
    3. Uncertainty

distinction to make is whether the decision is being made by an *individual* or a *group*. Some groups act as individuals; the difference is whether members of the group are expressing their own opinions from among which the group must choose. In this volume, we shall be almost exclusively concerned with the individual decisionmaker. For good summaries of the group decisionmaking problem, see Luce and Raiffa [1957], Sen [1970], or Fishburn [1972]. A classic reference is Arrow [1951]. More specifically for the decisionmaking problem involving elections, see Black [1958], Farquharson [1969], and Riker and Ordeshook [1973].

A second distinction between problems of decisionmaking involves how much certainty there is about outcomes of various actions. If we are trying to choose among alternative acts, we say that we are in a situation of *certainty* if for each act there is exactly one consequence. We are in a situation of *risk* if for each act there is a set of possible consequences, none of which occurs with certainty, but each of which occurs with a known probability. Finally, we are in a situation of *uncertainty* if the probabilities that consequences will occur are unknown. In practice many decisionmaking problems involve a mixture of risk and uncertainty (or even of certainty, risk, and uncertainty).

In this volume, we shall concentrate primarily on decisions involving certainty. However, in Chapter 7 we shall discuss decisionmaking under risk and under uncertainty. In Chapter 8, we shall discuss how to measure the probability of various outcomes, given subjective judgments about which of two outcomes is more probable.

4 Utility

If decisions are being made in a situation of certainty, then we often choose that act whose certain consequence maximizes (or minimizes) some index—for example, a measure of value, worth, satisfaction, or utility. The notion of *utility* goes back at least to the eighteenth century. Much of the original interest in this concept goes back to Jeremy Bentham, who defines utility in the first chapter of *The Principles of Morals and Legislation* (1789), as follows:

> By utility is meant that property in any object, whereby it tends to produce benefit, advantage, pleasure, good, or happiness (all this in the present case comes to the same thing), or (what comes again to the same thing) to prevent the happening of mischief, pain, evil, or unhappiness to the party whose interest is considered: if that party be the community in general, then the happiness of the community; if a particular individual, then the happiness of that individual.

Bentham formulated procedures for measuring utility, for he thought societies should strive for "the greatest good for the greatest number"— that is, maximum utility. In this volume, we shall usually look at the measurement of utility as a representation problem and list various axiom systems sufficient for the existence of *utility functions*, measures of utility. In the early applications of utility theory in economics, there was an emphasis by economists such as Walras and Jevons on obtaining utility functions that were additive in the sense that the utility of the combination of two objects was the sum of the individual utilities of the two objects. Such a utility function is called *cardinal*. (More generally, a utility function will be called *cardinal* if it is unique up to a positive linear transformation.) We shall discuss axioms for cardinal utility in Chapters 3 and 5. In the late nineteenth century, Edgeworth [1881] questioned the assumption of additivity. In the early twentieth century, the economist Pareto [1906] showed that much of economic theory depends only on the assumption that the utility function is *ordinal*—that is, that the utilities can be used only to decide which of two objects has a higher value, and addition in particular does not necessarily make sense. We shall discuss ordinal utility in Section 3.1.

Luce and Suppes [1965] classify theories of utility in several ways. (See Table 3.) A similar classification applies to theories of measurement in general. First, these theories can involve decisionmaking under certainty, risk, or uncertainty. We have already discussed these distinctions. A second distinction is whether the theories are *algebraic* and *deterministic* or *probabilistic*. Most of the traditional utility theories are algebraic, and the

Table 3. Classification of Theories of Utility

A. How Much Information about Consequences of Actions Is Known to the Decisionmakers?

1. Certainty
2. Risk
3. Uncertainty

B. What Kind of Representation Theorems?

1. Algebraic and deterministic
2. Probabilistic

C. What Judgments Is Measurement of Utility Based on?

1. Simple choices
2. Rankings or choices from a set

---

representation theorems state axioms that are very algebraic in nature. For
example, the axioms for cardinal utility functions stated in Section 3.2 say
that preference and combination of objects define a certain kind of
ordered semigroup. (Sometimes, if there is an interest in continuous utility
functions, topological axioms must be added.) Our basic formulation of
fundamental measurement is algebraic, starting with the idea of a relation
and an operation (Chapter 1), and then basing measurement on a relational system consisting of certain empirical relations and operations
(Section 1.8). Sometimes it is useful to modify algebraic theories. This is
the case when the fundamental judgments that representation theorems
describe are made inconsistently or made consistently, but according to
some statistical regularity. When data is inconsistent or there is no discern-
ible deterministic pattern available, the algebraic theories must be replaced
by probabilistic theories, which are built around this more random data.
Falmagne [1976] argues that we shall (almost) always have to replace
algebraic theories by probabilistic or random ones, at least in applications
to the behavioral sciences. He is concerned with developing probabilistic
analogues of the traditional algebraic theories of fundamental measurement. We discuss some probabilistic theories in Section 6.2, but otherwise
our basic approach is algebraic.

Both the algebraic and probabilistic theories of utility can take two
forms. The first form assumes that a utility function can be derived simply
on the basis of *simple choices* among pairs of alternatives or objects. The
second asks for a *ranking* among elements in each set of alternatives, or
choice of a best element from the set, before deriving a utility function. We
shall discuss only simple choice theories in this volume. Luce and Suppes
[1965, Section 6] and Krantz *et al.* [to appear] summarize some probabilistic
ranking theories.

Some excellent surveys of utility theory and decisionmaking are Fishburn [1968, 1970] and Luce and Suppes [1965] and, for utility with
multi-attributed alternatives, Farquhar [1977] and Keeney and Raiffa
[1976]. Aoki, Chipman, and Fishburn [1971] list many references on
preferences, utility, and demand. Other references on utility theory include
Adams [1960], Arrow [1951], Chipman [1960], Debreu [1959], Edwards
[1954, 1961], Fischer and Edwards [1973], Fishburn [1964], Luce and
Raiffa [1957], Majumdar [1958], Savage [1954], Slovic and Lichtenstein
[1971], Stigler [1950], and von Neumann and Morgenstern [1944]. For a
more complete list of references, see Luce and Suppes [1965] or Fishburn
[1970].

5 Mathematics

The study of measurement, decisionmaking, and utility has given rise to
interesting new mathematical results and questions, most of which are not

very well known among mathematicians. A major purpose of this book is to organize a class of these mathematical results and to introduce the mathematically trained reader to them.

Most of the mathematics in this book is distinctly algebraic in flavor. It deals with ordered algebraic systems and homomorphic mappings of one such system into another. Many of these systems are finite, so there is a discrete flavor to much of the mathematics. Fundamental properties of the real number system are used throughout, and many of the results are related to branches of logic such as set theory and foundations of geometry. A variety of results and tools of an analytic or topological nature are scattered throughout. For example, solutions of certain functional equations are studied in Chapter 4.

Finally, much of the mathematics described here has a flavor of its own. It is to be expected that, as more mathematicians become interested in problems of the social sciences, new forms of mathematics will have to be developed to solve these problems. Hopefully a work like this one will stimulate some mathematically trained individuals to become involved in such developments.

6 Organization of the Book

Since relational systems form a basis for much of the theory of measurement, I have chosen to include an introductory chapter on the theory of relations, which includes most of the relation-theoretic concepts needed later on. The reader familiar with this theory can skip much of Chapter 1, though he should read Theorems 1.2, 1.3, and 1.4 and Section 1.8.

Chapter 2 introduces fundamental and derived measurement and defines the two basic problems of fundamental measurement—the representation problem and the uniqueness problem. It introduces scale type and uses scale type to study what statements involving scales are meaningful. It applies the results to a variety of practical problems, such as the making of index numbers (consumer price, consumer confidence) and the measurement of air pollution. Chapter 3 illustrates fundamental measurement by giving three basic representation theorems, stating conditions under which measurement can be performed; one is for ordinal measurement, one for extensive measurement, and one for difference measurement. In the utility interpretation, the first theorem gives conditions for the existence of ordinal utility functions, and the second and third conditions for the existence of cardinal utility functions. Chapter 4 is an applications chapter. It presents the theory of psychophysical scaling, and applies the results about scale type and fundamental and derived measurement to this theory. It concentrates on the measurement of loudness, and shows how one might derive a measure of loudness from known physical scales of intensity of a sound.

Chapters 5 through 8 introduce various complications into the measurement picture. In Chapter 5, we study "complicated" or multidimensional alternatives. A new kind of fundamental measurement, called conjoint measurement, is introduced. The results have application to combinations of psychological factors such as drive and incentive, to binaural additivity of loudness, to mental testing problems, and to measuring discomfort due to different weather factors. They also have applications to problems of urban services, allocation of funds for education, treatment of medical problems, design of large public facilities such as airports, etc. In Chapter 6, we ask what to do when it is not even possible to perform ordinal measurement in the sense of Section 3.1—that is, when the necessary conditions for ordinal measurement are violated. We widen our scope, and introduce the idea of measurement without numbers or measurement when the basic data is inconsistent. We study Luce's fundamental idea of a semiorder and then give examples of probabilistic theories of measurement. A variety of applications to data from pair comparison experiments are presented.

Chapter 7 discusses the problem of decisionmaking under risk or uncertainty. It introduces the famous expected utility hypothesis, which goes back to Bernoulli [1738], and which says that a decisionmaker chooses that action which maximizes his expected utility. Accepting this as a *prescription*, we give applications to decisionmaking problems from such fields as transportation, medicine, and public health, and to calculation of utility. A number of *descriptive* utility theorists believe that although we do not consciously calculate expected utilities, we act *as if* we are maximizing expected utility. We present axiom systems which give conditions on choices among acts with risky or uncertain consequences sufficient to guarantee that these choices are made as if expected utility were being maximized.

In a decisionmaking situation under uncertainty, and in other situations as well, it is sometimes useful or necessary to be able to calculate probabilities that reflect our judgments that certain events are subjectively more probable than others. In Chapter 8, we discuss the measurement of subjective probability.

Chapters 1 and 2 and Section 3.1 form a groundwork for the rest of the book. Much of the rest of the book can be read in any order. For example, Sections 3.2 and 3.3 and Chapters 4, 5, 6, 7, and 8 are essentially independent, though Section 7.3 depends on Chapter 5 and parts of Chapter 8 depend on earlier material in the beginning of Chapter 7 and in Section 3.2. Even Sections 6.1 and 6.2 are essentially independent, with the exception that Section 6.2.4 depends on Section 6.1.

There are so many topics to be covered in the growing field of measurement theory that it is impossible to survey them all. However, references have been provided at the end of each chapter, and it is hoped they will lead the reader into the literature.

References

Adams, E. W., "Survey of Bernoullian Utility Theory," in H. Solomon (ed.), *Mathematical Thinking in the Measurement of Behavior*, The Free Press, Glencoe, Illinois, 1960.

Adams, E. W., "Elements of a Theory of Inexact Measurement," *Phil. Sci.*, **32** (1965), 205–228.

Aoki, M., Chipman, J. S., and Fishburn, P. C., "A Selected Bibliography of Works Relating to the Theory of Preferences, Utility, and Demand," in J. S. Chipman, L. Hurwicz, M. K. Richter, and H. F. Sonnenschein (eds.), *Preferences, Utility, and Demand*, Harcourt, Brace, Jovanovich, New York, 1971, pp. 437–492.

Arrow, K. J., *Social Choice and Individual Values*, Wiley, New York, 1951 (2nd ed., 1963).

Bentham, J., *The Principles of Morals and Legislation*, London, 1789.

Bernoulli, D., "Specimen Theoriae Novae de Mensura Sortis," *Comentarii Academiae Scientiarum Imperiales Petropolitanae*, **5** (1738), 175–192. Translated by L. Sommer in *Econometrica*, **22** (1954), 23–36.

Black, D., *The Theory of Committees and Elections*, Cambridge University Press, London, 1958.

Campbell, N. R., *Physics: The Elements*, Cambridge University Press, Cambridge, 1920. Reprinted as *Foundations of Science: The Philosophy of Theory and Experiment*, Dover, New York, 1957.

Campbell, N. R., *An Account of the Principles of Measurement and Calculation*, Longmans, Green, London, 1928.

Campbell, N. R., *Symposium: Measurement and Its Importance for Philosophy*, Aristotelian Society, Suppl. Vol. 17, Harrison, London, 1938.

Chipman, J. S., "The Foundations of Utility," *Econometrica*, **28** (1960), 193–224.

Cohen, M. R., and Nagel, E., *An Introduction to Logic and Scientific Method*, Harcourt, Brace, New York, 1934.

Debreu, G., *Theory of Value*, Wiley, New York, 1959.

Edgeworth, F. Y., *Mathematical Physics*, Kegan Paul, London, 1881.

Edwards, W., "The Theory of Decision Making," *Psychol. Bull.*, **51** (1954), 380–417.

Edwards, W., "Behavioral Decision Theory," in P. R. Farnsworth, O. McNemar, and Q. McNemar (eds.), *Ann. Rev. Psychol.* (1961), 473–498.

Ellis, B., *Basic Concepts of Measurement*, Cambridge University Press, London, 1966.

Falmagne, J. C., "Random Conjoint Measurement and Loudness Summation," *Psychol. Rev.*, **83** (1976), 65–79.

Farquhar, P. H., "A Survey of Multiattribute Utility Theory and Applications," in M. Starr and M. Zeleny (eds.), *TIMS/North Holland Studies in the Management Sciences: Multiple Criteria Decisionmaking*, **6** (1977), 59–89.

Farquharson, R., *Theory of Voting*, Yale University Press, New Haven, Connecticut, 1969.

Final Report of the British Association for the Advancement of Science, *Advan. Sci.*, **2** (1940), 331–349.

Fischer, G. W., and Edwards, W., "Technological Aids for Inference, Evaluation, and Decision-making: A Review of Research and Experience," Tech. Rept., Engineering Psychology Laboratory, University of Michigan, Ann Arbor, Michigan, 1973.

Fishburn, P. C., *Decision and Value Theory*, Wiley, New York, 1964.

Fishburn, P. C., "Utility Theory," *Management Sci.*, **14** (1968), 335–378.

Fishburn, P. C., *Utility Theory for Decisionmaking*, Wiley, New York, 1970.

Fishburn, P. C., *The Theory of Social Choice*, Princeton University Press, Princeton, New Jersey, 1972.

Guild, J., "Part III of Quantitative Estimation of Sensory Events (Interim Report)," *Rept. Brit. Assoc. Advan. Sci.*, (1938), 296–328.

Helmholtz, H. V., "Zählen und Messen," in *Philosophische Aufsätze*, Fues's Verlag, Leipzig, 1887, pp. 17–52. Translated by C. L. Bryan, "Counting and Measuring," Van Nostrand, Princeton, New Jersey, 1930.

Hempel, C. G., "Fundamentals of Concept Formation in Empirical Science," in O. Neurath *et al.* (eds.), *International Encyclopedia of Unified Science*, Vol. 2, No. 7, University of Chicago Press, Chicago, Illinois, 1952.

Keeney, R. L., and Raiffa, H., *Decisions with Multiple Objectives: Preferences and Value Tradeoffs*, Wiley, New York, 1976.

Krantz, D. H., "A Survey of Measurement Theory," in *Mathematics of the Decision Sciences*, Part 2, G. B. Dantzig and A. F. Veinott, Jr. (eds.), Vol. 12 of Lectures in Applied Mathematics, American Mathematical Society, Providence, Rhode Island, 1968, pp. 314–350.

Krantz, D. H., Luce, R. D., Suppes, P., and Tversky, A., *Foundations of Measurement*, Vol. I, Academic Press, New York, 1971.

Krantz, D. H., Luce, R. D., Suppes, P., and Tversky, A., *Foundations of Measurement*, Vol. II, Academic Press, New York, to appear.

Luce, R. D., and Raiffa, H., *Games and Decisions*, Wiley, New York, 1957.

Luce, R. D., and Suppes, P., "Preference, Utility, and Subjective Probability," in R. D. Luce, R. R. Bush, and E. Galanter (eds.), *Handbook of Mathematical Psychology*, Vol. III, Wiley, New York, 1965, pp. 249–410.

Majumdar, T., *The Measurement of Utility*, Macmillan, London, 1958.

Pareto, V., "Manuale di Economia Politica con una Introduzione ulla Scienza Sociale," Societa Editrice Libraria, Milan, Italy, 1906.

Pfanzagl, J., *Theory of Measurement*, Wiley, New York, 1968.

Reese, T. W., "The Application of the Theory of Physical Measurement to the Measurement of Psychological Magnitudes, with Three Experimental Examples," *Psychol. Monogr.*, **55** (1943), 1–89.

Riker, W. H., and Ordeshook, P. C., *Positive Political Theory*, Prentice-Hall, Englewood Cliffs, New Jersey, 1973.

Savage, L. J., *The Foundations of Statistics*, Wiley, New York, 1954.

Scott, D., and Suppes, P., "Foundational Aspects of Theories of Measurement," *J. Symbolic Logic*, **23** (1958), 113–128.

Sen, A. K., *Collective Choice and Social Welfare*, Holden-Day, San Francisco, California, 1970.

Slovic, P., and Lichtenstein, S., "Comparison of Bayesian and Regression Approaches to the Study of Information Processing in Judgement," *Organizational Behavior and Human Performance*, **6** (1971), 649–744.

Stevens, S. S., "On the Theory of Scales of Measurement," *Science*, **103** (1946), 677–680.

Stevens, S. S., "Mathematics, Measurement and Psychophysics," in S. S. Stevens (ed.), *Handbook of Experimental Psychology*, Wiley, New York, 1951, pp. 1–49.

Stevens, S. S., "Measurement, Psychophysics, and Utility," in C. W. Churchman and P. Ratoosh (eds.), *Measurement: Definitions and Theories*, Wiley, New York, 1959, pp. 18–63.

Stevens, S. S., "Measurement, Statistics, and the Schemapiric View," *Science*, **161** (1968), 849–856.

Stigler, G. J., "The Development of Utility Theory," *J. Polit. Econ.*, **58** (1950), 307–327, 373–396.

Suppes, P., and Zinnes, J., "Basic Measurement Theory," in R. D. Luce, R. R. Bush, and E. Galanter (eds.), *Handbook of Mathematical Psychology*, Vol. I, Wiley, New York, 1963, pp. 1–76.

Torgerson, W. S., *Theory and Methods of Scaling*, Wiley, New York, 1958.

von Neumann, J., and Morgenstern, O., *Theory of Games and Economic Behavior*, Princeton University Press, Princeton, New Jersey, 1944, 1947, 1953.

CHAPTER 1

Relations

1.1 Notation and Terminology

In this chapter, we present a mathematical topic, the theory of relations. The concepts and techniques presented here will be used throughout the rest of the volume. The notation we shall use is summarized in Table 1.1, fundamental properties of relations are defined in Table 1.2, and types of relations are defined in Table 1.3. Many readers will be familiar with most of the material of this chapter. The reader may simply want to glance at Tables 1.1, 1.2, and 1.3. He should, however, familiarize himself with Theorems 1.2, 1.3, and 1.4 of Section 1.5 and with the material of Section 1.8.

1.2 Definition of a Relation

Suppose A and B are sets. The *Cartesian product* of A with B, denoted $A \times B$, is the set of all *ordered* pairs (a, b) so that a is in A and b is in B. More generally, if $A_1, A_2, \ldots, A_n$ are sets, the *Cartesian product*

$$A_1 \times A_2 \times \ldots \times A_n$$

is the set of all ordered n-tuples $(a_1, a_2, \ldots, a_n)$ such that $a_1 \in A_1$, $a_2 \in A_2, \ldots, a_n \in A_n$. The notation A^n denotes the Cartesian product of A with itself n times.

A *binary relation* R on the set A is a subset of the Cartesian product $A \times A$, that is, a set of ordered pairs (a, b) such that a and b are in A. If A is the set $\{1, 2, 3, 4\}$, then examples of binary relations on A are given by

$$R = \{(1, 1), (1, 2), (2, 1), (3, 3), (3, 4), (4, 3)\}, \tag{1.1}$$

$$S = \{(1, 1), (1, 2), (2, 1), (2, 2), (3, 3), (3, 4), (4, 3), (4, 4)\}, \tag{1.2}$$

ENCYCLOPEDIA OF MATHEMATICS and Its Applications, Gian-Carlo Rota (ed.). Vol. 7: Fred S. Roberts, Measurement Theory

Table 1.1. Notation

Set-Theoretic Notation

∪	union
∩	intersection
⊆	subset (contained in)
⊊	proper subset
⊄	is not a subset
⊇	contains (superset)
∈	member of
∉	not a member of
∅	empty set
$\{\ldots\}$	the set ...
$\{\ldots : \ldots\}$	the set of all ... such that ...
A^c	complement of A
$A - B$	$A \cap B^c$
$|A|$	cardinality of A, the number of elements in A

Logical Notation

~	not
⇒	implies
⇔	if and only if (equivalence)
∀	for all
∃	there exists
iff	if and only if

Sets of Numbers

Re	the real numbers
Re^+	the positive real numbers
Q	the rational numbers
Q^+	the positive rational numbers
N	the positive integers
Z	the integers

Miscellaneous

$f \circ g$	composition of the two functions f and g
$f(A)$	the image of the set A under the function f; i.e., $\{f(a): a \in A\}$.
≈	approximately equal to
≡	congruent to
Π	product
Σ	sum
$\int$	integral sign

Table 1.2. Properties of Relations

A Binary Relation (A, R) Is:	Provided That:
Reflexive	aRa, all $a \in A$
Nonreflexive	it is not reflexive
Irreflexive	$\sim aRa$, all $a \in A$
Symmetric	$aRb \Rightarrow bRa$, all $a, b \in A$
Nonsymmetric	it is not symmetric
Asymmetric	$aRb \Rightarrow \sim bRa$, all $a, b \in A$
Antisymmetric	aRb & $bRa \Rightarrow a = b$, all $a, b \in A$
Transitive	aRb & $bRc \Rightarrow aRc$, all $a, b, c \in A$
Nontransitive	it is not transitive
Negatively transitive	$\sim aRb$ & $\sim bRc \Rightarrow \sim aRc$, all $a, b, c \in A$; equivalently: $xRy \Rightarrow xRz$ or zRy, all $x, y, z \in A$
Strongly complete	for all $a, b \in A$, aRb or bRa
Complete	for all $a \neq b \in A$, aRb or bRa
Equivalence relation	it is reflexive, symmetric and transitive

Table 1.3. Order Relations*

Property	Quasi Order	Weak Order	Simple Order	Strict Simple Order	Strict Weak Order	Partial Order	Strict Partial Order
Reflexive	√					√	
Symmetric							
Transitive	√	√	√	√		√	√
Asymmetric				√	√		√
Antisymmetric			√			√	
Negatively transitive					√		
Strongly complete		√	√				
Complete				√			

*A given type of relation can satisfy more of these properties than those indicated. Only the defining properties are indicated.

and

$$T = \{(1, 2), (1, 3), (1, 4), (2, 3), (2, 4), (3, 4)\}. \tag{1.3}$$

The binary relation T is the "less than" relation on A; an ordered pair (a, b) is in the binary relation T if and only if $a < b$. Similarly, "less than" defines a binary relation on the set A of all real numbers, as does "greater than," "equals," and so on. Of course, the set A does not have to be a set of numbers. If A is the set $\{SO_2, DDT, NO_x\}$, then examples of binary relations on A are given by

$$U = \{(SO_2, DDT), (DDT, NO_x), (SO_2, NO_x)\} \tag{1.4}$$

and

$$V = \{(SO_2, NO_x), (NO_x, SO_2)\}. \tag{1.5}$$

In the case of a binary relation R on a set A, we shall usually write aRb to denote the statement that $(a, b) \in R$. Thus, for example, if S is the relation defined by Eq. (1.2), then $3S3$ and $2S1$, but not $3S1$. If U is the relation defined by Eq. (1.4), then $SO_2 U DDT$.

Binary relations arise very frequently from everyday language. For example, if A is the set of people in the world, then the set

$$F = \{(a, b): a \in A \text{ and } b \in A \text{ and } a \text{ is the father of } b\} \tag{1.6}$$

defines a binary relation on A, which we may call, by a slight abuse of language, "father of." To give another example, suppose A is any collection of alternatives among which you are choosing, for example, a collection of designs for a regional transportation system, and suppose

$$P = \{(a, b) \in A \times A: \text{you (strictly) prefer } a \text{ to } b\}. \tag{1.7}$$

Then P may be called your relation of "strict preference."* Someone else's relation of preference might be quite different, and that of course is where problems arise. To give yet another example, suppose A is a set of airplane engines and L is the relation "sounds louder than when heard at a horizontal distance of 500 feet." The relation L on A can play a role in the design of airplanes. It is hoped that the relation L is related to some physical characteristics of the engine design and of the sounds engines emit. The study of the relationship between the physical properties of these sounds and the psychological ones, such as perceived loudness, is the subject matter of the field called psychophysics. (We return to this study in Chapter 4.)

The properties of a relation are not clearly defined without giving its underlying set. Thus, if A is the set of all people in the United States and B is the set of all males in the United States, then the relation

$$R = \{(a, b) \in A \times A: a \text{ is the brother of } b\} \tag{1.8}$$

is different from the relation

$$R' = \{(a, b) \in B \times B: a \text{ is the brother of } b\}. \tag{1.9}$$

For example, R' has certain symmetry properties; that is, if $aR'b$, then $bR'a$. These properties are not shared by R. Moreover, R' has different properties if it is thought of as a relation on the set B or as a relation on the set A (even though R' contains no ordered pairs with elements not in B). To make sure that the set A on which a relation R is defined is given explicitly, it is necessary to speak formally of a *relational system* (A, R)

---

*Strict preference is to be distinguished from weak preference: the former means "better than," and the latter "at least as good as." See Section 1.5.

rather than just a relation R. By an abuse of language, we shall simply call (A, R) a relation. We shall see more precisely below why specification of the underlying set is important. In general, if $B \subseteq A$ and R is a binary relation on A, we shall refer to

$$S = \{(a, b) \in B \times B : aRb\}$$

as the *restriction of R to B* or the *subrelation generated by B*. Thus the relation (B, R') defined by Eq. (1.9) is the restriction of the relation (A, R) of Eq. (1.8) to the set B of all males in the United States.

We may also speak of n-ary relations, where n is a positive integer. An *n-ary relation R* on a set A is a subset of the Cartesian product A^n. (We shall frequently speak of a relation, rather than an n-ary relation, if n is understood.) For example, if $A = \{1, 2, 6\}$, then a 3-ary or *ternary* relation R on A is given by

$$R = \{(1, 2, 6), (6, 2, 1), (6, 6, 6)\}. \tag{1.10}$$

If A is the set of all lines in the plane, we might define a ternary relation R on A as follows:

$$(a, b, c) \in R \Leftrightarrow a, b, \text{ and } c \text{ are parallel and } b \text{ is} \tag{1.11}$$

strictly between a and c.

A similar ternary relation R might be defined on the set A of all students in a given school. Then we would take

$$(a, b, c) \in R \Leftrightarrow b\text{'s grade point average is strictly} \tag{1.12}$$

between those of a and c.

If $A = \{1, 2, 6\}$, then a 4-ary or *quaternary* relation R on A is given by

$$R = \{(1, 2, 2, 6), (1, 2, 1, 6), (6, 6, 6, 6)\}. \tag{1.13}$$

If A is once again a set of alternatives such as designs for alternative regional transportation systems, you might make statements like S: "I prefer a to b at least as much as I prefer c to d." A quaternary relation on A is given by the collection D of all ordered 4-tuples (a, b, c, d) such that a, b, c, d are in A and for which you assert such a statement S of comparative preference. In such a case, we shall use either the notation $D(a, b, c, d)$ or the notation $abDcd$ to mean that $(a, b, c, d) \in D$.

In what follows, we shall most frequently deal with binary relations, and shall often speak just of a *relation* when we mean a binary one.

Exercises

1. If (A, R) is a binary relation, the *converse* is the relation R^{-1} on A defined by

$$aR^{-1}b \quad \text{iff} \quad bRa.$$

(The notation $\breve{R}$ is sometimes used in place of R^{-1}.) For example, if A is the set of all males in the United States and (A, R) is "father of," then (A, R^{-1}) is "son of." Identify the converse of the following relations:

    (a) "Sister of" on the set of all people in the United States.

    (b) "Uncle of" on the set of all people in the Soviet Union.

    (c) "Greater than" on the set Re.

    (d) $(Re, =)$.

    (e) "As tall as" on the set of all men in New Jersey.

  2. If (A, R) and (A, S) are binary relations, the *intersection* $R \cap S$ on A is defined by

$$R \cap S = \{(a, b): aRb \text{ and } aSb\}.$$

For example, if (A, R) is "brother of" and (A, S) is "sibling of," then $(A, R \cap S)$ is "brother of." Identify $(A, R \cap S)$ in the following cases:

    (a) $A = $ a set of people, $R = $"father of," $S = $"relative of."

    (b) $A = Re$, $R = \geqq$, $S = \neq$.

    (c) $A = $ a set of people, $R = $"older than," $S = $"father of."

    (d) $A = $ a set of sets, $R = \subsetneqq$, $S = $"are disjoint."

  3. If (A, R) and (A, S) are binary relations, the *union* $R \cup S$ on A is defined by

$$R \cup S = \{(a, b): aRb \text{ or } aSb\}.$$

For example, if (A, R) is "brother of" and (A, S) is "sister of," then $(A, R \cup S)$ is "sibling of." Identify $(A, R \cup S)$ in the examples of Exer. 2.

  4. If (A, R) and (A, S) are binary relations, then the *relative product* $R \circ S$ on A is defined by

$$R \circ S = \{(a, b): \text{for some } c, aRc \text{ and } cSb\}.$$

For example, if (A, R) is "father of" and (A, S) is "parent of," then $a(R \circ S)b$ holds if and only if, for some c, a is father of c and c is parent of b—that is, if and only if a is grandfather of b. Identify $R \circ S$ in the following examples:

    (a) $A = $ a set of people, $R = $"father of," $S = $"mother of."

    (b) $A = $ a set of people, $R = $"older than," $S = $"older than."

    (c) $A = Re$, $R = >$, $S = >$.

    (d) $A = Re$, $R = >$, $S = <$.

  5. Show the following:

    (a) $(R \cup S) \circ T = (R \circ T) \cup (S \circ T)$.

    (b) $(R \cap S) \circ T$ may not be $(R \circ T) \cap (S \circ T)$.

    (c) $R \circ (S \circ T) = (R \circ S) \circ T$.

    (d) $(R \circ S)^{-1}$ may not be $R^{-1} \circ S^{-1}$.

  6. Show that $(A, S \circ R)$ can be different from $(A, R \circ S)$.

  7. Suppose $A = \{1, 2, 6\}$ and (A, R) is the quaternary relation of Eq. (1.13). Then we may define a ternary relation S on A and a binary relation T on A by

$$(a, b, c) \in S \Leftrightarrow (\exists x)[R(a, b, c, x)]$$

and
$$aTb \Leftrightarrow (\exists x)(\exists y)[R(a, b, x, y)].$$

(a) Write out (A, S) and (A, T) as sets of ordered n-tuples.

(b) Let U be the quaternary relation defined by the restriction of R to $B = \{1, 6\}$. Write out (B, U).

1.3 Properties of Relations

There are certain properties that are common to many naturally occurring relations. We discuss some of these properties in this section, and they are summarized in Table 1.2 at the beginning of the chapter. Let us say that a binary relation (A, R) is *reflexive* if, for all $a \in A$, aRa. Thus, for example, if A is a set of real numbers and R is the relation "equality" on A, then (A, R) is reflexive because a number is always equal to itself. If $A = \{1, 2, 3, 4\}$, then the relation (A, R) defined by Eq. (1.1) is not reflexive, since $2R2$ does not hold. On the other hand, the restriction of this relation to the set $B = \{1, 3\}$ is reflexive. This observation demonstrates again why it is important to refer to the underlying set when speaking of a relation.

If A is the set of people in the world and F is the relation "father of" on A, then (A, F) is not reflexive, since a person is not his own father. This relation is not reflexive in a very strong way, since the condition of reflexivity is violated for every a in A. We shall say that a binary relation (A, R) is *irreflexive* if, for all $a \in A$, $\sim aRa$. Thus the relation "father of" is irreflexive. So is the relation (A, T) where $A = \{1, 2, 3, 4\}$ and T is defined by Eq. (1.3), and the relation (A, U) where $A = \{SO_2, DDT, NO_x\}$ and U is defined by Eq. (1.4). This terminology should be distinguished from the terminology *nonreflexive*, which means simply "not reflexive."

A binary relation (A, R) is called *symmetric* if, for all $a, b \in A$,
$$aRb \Rightarrow bRa.$$
That is, (A, R) is symmetric if, whenever $(a, b) \in R$, then $(b, a) \in R$. The relation "equality" on any set of numbers is symmetric. So are the relations (A, R) where $A = \{1, 2, 3, 4\}$ and R is defined by Eq. (1.1), and (A, V) where $A = \{SO_2, DDT, NO_x\}$ and V is defined by Eq. (1.5). The relation (A, T) where $A = \{1, 2, 3, 4\}$ and T is given by Eq. (1.3), is not symmetric, for $1T2$ but $\sim 2T1$. The relation "brother of" on the set of all people in the United States is not symmetric, for if a is the brother of b, it does not follow that b is the brother of a. However, the relation "brother of" on the set of all males is symmetric.

Other examples of *nonsymmetric* (not symmetric) relations are the relation "father of" on the set of people in the world, the relation P of strict preference on a set of alternatives, and the relation "sounds louder than" on a set of sounds. These three relations are highly nonsymmetric. They are *asymmetric*, i.e., they satisfy the rule
$$aRb \Rightarrow \sim bRa.$$

Other asymmetric relations are the relation "greater than," $>$, on the set of real numbers, the relation "strictly contained in," $\subsetneqq$, on any collection of sets, and the relation (A, U) where $A = \{SO_2, DDT, NO_x\}$ and U is given by Eq. (1.4).

Some relations (A, R) are not quite asymmetric, but are almost asymmetric in the sense that aRb and bRa holds only if $a = b$. Let us say that (A, R) is *antisymmetric* if, for all $a, b \in A$,

$$aRb \ \& \ bRa \Rightarrow a = b.$$

An example of an antisymmetric relation is the relation "greater than or equal to," $\geq$, on the set of real numbers. Another example is "contained in," $\subseteq$, on any collection of sets. Every asymmetric binary relation (A, R) is antisymmetric. But the converse is false: the relation $\geq$ is antisymmetric but not asymmetric.

A relation (A, R) is called *transitive* if, for all $a, b, c \in A$, whenever aRb and bRc, then aRc. In symbols, (A, R) is transitive if

$$aRb \ \& \ bRc \Rightarrow aRc.$$

Examples of transitive relations are the relations "equality" and "greater than" on the set of real numbers and "implies" on a set of statements. It is left to the reader to verify that the relations (A, S) where $A = \{1, 2, 3, 4\}$ and S is defined by Eq. (1.2), and (A, U) where $A = \{SO_2, DDT, NO_x\}$ and U is defined by Eq. (1.4), are transitive. It seems reasonable to assume that the relation of strict preference among alternative designs of transportation systems is transitive, for if you prefer a to b and b to c, you should be expected to prefer a to c. We shall discuss this point further in later chapters. Similarly, it seems reasonable to assume that the relation L, "sounds louder than," on a set of airplane engines is transitive, though this must be left to empirical data to verify. If $A = \{SO_2, DDT, NO_x\}$ and V is defined by Eq. (1.5), then (A, V) is not transitive. For $SO_2 V NO_x$ and $NO_x V SO_2$, but not $SO_2 V SO_2$. Another relation that is not transitive is the relation "father of."

In studying binary relations (A, R), it will often be convenient to use the abbreviation $aRbRc$ for the statement $aRb \ \& \ bRc$. Thus, (A, R) is transitive if $aRbRc$ implies aRc. Similarly, $aRbRcRd$ will abbreviate $aRb \ \& \ bRc \ \& \ cRd$. And so on. If transitivity holds, then $aRbRcRd$ implies aRd. More generally, $a_1 Ra_2 Ra_3 \ldots Ra_n$ implies $a_1 Ra_n$. The proof is easily accomplished by mathematical induction.

A binary relation (A, R) is called *negatively transitive* if, for all $a, b, c \in A$, not aRb and not bRc imply not aRc. A binary relation (A, R) is negatively transitive if the relation "not in the relation R," defined on the set A, is transitive. To give an example, the relation $R =$ "greater than" on a set of real numbers is negatively transitive, for "not in R" is the relation "less than or equal to," which is transitive. It is easy to show that if $A = \{SO_2, DDT, NO_x\}$ and U is defined by Eq. (1.4), then (A, U) is negatively transitive. Similarly, strict preference on a set of alternatives and

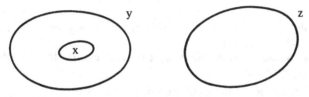

Figure 1.1. The relation "contained in" is not necessarily negatively transitive.

"sounds louder than" on a set of sounds are probably negatively transitive. Verifying negative transitivity can be annoyingly confusing; it is often easier to test the following equivalent condition: For all $x, y, z \in A$, if xRy, then xRz or zRy. To prove that these two conditions are equivalent, we observe that the equivalent version is just the contrapositive of the condition in negative transitivity.* Using this notion, we see easily that the relation "greater than" is negatively transitive, for if $x > y$, then for all z, either $x > z$ or $z > y$. Similarly, one sees that "contained in" is not negatively transitive, for if x is contained in y, there may very well be a z so that x is not contained in z and z is not contained in y. (See Fig. 1.1 for an example.) The relation "father of" on the set of all people in the world is not negatively transitive, nor is the relation (A, R) where $A = \{1, 2, 3, 4\}$ and R is given by Eq. (1.1). To see the latter, note that $1R2$, but not $1R3$ and not $3R2$.

Exercises

1. (a) Show that if (A, R) is a binary relation and (A, R^{-1}) is its converse (Exer. 1, Section 1.2), then (A, R^{-1}) is reflexive if and only if (A, R) is.
    (b) Is a similar statement true for the property irreflexive?
    (c) Symmetric?
    (d) Asymmetric?
    (e) Antisymmetric?
    (f) Transitive?
    (g) Negatively transitive?

2. Suppose (A, R) and (A, S) are binary relations.
    (a) If both relations (A, R) and (A, S) are reflexive, is the intersection $(A, R \cap S)$ reflexive?
    (b)–(g) Repeat for the properties in (b) through (g) of Exer. 1.

3. Repeat Exer. 2 for the union $(A, R \cup S)$.

4. Repeat Exer. 2 for the relative product $(A, R \mathbf{o} S)$.

5. Which of the properties in (a) through (g) of Exer. 1 hold for the relational system $(A, \varnothing)$?

6. Which of the properties in (a) through (g) of Exer. 1 hold for the relational system $(A, A \times A)$?

---

*If $\mathcal{S}$ is the statement A implies B, then the *contrapositive* of $\mathcal{S}$ is the statement "not B" implies "not A."

7. (a) Show that it is not possible for a binary relation to be both symmetric and asymmetric.

(b) Show that it is possible for a binary relation to be both symmetric and antisymmetric.

8. Show that there are binary relations that are
(a) transitive but not negatively transitive;
(b) negatively transitive but not transitive;
(c) neither negatively transitive nor transitive;
(d) both negatively transitive and transitive.

9. (a) Consider the relation x divides y on the set of positive integers. Which of the properties in (a) through (g) of Exer. 1 does this relation have?

(b) Repeat part (a) for the relation "uncle of" on a set of people.

(c) For the relation of "having the same weight as" on a set of mice.

(d) For the relation "feels smoother than" on a set of objects.

(e) For the relation "admires" on a set of people.

10. Suppose $A = Re$ and

$$aRb \Leftrightarrow a > b + 1.$$

This relation will arise in our study of preference in Chapter 6. Which of the properties in (a) through (g) of Exer. 1 hold for (A, R)?

11. Consider the binary relation (A, S) where $A = Re$ and

$$aSb \Leftrightarrow |a - b| \leq 1.$$

This relation is closely related to the binary relation (A, R) of Exer. 10, and will arise in our study of indifference in Chapter 6. Which of the properties in (a) through (g) of Exer. 1 hold for (A, S)?

12. If (A, R) is a binary relation, the *symmetric complement* is the binary relation S on A defined by

$$aSb \Leftrightarrow (\sim aRb \ \& \ \sim bRa).$$

Note that if R is strict preference, then S is indifference: you are indifferent between two alternatives if and only if you prefer neither.

(a) Show that the symmetric complement is always symmetric.

(b) Show that if (A, R) is negatively transitive, then the symmetric complement is transitive.

(c) Show that the converse of (b) is false.

(d) Show that if $A = Re$ and R is as defined in Exer. 10, then S as defined in Exer. 11 is the symmetric complement.

(e) Identify the symmetric complement of the following relations:
(i) $(Re, >)$.
(ii) $(Re, =)$.
(iii) (N, R), where xRy means that x does not divide y.

13. The data of Table 1.4 shows the (consensus) preferences among composers of the members of an orchestra. The i,j entry is 1 if and only if composer i is (strictly) preferred to composer j. Which of the properties in

(a) through (g) of Exer. 1 hold for the orchestra members' relation of preference on the set

$$A = \{\text{Beethoven, Brahms, Mozart, Wagner}\}?$$

Table 1.4. Preferences of Orchestra Members Among Composers*
(The i,j entry is 1 iff composer i is (strictly) preferred to composer j.)

	Beethoven	Brahms	Mozart	Wagner
Beethoven	0	1	1	1
Brahms	0	0	1	1
Mozart	0	0	0	1
Wagner	0	0	0	0

*Data based on an experiment of Folgmann [1933], with strict preference for the orchestra taken to mean a majority of the orchestra members have that preference.

14. The data of Table 1.5 represent taste preferences for vanilla puddings by a group of judges, with entry i, j equal to 1 if and only if the group (strictly) prefers pudding i to pudding j. Which of the properties in (a) through (g) of Exer. 1 hold for this relation of preference on the set $\{1, 2, 3, 4, 5\}$?

Table 1.5. Taste Preference for Vanilla Puddings*
(Entry i,j is 1 iff pudding i is (strictly) preferred to pudding j by a group of judges.)

	1	2	3	4	5
1	0	0	1	1	0
2	1	0	0	1	0
3	0	1	0	0	0
4	0	0	1	0	0
5	1	1	0	1	0

*Data obtained from an experiment of Davidson and Bradley [1969].

15. The data of Table 1.6 state judgments of relative loudness of different sounds, with entry i, j taken to be 1 if and only if sound i is judged (definitely) louder than sound j. Which of the properties in (a) through (g) of Exer. 1 hold for the relation "louder than" on the set of sounds $\{1, 2, 3, 4, 5\}$?

Table 1.6. Judgments of Relative Loudness.
(Entry i,j is 1 iff sound i is judged (definitely) louder than sound j.)

	1	2	3	4
1	0	1	0	1
2	0	0	1	1
3	1	0	0	0
4	0	0	0	0

16. The data of Table 1.7 present the judgments of "sameness" among different sounds by individuals in a group, with entry i,j taken to be 1 if and only if sound i is judged to be the same as sound j a sufficiently large percentage of the time. Which of the properties in (a) through (g) of Exer. 1 hold for this relation of sameness on the set $\{B, C, F, J\}$?

Table 1.7. Judgments of Sameness of Sounds*
(Entry i,j is taken to be 1 iff sound i
is judged to be the same as sound j
at least 25% of the time.)

	B	C	F	J
B	1	1	1	0
C	1	1	1	1
F	1	1	1	1
J	0	1	1	1

*Data from an experiment of Rothkopf [1957].

17. The data of Table 1.8 present judgments of relative importance of different objectives for a library system in Dallas, with entry i,j equal to 1 if and only if objective i is judged more important than objective j. Which of the properties in (a) through (g) of Exer. 1 hold for this relation of relative importance on the set $\{a, b, c, d, e, f\}$?

Table 1.8. Judgments of Relative Importance of Objectives for a Library System in Dallas*
(Entry i,j is 1 iff i is judged more important than j.)

	a	b	c	d	e	f
a	0	1	1	1	1	1
b	0	0	0	0	0	0
c	0	0	0	0	0	1
d	0	1	0	0	1	0
e	0	0	0	0	0	0
f	0	0	0	0	0	0

*Data from Farris [1975].

Key

a Convenient, accessible library facilities
b Convenient operating hours
c Efficient inter-library system
d Good local libraries
e Good reproduction facilities
f Rapid inter-library response time

18. The data of Table 1.9 present judgments of relative importance of different objectives for a state environmental agency in Ohio, with entry

Table 1.9. Judgments of Relative Importance of Goals for a
State Environmental Agency in Ohio*
(Entry i, j is 1 iff i is judged more important than j.)

	a	b	c	d	e	f
a	0	1	1	1	1	1
b	0	0	0	0	0	0
c	0	1	0	0	1	0
d	0	1	0	0	1	0
e	0	1	0	0	0	0
f	0	1	0	0	1	0

*Data from Hart [1975].

Key

a   Enhance and protect State's environment
b   Improve and insure the quality of the air
c   Develop a comprehensive program of environmental quality
    planning
d   Protect and promote the State's natural attractions
e   Prevent the future occurrence of pollution emergencies
f   Promote an environment that is beneficial to human health
    and welfare

i, j taken to be 1 if and only if objective i is judged more important than
objective j. Which of the properties in (a) through (g) of Exer. 1 hold for this
relation of relative importance on the set $\{a, b, c, d, e, f\}$?

1.4 Equivalence Relations

Many binary relations satisfy the three properties reflexivity, symmetry,
and transitivity. Such relations are called *equivalence relations*. The relation
equality on any set of numbers is an equivalence relation. So is the relation
(A, S), where A is $\{1, 2, 3, 4\}$ and S is defined by Eq. (1.2). If A is a set of
lines and aRb holds if and only if a and b are parallel, then (A, R) is an
equivalence relation provided that we say a line is parallel to itself. Other
examples are the following: A is the set of integers $\{0, 1, 2, \ldots, 26\}$, and
aRb iff $a \equiv b(\mod 3)$; A is a set of people who have been blood-typed, and
aRb iff a and b have the same blood type; A is a set of people, and aRb iff
a has the same height as b; A is a set of children, and aRb iff a and b have
the same IQ; A is a set of animals, and aRb iff a and b are in the same
species.

If (A, R) is an equivalence relation and $a \in A$, let a^* denote

$$\{b \in A: aRb\}.$$

This set will be called the *equivalence class containing a*. The element a will
be called a *representative* of the equivalence class a^*. To give an example,

if $A = \{1, 2, 3, 4\}$ and S is defined by Eq. (1.2), then

$$1^* = \{1, 2\},$$
$$2^* = \{1, 2\},$$
$$3^* = \{3, 4\},$$
$$4^* = \{3, 4\}.$$

Here, there are two different equivalence classes, $\{1, 2\}$ and $\{3, 4\}$. If $A = \{0, 1, 2, \ldots, 26\}$ and aRb iff $a \equiv b$ (mod 3), then

$$0^* = \{0, 3, 6, 9, 12, 15, 18, 21, 24\},$$
$$1^* = \{1, 4, 7, 10, 13, 16, 19, 22, 25\},$$
$$2^* = \{2, 5, 8, 11, 14, 17, 20, 23, 26\},$$
$$3^* = \{0, 3, 6, 9, 12, 15, 18, 21, 24\},$$
$$\cdots$$

Here, there are three different equivalence classes, 0^*, 1^*, and 2^*.

The most important properties of equivalence relations are summarized in the following theorem.

THEOREM 1.1. *Suppose (A, R) is an equivalence relation. Then:*
(a) Any two equivalence classes are either disjoint or identical.
(b) The collection of (distinct) equivalence classes partitions A; that is, every element of A is in one and only one (distinct) equivalence class.

Proof. To prove (a), we shall show that for all $a, b \in A$, either $a^* = b^*$ or $a^* \cap b^* = \varnothing$. In particular, we shall show that

$$aRb \Rightarrow a^* = b^* \tag{1.14}$$

and

$$\sim aRb \Rightarrow a^* \cap b^* = \varnothing. \tag{1.15}$$

To demonstrate (1.14), we assume aRb and show that aRc holds if and only if bRc holds. Thus, suppose aRc holds. Then cRa follows, since (A, R) is symmetric. Now we have cRa and aRb, so we conclude cRb from transitivity of (A, R). Finally, cRb implies bRc, by symmetry. A similar proof, left to the reader, shows that bRc implies aRc. Thus, we have established (1.14). To prove (1.15), suppose $a^* \cap b^* \neq \varnothing$ and let $c \in a^* \cap b^*$. Then $c \in a^*$ implies aRc, and $c \in b^*$ implies bRc. By symmetry, we have cRb. Finally, by transitivity, aRc and cRb imply aRb. This proves (1.15), and completes the proof of part (a).

To prove part (b), note that every element a in A is in some equivalence class, namely a^*, and in no more than one, by part (a).   ∎

In general, whenever (A, R) is an equivalence relation, we can gain much in the way of economy by dealing with the equivalence classes rather than the objects of A themselves. There are in general many fewer such. We shall see this clearly in the next section, when we discuss the process of reduction.

Exercises

1. Suppose A is the set of sequences of 0's and 1's of length 10, and aRb holds if and only if sequences a and b have the same number of 1's.
  (a) Show that (A,R) is an equivalence relation.
  (b) Identify the equivalence classes.

2. (a) Show that if A is a set of sounds and aRb holds if and only if a and b *sound* equally loud, then (A,R) may not be an equivalence relation.
  (b) If A is a set of sounds, and aRb holds if and only if a and b are measured to have the same decibel level (a measure of sound intensity), is (A,R) an equivalence relation?
  (c) If A is a set of people and aRb holds if and only if a and b *look* equally tall, is (A,R) an equivalence relation?
  (d) If A is a set of people and aRb holds if and only if a and b are measured to have the same height, is (A,R) an equivalence relation?

3. Suppose (A,S) is the binary relation of Exer. 11, Section 1.3. Is (A,S) an equivalence relation?

4. Show that all the properties in the definition of an equivalence relation are needed. In particular, show that there are binary relations that are
  (a) reflexive, symmetric, and not transitive;
  (b) reflexive, transitive, and not symmetric;
  (c) symmetric, transitive, and not reflexive.

5. Suppose (A, R) and (A, S) are equivalence relations.
  (a) Show that $(A, R \cap S)$ is an equivalence relation.
  (b) Show that $(A, R \cup S)$ does not have to be an equivalence relation.
  (c) What about $(A, R \text{ o } S)$?

6. Show that if (A, R) is an equivalence relation, then

$$a^* = b^* \quad \text{iff} \quad a \in b^*.$$

7. If (A,R) is an equivalence relation, we may define a binary relation R^* on the set A^* of equivalence classes as follows: if α and β are equivalence classes, and a is in α and b is in β, then

$$\alpha R^* \beta \quad \text{iff} \quad aRb.$$

  (a) Show that R^* is well-defined in the sense that if a' is in α and b' is in β, then
$$aRb \quad \text{iff} \quad a'Rb'.$$

  (b) Moreover, show that (A^*, R^*) is an equivalence relation.

8. Suppose A is finite and R is a binary relation on A. Let R^2 be the binary relation $R \text{ o } R$ on A, and define R^n to be $R^{n-1} \text{ o } R$.

(a) Show that if $n \geq 1$, then aR^nb if and only if there are $a_1, a_2, \ldots, a_{n-1}$ so that $aRa_1, a_1Ra_2, \ldots, a_{n-2}Ra_{n-1}, a_{n-1}Rb$.

(b) Define R^0 on A to be $\{(a, a): a \in A\}$. If aR^nb for some $n \geq 0$, we say that there is a (finite) *path* from a to b of *length* n. If A is finite, show that there is a number k so that if there is a path from a to b, there is one of length at most k.

(c) Let S be defined on A as follows: aSb if and only if there is a path from a to b. Show that (A, S) is transitive.

(d) Let T be defined on A as follows: $T = S \cap S^{-1}$. Show that aTb if and only if there is a path from a to b and a path from b to a.

(e) Show that (A, T) is an equivalence relation. (The equivalence classes are called *strong components*.)

(f) (A, R) is called *strongly connected* if there is just one equivalence class under T. Is the relation $<$ on a finite set of numbers strongly connected?

(g) If A is finite, the *transitive closure* of (A, R) is the transitive relation on A containing all ordered pairs in R and the smallest possible number of ordered pairs. Show that the notion of transitive closure is well-defined.

(h) How is transitive closure related to the relation S defined in (c)?

(i) Identify R^2, R^3, S, T, the strong components, and the transitive closure in the following examples. Determine which of these examples is strongly connected.

    (i) $A = Re$, $R = \geq$.

    (ii) $A = \{1, 2, 3, 4\}$, $R = \{(1, 2), (2, 3), (3, 4)\}$.

    (iii) $A = \{1, 2, 3\}$, $R = \{(1, 2), (2, 1), (1, 3)\}$.

    (iv) $A = \{+, *, \#, \$\}$, $R = \{(+, \#), (*, *), (+, *)\}$.

9 (a) Does the data of Table 1.4 define an equivalence relation?

  (b) What about the data of Table 1.5?

  (c) Table 1.6?

  (d) Table 1.7?

  (e) Table 1.8?

  (f) Table 1.9?

1.5 Weak Orders and Simple Orders

In this section, we define various order relations. The results are summarized at the beginning of the chapter in Table 1.3, which the reader is urged to consult for reference.

Suppose (A, P) is the binary relation of (strict) preference defined by Eq. (1.7), where A is a set of alternatives among which you are choosing—for example, alternative designs for a regional transportation system. In general, we can suppose that for each pair of alternatives a and b, you do one of three things: You prefer a to b, you prefer b to a, or you are indifferent between a and b. Let us say you *weakly prefer* a to b if either you (strictly) prefer a to b or you are indifferent between a and b. We denote the binary relation of weak preference on the set A by W, and the binary relation of indifference on A by I. Then we have

$$aWb \Leftrightarrow (aPb \text{ or } aIb).$$

It is reasonable to assume that the relation (A, W) is both reflexive and transitive, though later we shall question transitivity. A binary relation that satisfies these two properties is called a *quasi order* or pre-order. The relation $\geq$ on the set of real numbers is another example of a quasi order; so is the relation $\subseteq$ on any collection of sets; so is the relation "at least as tall as"; and so is any equivalence relation.

The relation (A, W) presumably also has the property that for every a and b in A, including $a = b$, either aWb or bWa. A binary relation with this property is called *strongly complete* (sometimes the terms connected or strongly connected are used). A binary relation is called a *weak order* if it is transitive and strongly complete. Thus, weak preference is a weak order. So is the relation $\geq$ on the set Re of real numbers. The relation $>$ on Re is not a weak order, for it is not strongly complete; it is not the case that $1 > 1$. Similarly, $\subseteq$ is not weak, because it is not strongly complete. Any weak order (A, R) is a quasi order. It suffices to prove reflexivity, which follows by strong completeness. A quasi order is not necessarily a weak order. An example of a quasi order that is not weak is given by any equivalence relation with more than one equivalence class, as, for example, the relation (A, S), where $A = \{1, 2, 3, 4\}$ and S is defined by Eq. (1.2). One of the most helpful examples of a weak order is the following, which we shall use as an example frequently. Let $A = \{0, 1, 2, \ldots, 26\}$. Every number a in A is congruent (mod 3) to one of the numbers 0, 1, or 2. Let us call this number a mod 3. Thus, 8 mod 3 is 2, 10 mod 3 is 1, etc. We define R on A as follows:

$$aRb \Leftrightarrow a \bmod 3 \geq b \bmod 3. \tag{1.16}$$

Thus, $2R1$, $8R10$, etc. It is left to the reader to prove that (A, R) is a weak order. We may think of (A, R) as follows: Elements of A are listed in vertical columns above the number 0, 1, or 2 to which they are congruent. Then aRb if and only if a is at least as far to the right as b. (See Fig. 1.2.) It will follow from Theorem 1.2 below that essentially every weak order on a finite set can be thought of as an ordering "weakly to the right of" on a comparable array of points arranged in vertical columns. A sample array is shown in Fig. 1.3.

A weak order (A, R) that is also antisymmetric is called a *simple order*. (The terms *linear order* and *total order* are also used.) The prototype of simple orders is the relation $\geq$. In a simple order R on a finite set A, the elements of A may be laid out on the line with aRb holding if and only if a is to the right of b or equal with b, as shown in Fig. 1.4. (We prove this formally in Section 3.1.)

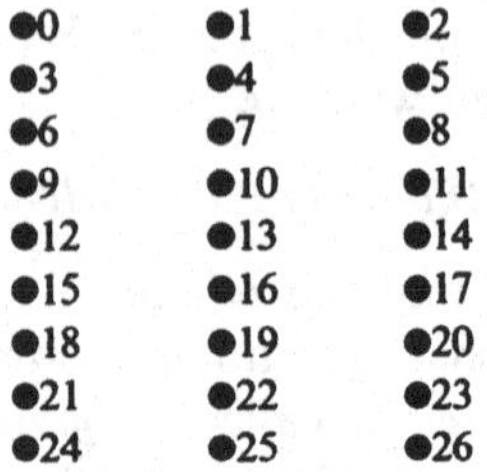

```
●0        ●1        ●2
●3        ●4        ●5
●6        ●7        ●8
●9        ●10       ●11
●12       ●13       ●14
●15       ●16       ●17
●18       ●19       ●20
●21       ●22       ●23
●24       ●25       ●26
```

Figure 1.2. The binary relation $\geqq$ (mod 3) on $A = \{0, 1, 2, \ldots, 26\}$.

Figure 1.3. A weak order (A, R); aRb iff a is to the right of b or in the same vertical column as b.

●   ●   ●   ●●●●   ●   ●●   ●●●   ●   ●●

Figure 1.4. A simple order.

If (A, R) is a binary relation, suppose we define a binary relation E on A by

$$aEb \Leftrightarrow aRb \ \& \ bRa. \tag{1.17}$$

If we think of an array like that of Fig. 1.3, the relation E can be interpreted as "being in the same vertical column." If (A, R) is a simple order, then antisymmetry implies that E is equality. More generally, if (A, R) is a quasi order, then E is an equivalence relation. To see this, suppose (A, R) is a quasi order. We verify first that (A, E) is symmetric. If aEb holds, then aRb and bRa hold; hence bRa and aRb hold; hence bEa holds. (Notice that this proof does not use any of the properties of a quasi order.) Proof that (A, E) is reflexive and transitive is left to the reader.

If (A, R) is a weak order, then it is a quasi order, and so E is an equivalence relation. The relation R tells how to simply order the equivalence classes under E. To be precise, let A^* be the collection of equivalence classes, and define R^* on A^* as follows: If α and β are equivalence classes, pick $a \in \alpha$ and $b \in \beta$, and let $\alpha R^* \beta$ hold if and only if aRb holds. Of course, this process may lead to ambiguities, because whether we take $\alpha R^* \beta$ may depend on which a and b we chose. We shall show below that this is not the case, that is, that R^* is well-defined. Assuming this for now, we can summarize the definition of R^* as follows:

$$a^* R^* b^* \Leftrightarrow aRb. \qquad (1.18)$$

Then (A^*, R^*) is called the *reduction* or *quotient* of (A, R).†

To give an example, suppose $A = \{0, 1, 2, \ldots, 26\}$ and aRb holds iff $a \bmod 3 \geq b \bmod 3$. Then aEb holds iff $a \equiv b \pmod 3$. There are three equivalence classes: 0^*, 1^*, and 2^*. We have $2^* R^* 1^*$, since $2R1$; and similarly $2^* R^* 0^*$, $2^* R^* 2^*$, etc. We see that here the reduction (A^*, R^*) is a simple order on the set A^* of equivalence classes; it is like the usual simple order $\geq$ on $\{0, 1, 2\}$. That this example is not a special case is summarized in the following theorem.

THEOREM 1.2. *Suppose (A, R) is a weak order. Then the reduction (A^*, R^*) is well-defined, and it is a simple order.*

Proof. To show that R^* is well-defined, we need to show that its definition does not depend on the particular choice of elements $a \in \alpha$ and $b \in \beta$ of the equivalence classes α and β. Put in other words, if we choose $c \in \alpha$ and $d \in \beta$, then we need to show that aRb iff cRd. To show this, suppose aRb. Since a and c are in α, we conclude that aEc. Similarly, bEd. Now aEc implies aRc and cRa, and bEd implies bRd and dRb. Now, if aRb, then using cRa, we conclude cRb by transitivity. From cRb and bRd we conclude cRd, again by transitivity. The proof that cRd implies aRb is analogous. Thus, R^* is well-defined.

To prove that (A^*, R^*) is a simple order, one must verify that it is transitive, strongly complete, and antisymmetric. To show that it is transitive, suppose that α, β, and γ are in A^* and that $\alpha R^* \beta R^* \gamma$. To show $\alpha R^* \gamma$, pick a in α and c in γ and show aRc. Now given $b \in \beta$, we have aRb, since $\alpha R^* \beta$, and we have bRc, since $\beta R^* \gamma$. Since (A, R) is transitive, aRc follows. This implies $\alpha R^* \gamma$. The rest of the proof is left to the reader. ∎

COROLLARY *The reduction is well-defined even if (A, R) is only a quasi order.*

Proof. The proof of well-definedness uses only this fact. ∎

If we accept the fact that every simple order on a finite set can be realized as the ordering $\geq$ on a set of real numbers, then we may interpret Theorem 1.2 as follows: Every weak order (A, R) on a finite set A may be

---

†(A^*, R^*) is sometimes denoted $(A/E, R/E)$, and the process of reduction is sometimes called "canceling out the equivalence relation." This is similar to what we do in group theory when we pass from a group to its cosets.

realized on the real line with equivalent elements in the same vertical column and so that aRb holds if and only if a is to the right of b or equal with b (cf. Fig. 1.3).

The relation $>$ on the real numbers is not a simple order. It is not strongly complete, because it is not reflexive. But it is *complete* in the sense that for all $a \neq b$, aRb or bRa. The relation $(Re, >)$ has many properties, among them transitivity, completeness, asymmetry, antisymmetry, and negative transitivity. We would like to characterize this ordering $>$ just as we characterized the ordering $\geq$ by listing properties or axioms that essentially determined it—namely, the axioms for a simple order. In axiom-building, one tries to be as frugal as possible, and list only those properties that are needed. Of the above list, some are superfluous. For asymmetry implies antisymmetry. Similarly, transitivity and completeness imply negative transitivity. (The proofs are left to the reader.) These observations lead us to adopt the following definition: A binary relation (A, R) is called a *strict simple order* if it is asymmetric, transitive, and complete. (It is left to the reader to show that none of the conditions in this definition is superfluous.) It is often assumed that strict preference is a strict simple order, though it probably violates at least completeness: We can be indifferent between two alternatives. The relation of strict containment $\subsetneq$ is not a strict simple order, because it violates completeness. The relation "beats" in a round-robin tournament usually does not determine a strict simple order, for transitivity is violated. Naturally, $>$ on the set of reals is a strict simple order. We shall prove in Section 3.1 that in every strict simple order R on a finite set A, the elements of A may be laid out on the line with aRb holding if and only if a is strictly to the right of b. Thus, $>$ is indeed the prototype of strict simple orders. It is also easy to show, and we shall ask the reader to show it, that the strict simple orders and simple orders are related to each other in the same way that $>$ is related to $\geq$. Namely, suppose (A, R) is an irreflexive relation and S is defined on A by

$$aSb \Leftrightarrow aRb \text{ or } a = b.$$

Then (A, R) is a strict simple order if and only if (A, S) is a simple order.

To complete our discussion of order relations, we would like to define a type of order relation, called *strict weak*, which corresponds to strict simple orders in the same way that weak orders correspond to simple orders. The paradigm example will again be an ordering of an array of vertical columns, but now an element a is in the relation R to an element b if and only if a is strictly to the right of b. An example can be defined as follows: Let $A = \{0, 1, 2, \ldots, 26\}$, and let aRb hold if and only if $a \bmod 3 > b \bmod 3$. Such a relation will clearly be asymmetric and transitive, but no longer complete: of two elements a and b in the same vertical column, neither aRb nor bRa. Two elements a and b are in the same vertical column if and only if $\sim aRb$ and $\sim bRa$, and this suggests that we should study the following binary relation:

$$aEb \Leftrightarrow \ \sim aRb \text{ and } \sim bRa.* \tag{1.19}$$

If (A, R) is strict simple, then (A, E) is equality. In general, we would like (A, E) to be an equivalence relation. We could define (A, R) to be a strict weak order if it is asymmetric and transitive and (A, E) is an equivalence relation. (The reader might wish to check that our relation $>$ (mod 3) on $\{0, 1, 2, \ldots, 26\}$ satisfies these properties.) However, there is something unsatisfactory about this definition. Specifically, all our definitions of order relations (A, R) so far have been stated in terms of properties of (A, R), and not in terms of properties of relations like (A, E) which are defined from (A, R). Any definition of strict weak order should turn out to be equivalent to this potential definition, and that will be the case with the definition we adopt. We shall say that (A, R) is by definition a *strict weak order* if (A, R) is asymmetric and negatively transitive. To justify this definition, we prove the following theorem.

THEOREM 1.3. *(A, R) is a strict weak order if and only if*
(a) *(A, R) is asymmetric,*
and
(b) *(A, R) is transitive,*
and
(c) *(A, E) is an equivalence relation, where E is defined by Eq. (1.19).*

Proof. Assume (A, R) is strict weak. Then it is asymmetric by definition. To show that it is transitive, suppose aRb and bRc. To show aRc, suppose by way of contradiction that $\sim aRc$. By asymmetry, bRc implies $\sim cRb$. Now by negative transitivity, $\sim aRc$ and $\sim cRb$ imply $\sim aRb$, which is a contradiction. It is left to the reader to prove that if (A, R) is strict weak, then (A, E) is an equivalence relation.

To complete the proof of Theorem 1.3, let us assume that (A, R) satisfies conditions (a), (b), and (c). To show that (A, R) is strict weak, it is sufficient to show that (A, R) is negatively transitive. To demonstrate negative transitivity, we assume that $\sim aRb$ and $\sim bRc$, and show $\sim aRc$. We argue by cases.

CASE 1: bRa. In this case, if aRc, we conclude bRc by transitivity. Thus, $\sim aRc$.

CASE 2: $\sim bRa$. Here, there are two subcases.

CASE 2a: cRb. In this case, if aRc, we conclude aRb by transitivity. Thus, $\sim aRc$.

---

*In Exer. 12 of Section 1.3, (A, E) is called the symmetric complement of (A, R).

CASE 2b: $\sim cRb$. Here, we have aEb and bEc, since $\sim aRb$, $\sim bRa$, $\sim bRc$, and $\sim cRb$. Since (A, E) is an equivalence relation, we conclude aEc, from which $\sim aRc$ follows. ■

Even if indifference is allowed, it is often assumed that the relation of strict preference is an example of a strict weak order, though later we shall question this assumption. Another example is the binary relation "weighs more than" on a set of people, if weight is measured on a precise scale. A third example is the binary relation "warmer than" on a set of objects, if warmer than is based on temperature and temperature is measured on a precise scale. We shall return to these examples in Chapter 3.

The process of reduction is the same for strict weak orders as it is for weak orders. If (A, R) is a strict weak order, let the equivalence relation E be defined by Eq. (1.19) and let A^* be the collection of equivalence classes under E. Then define the *reduction* R^* on A^* as before, by

$$a^* R^* b^* \Leftrightarrow aRb.$$

THEOREM 1.4 *Suppose* (A, R) *is a strict weak order. Then the reduction* (A^*, R^*) *is well-defined and it is a strict simple order.*

Proof. The proof is left to the reader. ■

Exercises

1. (a) Is the converse R^{-1} of a strict weak order R necessarily strict weak?
    (b) Is the converse of a weak order necessarily weak?
    (c) Is the converse of a strict simple order necessarily strict simple?
    (d) Is the converse of a simple order necessarily simple?

2. Is every quasi order a simple order?

3. Is every strict weak order strict simple?

4. Let $A = \{(a, b): a, b \in \{1, 2, 3, 4\}\}$, and suppose

$$(a, b)R(c, d) \quad \text{iff} \quad a > c.$$

Show that (A, R) is a strict weak order and calculate the reduction (A^*, R^*).

5. Suppose $A = \{a, b, x, y, \alpha, \beta, \gamma\}$, and R consists of the following ordered pairs:

$$(a, a), (b, b), (x, x), (y, y), (\alpha, \alpha), (\beta, \beta), (\gamma, \gamma),$$
$$(y, x), (x, y), (x, a), (x, b), (y, a), (y, b), (\alpha, \beta),$$
$$(\alpha, \gamma), (\beta, \alpha), (\beta, \gamma), (\gamma, \alpha), (\gamma, \beta), (\alpha, x), (\alpha, y),$$
$$(\alpha, a), (\alpha, b), (\beta, x), (\beta, y), (\beta, a), (\beta, b), (\gamma, x),$$
$$(\gamma, y), (\gamma, a), (\gamma, b).$$

Note that (A, R) is a weak order and calculate the reduction.

6. Suppose $A = Re \times Re$, and suppose P is defined on A by

$$(a, b)P(s, t) \quad \text{iff} \quad a \geqq s \ \& \ b \geqq t \ \& \ (a > s \text{ or } b > t).$$

Show that (A, P) is not a strict weak order or a weak order.

7. Suppose $A = Re \times Re$, and suppose P is defined on A by

$$(a, b)P(s, t) \quad \text{iff} \quad a > s \text{ or } (a = s \text{ and } b > t).$$

    (a) Show that (A, P) is a strict weak order—it is called the *lexicographic ordering of the plane.*
    (b) Is (A, P) strict simple?
    (c) Is (A, P) weak?

8. Suppose (A, R) is the binary relation of Exer. 10, Section 1.3.
    (a) Is (A, R) a strict weak order?
    (b) A weak order?

9. Suppose A is a set of students and aRb holds if and only if a has a higher grade point average than b, or a and b have the same grade point average and a has a smaller number of absences.
    (a) Is (A, R) strict weak?
    (b) If so, what is its reduction?

10. Suppose (A, R) and (A, S) are weak orders.
    (a) Is $(A, R \cap S)$ necessarily weak?
    (b) What about $(A, R \cup S)$?
    (c) What about $(A, R \circ S)$?

11. Show that the axioms for a simple order are all needed by giving examples of binary relations that are
    (a) transitive, antisymmetric, and not strongly complete;
    (b) transitive, strongly complete, and not antisymmetric;
    (c) antisymmetric, strongly complete, and not transitive.

12. Show that the axioms for a strict simple order are all needed by giving examples of binary relations that are
    (a) asymmetric, transitive, and not complete;
    (b) asymmetric, complete, and not transitive;
    (c) transitive, complete, and not asymmetric.

13. Prove that transitivity and completeness imply negative transitivity.

14. Prove that asymmetry implies antisymmetry.

15. If (A, R) is a simple order, prove that (A^*, R^*) is
    (a) strongly complete;
    (b) antisymmetric.

16. Show that if (A, R) is a strict weak order and E is defined on A by Eq. (1.19), then (A, E) is an equivalence relation.

17. If (A, R) is a quasi order and E is defined on A by Eq. (1.17), show that (A, E) is
   (a) reflexive;
   (b) transitive.

18. Prove Theorem 1.4.

19. Suppose (A, R) is irreflexive. Define S on A by

$$aSb \quad \text{iff} \quad (aRb \text{ or } a = b).$$

Show that (A, R) is strict simple if and only if (A, S) is simple.

20. Suppose (A, R) is strict weak and E is defined on A by Eq. (1.19). Define S on A by

$$aSb \quad \text{iff} \quad (aRb \text{ or } aEb).$$

Show that (A, S) is weak.

21. Suppose (A, R) is weak. Define S on A by

$$aSb \quad \text{iff} \quad (aRb \,\&\, \sim bRa).$$

Show that (A, S) is strict weak.

22. (a) Does the data of Table 1.4 define a strict weak order?
   (b) A weak order?
   (c) A strict simple order?
   (d) A simple order?

23. Repeat Exer. 22 for the data of Table 1.5.

24. Repeat Exer. 22 for the data of Table 1.6.

25. Repeat Exer. 22 for the data of Table 1.7.

26. Repeat Exer. 22 for the data of Table 1.8.

27. Repeat Exer. 22 for the data of Table 1.9.

1.6 Partial Orders

Very often an ordering relation violates completeness, as we have seen. For example, the relation $>(\text{mod}\,3)$ does; the relation "father of" does; most probably, the relation "louder than" on a set of airplane engines does, since very possibly two different engines will sound equally loud; and the relation $\subseteq$ of containment on a collection of sets does. The latter relation has the properties of reflexivity and transitivity and also antisymmetry: $X \subseteq Y$ and $Y \subseteq X$ implies $X = Y$. A binary relation satisfying these three properties is called a *partial order*. Thus, a partial order is an antisymmetric quasi order. Each simple order is a partial order, but not conversely. The relations "father of" and "louder than" are not partial orders, since they are not reflexive.

Partial orders often arise if we are stating preferences among alternatives that have several aspects or dimensions. Thus, suppose A is a set of alternative designs for a transportation system. Suppose we judge these designs on the basis of two aspects: cost and number of people served. Let us suppose we (weakly) prefer design a to design b if and only if a costs no more than b and a serves at least as many people as b. Let us denote this weak preference relation as W. Then (A, W) is certainly not complete: if a costs less than b and serves fewer people, then $\sim aWb$ and $\sim bWa$. On the other hand, (A, W) is transitive and, if we assume that no two designs both cost the same and serve the same number of people, it is also antisymmetric. Thus, (A, W) is a partial order.

More generally, suppose A is a collection of alternatives each of which has n dimensions or aspects, and suppose the real number $f_i(x)$ measures the "worth" of x on the ith aspect. It is not unreasonable to define a (weak) preference relation W on A by

$$aWb \quad \text{iff} \quad [\, f_i(a) \geqq f_i(b) \quad \text{for each } i \,].$$

The binary relation (A, W) defines a partial order.

To give yet another example of a partial order, let A be a set of points in the plane, some of which are joined by straight lines, as in the diagram of Fig. 1.5. If $a, b \in A$, let aRb hold if and only if $a = b$ or there is a continually descending path from a to b, following lines of the diagram. For example, in Fig. 1.5, we have $1R2$, $1R3$, $1R4$, $1R5$, $2R4$, $3R5$, and aRa for every a. It is fairly easy to prove that (A, R) is a partial order. (Proof is left to the reader.) What is not so obvious is that every partial order R on a finite set A arises from such a diagram, called a *Hasse diagram* of the partial order. To give an example, let us consider the partial order $\subseteq$ on $A =$ the set of subsets of $\{1, 2, 3\}$. A Hasse diagram corresponding to $(A, \subseteq)$ is shown in Fig. 1.6. The reader will note that the Hasse diagram of a simple order is a "chain," as shown in Fig. 1.7. Usually it is convenient in Hasse diagrams to omit a line from point a to point b if there is a continuously descending path of more than one link from a to b. We shall follow this procedure in our diagrams.

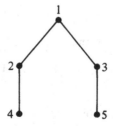

Figure 1.5. Hasse diagram of the partial order (A, R) defined by $A = \{1, 2, 3, 4, 5\}$, $R = \{(1, 1), (2, 2), (3, 3), (4, 4), (5, 5), (1, 2), (1, 3), (1, 4), (1, 5), (2, 4), (3, 5)\}$.

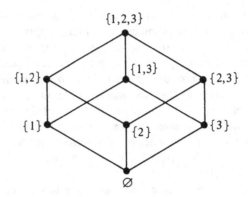

Figure 1.6. Hasse diagram of the partial order of inclusion $\subseteq$ on the set of subsets of {1, 2, 3}.

Figure 1.7. Hasse diagram of a simple order.

Analogously to strict simple orders, we can speak of strict partial orders. A binary relation is a *strict partial order* if it is asymmetric and transitive. Partial orders and strict partial orders are related in the same way that simple orders and strict simple orders are. Namely, suppose (A, R) is an irreflexive binary relation and we define

$$aSb \quad \text{iff} \quad (aRb \text{ or } a = b).$$

Then (A, R) is a strict partial order if and only if (A, S) is a partial order. Again, a strict partial order R corresponds to a Hasse diagram, except that now an element is not in the relation R to itself.

Exercises

1. Suppose $A = \{1, 2, 3, 4\}$ and

$$R = \{(1, 1), (2, 2), (3, 3), (4, 4), (1, 3), (1, 4), (2, 3), (2, 4), (3, 4)\}.$$

Show that
      (a) (A, R) is a partial order.
      (b) Draw the Hasse diagram.

2. Which of the following are strict partial orders?
   (a) $\subsetneq$ on the collection of subsets of $\{1,2,3,4\}$.
   (b) (A,P) where $A = Re \times Re$ and

$$(a,b)P(s,t) \quad \text{iff} \quad (a > s \text{ and } b > t).$$

   (c) (A, Q) where A is a set of n-dimensional alternatives, $f_1, f_2, \ldots, f_n$ are real-valued scales on A, and Q is defined by

$$aQb \Leftrightarrow [f_i(a) > f_i(b) \quad \text{for each } i].$$

   (d) The relation (A, Q) where A is as in part (c) and

$$aQb \Leftrightarrow [f_i(a) \geq f_i(b) \text{ for each } i \text{ and } f_i(a) > f_i(b) \text{ for some } i].$$

   (e) (A,R) of Exer. 10, Section 1.3.

3. (a) Is the converse R^{-1} of a strict partial order necessarily a strict partial order?
   (b) Is the converse of a partial order necessarily a partial order?

4. (a) Is every weak order a partial order?
   (b) Is every partial order a weak order?

5. Is every quasi order a partial order?

6. Show that there are quasi orders that are not partial orders, not weak orders, and not equivalence relations.

7. Show that every strict weak order is a strict partial order.

8. Draw the Hasse diagram of the general strict weak order.

9. If (A,R) is strict weak, define S on A by

$$aSb \quad \text{iff} \quad (aRb \text{ or } a = b).$$

Show that (A,S) is a partial order.

10. Show that none of the axioms for a partial order is superfluous by giving examples of binary relations that are
   (a) reflexive, transitive, and not antisymmetric;
   (b) reflexive, antisymmetric, and not transitive;
   (c) transitive, antisymmetric, and not reflexive.

11. Suppose (A, R) is irreflexive, and define S on A as in Exer. 9. Show that (A, R) is a strict partial order if and only if (A, S) is a partial order.

12. Exercises 12 through 18 introduce the notion of dimension of a strict partial order. For further reference on this subject, see Baker, Fishburn, and Roberts [1971] or Trotter [1978]. Also, see Exer. 2 of Section 5.1 and Exers. 21 and 31 of Section 6.1. Suppose (A, P) is a strict partial order. A strict simple order R on A such that $P \subseteq R$ is called a *strict simple extension* of (A, P). For example, the strict partial order defined from the Hasse

diagram of Fig. 1.5 has as one strict simple extension the ordering in which 1 comes first, 2 second, 3 third, 4 fourth, and 5 fifth. List all the strict simple extensions of this strict partial order.

13. Szpilrajn's Extension Theorem [1930] states that if (A, P) is a strict partial order, if $a \neq b$, and if $\sim aPb$ and $\sim bPa$, then there is a strict simple extension (A, R) of (A, P) so that aRb. Show from this that there is a family $\mathcal{F}$ of strict simple extensions so that

$$P = \cap \{R \colon R \in \mathcal{F}\}.$$

14. Dushnik and Miller [1941] define the *dimension* of a strict partial order (A, P) as the smallest cardinal number m so that (A, P) is the intersection of m strict simple extensions. (By Exer. 13, dimension is well-defined.) As an example, the strict partial order (A, P) defined from Fig. 1.5 has dimension 2. Show this by observing that (A, P) is not strict simple and writing (A, P) as the intersection of two strict simple extensions.

15. If A is the set of all subsets of $\{1, 2, 3\}$ and P is the strict partial order $\subsetneq$, show that (A, P) has dimension 3. (Komm [1948] proves that the strict partial order $\subsetneq$ on the set of subsets of a set S has dimension $|S|$.)

16. Show that every strict weak order has dimension at most 2.

17. Hiraguchi [1955] shows that if (A, P) is a strict partial order with $|A|$ finite and at least 4, then (A, P) has dimension at most $|A|/2$. For a simple proof of this result, see Trotter [1975]. Show that dimension can be less than $[|A|/2]$, where $[a]$ is the greatest integer less than or equal to a.

18. (a) Use Hiraguchi's Theorem (Exer. 17) to obtain upper bounds for dimension of the strict partial orders whose Hasse diagrams are shown in Fig. 1.8.

(b) Use Komm's Theorem (Exer. 15) in one case and a specific construction in the other case to determine the exact dimensions.

19. (a) Does the data of Table 1.4 define a partial order?

(b) A strict partial order?

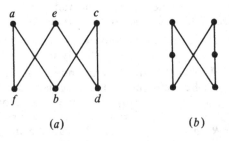

(a)   (b)

Figure 1.8. Hasse diagrams of two strict partial orders.

20. Repeat Exer. 19 for the data of Table 1.5.
21. Repeat Exer. 19 for the data of Table 1.6.
22. Repeat Exer. 19 for the data of Table 1.7.
23. Repeat Exer. 19 for the data of Table 1.8.
24. Repeat Exer. 19 for the data of Table 1.9.

1.7 Functions and Operations

Suppose A is a set. A function $f:A \to A$ can be thought of as a binary relation (A, R) with the following properties:

$$(\forall a \in A)(\exists b \in A)(aRb), \tag{1.20}$$

$$(\forall a, b, c \in A)(aRb \ \& \ aRc \Rightarrow b = c). \tag{1.21}$$

Conversely, any binary relation (A, R) satisfying (1.20) and (1.21) can be thought of as a function from A into A. More generally, a function $f:A^n \to A$ can be thought of as an $(n+1)$-ary relation (A, R) satisfying properties analogous to (1.20) and (1.21). For example, if $n=2$, these properties are as follows:

$$(\forall a, b \in A)(\exists c \in A)[(a, b, c) \in R], \tag{1.22}$$

$$(\forall a, b, c, d \in A)[(a, b, c) \in R \ \& \ (a, b, d) \in R \Rightarrow c = d]. \tag{1.23}$$

Such functions from $A \times A$ into A are sometimes called *binary operations*, or just *operations* for short. Abstractly, a ternary relation defines a binary operation if and only if it satisfies Eqs. (1.22) and (1.23).

Let us give some examples. Consider the operation $+$ of addition of real numbers. Given a pair of real numbers a and b, $+$ assigns a third real number c so that $c = a + b$. The corresponding ternary relation $\oplus$ on *Re* is defined as follows:

$$(a, b, c) \in \oplus \quad \text{iff} \quad c = a + b.$$

Thus, $(1, 2, 3) \in \oplus$ and $(1, 3, 4) \in \oplus$, but $(2, 5, 3) \notin \oplus$. The operation $\times$ of multiplication corresponds to the ternary relation $\otimes$ on *Re* defined as follows:

$$(a, b, c) \in \otimes \quad \text{iff} \quad c = a \times b.$$

To give yet another example, suppose $A = Re$ and

$$o(a, b, c) \Leftrightarrow c = a/b.$$

Then the ternary relation (A, o) is not an operation because there is no c

so that $o(1,0,c)$. Next, suppose $A = Re$ and

$$o(a, b, c) \Leftrightarrow c = \sqrt{ab} \, .$$

Then (A, o) is not an operation because there is no c so that $o(2, -2, c)$. If we again take $A = Re$ and now define

$$o(a, b, c) \Leftrightarrow c = \sqrt{|ab|} \, ,$$

then (A, o) is still not an operation, for $o(2, 2, 2)$ and $o(2, 2, -2)$. However, (A, o) is an operation if we only allow the positive square root. Another operation o on $A = Re$ can be defined by taking

$$o(a, b, c) \Leftrightarrow c = a + 2b.$$

If (A, o) is an operation and $o(a,b,c)$ holds, we usually write $c = a o b$. Thus, in our present example, $5 = 1 o 2$ and $8 = 2 o 3$.

To give two further examples, suppose $A = \{1,2,3\}$ and we define

$$R = \{(1, 1, 1), (1, 2, 2), (1, 3, 1), (2, 1, 2), (2, 2, 1),$$
$$(2, 3, 2), (3, 1, 1), (3, 2, 2), (3, 3, 1)\}$$

and

$$S = \{(1, 1, 1), (1, 2, 1), (1, 3, 1), (2, 1, 2), (2, 2, 2),$$
$$(2, 3, 2), (3, 1, 3), (3, 2, 3), (1, 3, 2)\}.$$

Then (A, R) is a binary operation: it is the operation that assigns to a and b the number 1 if $a + b$ is even and the number 2 if $a + b$ is odd. (A, S) is not a binary operation, since both $S(1, 3, 1)$ and $S(1, 3, 2)$.

To give still another example, which we shall encounter later, suppose A is the set of aircraft engines discussed above, among which we are interested in comparing subjective loudness. If a and b are two engines, let us think of $a o b$ as the object consisting of both engines placed next to each other. As far as loudness is concerned, the loudness of $a o b$ is compared to other loudnesses by running both engines at once. We might call o an operation of "combination" and speak of comparing combined loudness. Unfortunately, o does not allow us to define an operation on the set A in the precise sense we have defined. We would think of defining the ternary relation $o(a, b, c)$ which holds if and only if c is $a o b$. Unfortunately, if a and b are in A, $a o b$ is not necessarily in A, and so there is no c for which $o(a, b, c)$ holds, violating condition (1.22). Moreover, the combination $a o a$ does not make sense, and yet an operation must be defined on all pairs in $A \times A$ including pairs (a, a).

(To rectify this situation, we can think in terms of a hypothetical set B consisting of infinitely many copies of each element a of A, and to speak of a set C consisting of finite sets of elements from B. Two sets in C are considered equivalent if for each element a in A the two sets have the same number of copies of a. Then equivalence is indeed an equivalence relation. Let C^* be the collection of equivalence classes. If α and β are in C^*, pick disjoint representatives x in α and y in β and define $\alpha \circ \beta$ to be $x \cup y$. It is not hard to show that $\circ$ is well-defined and defines an operation on C^*.)

Exercises

1. Suppose A is the set of positive integers.
    (a) Show that $R(a, b, c)$ iff $a + b = c$ defines an operation on A.
    (b) Show that $S(a, b, c)$ iff $a - b = c$ does not define an operation on A.
    (c) Show that $T(a, b, c)$ iff $a + b + c = 0$ does not define an operation on A.

2. Suppose A is all of the integers.
    (a) Is the relation (A, R) of Exer. 1 an operation?
    (b) What about the relation (A, S) of Exer. 1?
    (c) What about the relation (A, T) of Exer. 1?

3. Which of the following relations (A, U) define operations?
    (a) $A = Re$, $U(a, b, c)$ iff $a + b + c = 0$.
    (b) $A = Re$, $U(a, b, c)$ iff $abc = 0$.
    (c) $A = Re$, $U(a, b, c)$ iff $a > b > c$.

4. If $A = \{0, 1, 2\}$ and

$$R = \{(0, 0, 0), (0, 1, 0), (0, 2, 0), (1, 0, 0), (1, 1, 1),$$
$$(1, 2, 2), (2, 0, 0), (2, 1, 2), (2, 2, 1)\},$$

show that (A, R) is an operation. (What operation is it?)

5. (a) If $A = \{$Los Angeles, Chicago, New York$\}$, show that the following relation on A is not a binary operation:

$$R = \{(\text{Los Angeles, Chicago}), (\text{Los Angeles, New York}),$$
$$(\text{Chicago, New York})\}.$$

(b) Which of the following relations on A is a binary operation?
    (i) S given by the following set of triples:
    (Los Angeles, Los Angeles, Chicago)
    (Los Angeles, Chicago, Chicago)
    (Los Angeles, New York, Chicago)
    (Chicago, Chicago, Chicago)
    (Chicago, New York, Chicago)
    (New York, New York, Chicago).

    (ii) $T = \{(x, y, \text{NewYork}): x, y \in A\} \cup$
                                $\{(\text{Chicago, Chicago, Chicago})\}$.

6. (a) Show that the following binary relations (A, R) are functions:
      (i) $A = Re$, $R = \{(a, a + 1): a \in A\}$.
      (ii) $A = Re$, aRb iff $b = a^2$.
      (iii) $A = Re$, aRb iff $a = b$.
   (b) Show that the following binary relations are not functions:
      (i) $A = Re$, aRb iff $a = b^2$.
      (ii) $A = Re$, aRb iff $a > b$.
   (c) Which of the following binary relations are functions?
      (i) $A = Re$, aRb iff $6a + 2b = 0$.
      (ii) $A = Re$, aRb iff $a > b + 1$.
      (iii) $A = Re$, aRb iff a divides b.

7. Which of the following quaternary relations (A, R) correspond to functions from $A \times A \times A$ to A?
   (a) $A = \{SO_2, DDT\}$
      $R = \{(SO_2, SO_2, SO_2, NO_x), (NO_x, NO_x, NO_x, SO_2)\}$.
   (b) $A = Re$, $R = \{(a, b, c, d): d = \sqrt{abc}\ \}$.
   (c) $A = Re$, $R = \{(a, b, c, d): d > a + b + c\}$.

8. Suppose R is a function on A and S is a function on A.
   (a) Is the relative product $R \circ S$ necessarily a function on A?
   (b) How does the notion of relative product compare to the usual notion of composition of two functions?

1.8 Relational Systems and the Notion of Reduction

Suppose $R_1, R_2, \ldots, R_p$ are (not necessarily binary) relations on the same set A and $\mathbf{o}_1, \mathbf{o}_2, \ldots, \mathbf{o}_q$ are binary operations on A. We shall call the $(p + q + 1)$-tuple $\mathfrak{A} = (A, R_1, R_2, \ldots, R_p, \mathbf{o}_1, \mathbf{o}_2, \ldots, \mathbf{o}_q)$ a *relational system*. Of course, we could treat binary operations as relations, and so simply speak of a relational system as a $(p + 1)$-tuple $(A, R_1, R_2, \ldots, R_p)$. However, in what follows, it will be convenient to single out the binary operations. On the other hand, we do not single out any other functions.

It will be useful to generalize the reduction or quotient procedure of Section 1.5 to relational systems $\mathfrak{A}$. The procedure is called by Scott and Suppes [1958] the *method of cosets*. Let us start with a relational system $(A, R_1, R_2, \ldots, R_p)$ having no operations. We define a relation of equivalence E on A by saying that aEb holds if and only if a and b are "perfect substitutes" for each other with respect to all the relations R_i. (Formally, a and b are *perfect substitutes* for each other with respect to an m-ary relation R_i if the following condition holds: given sequences $(a_1, a_2, \ldots, a_m)$ and $(b_1, b_2, \ldots, b_m)$ from A, if for $j = 1, 2, \ldots, m$, $a_j \neq b_j$ implies $\{a_j, b_j\} = \{a, b\}$, then

$$R_i(a_1, a_2, \ldots, a_m) \quad \text{iff} \quad R_i(b_1, b_2, \ldots, b_m).$$

This definition, and the notion of reduction defined below, appear in Scott and Suppes [1958].) It is not hard to show that if a^* is the equivalence class containing a, then the following relations R_i^* are well-defined on the collection A^* of equivalence classes.

$$R_i^*(a_1^*, a_2^*, \ldots, a_m^*) \quad \text{iff} \quad R_i(a_1, a_2, \ldots, a_m).$$

The relational system $\mathfrak{A}^* = (A^*, R_1^*, R_2^*, \ldots, R_p^*)$ is called the *reduction* or *quotient* of $\mathfrak{A}$, and is often denoted as $\mathfrak{A}/E$. The reader should verify that if (A, R) is a strict weak order, then the perfect substitutes relation E is the same as the tying relation E defined in Eq. (1.19) of Section 1.5.

Handling the general case, suppose we are given a relational system $\mathfrak{A} = (A, R_1, R_2, \ldots, R_p, o_1, o_2, \ldots, o_q)$, and aEb holds for a, b in A if and only if a and b are perfect substitutes for each other with respect to all the relations R_i. Let us define binary operations o_i^* on A^* as follows:

$$a^* \, o_i^* \, b^* = (a \, o_i \, b)^*.$$

The operation o_i^* is well-defined provided the following condition holds:

$$(aEa' \, \& \, bEb') \Rightarrow (a \, o_i \, b)E(a' \, o_i \, b'). \tag{1.24}$$

If (1.24) holds for all i, we say the relational system $\mathfrak{A}$ is *shrinkable*, and we call the relational system $\mathfrak{A}^* = (A^*, R_1^*, R_2^*, \ldots, R_p^*, o_1^*, o_2^*, \ldots, o_q^*)$ the *reduction* of $\mathfrak{A}$. We say that a relational system $\mathfrak{A}$ is *irreducible* if every equivalence class with respect to E has exactly one element.

We illustrate these ideas with an example. Let $A = \{0, 1, 2, \ldots, 26\}$, and define R and o on A by

$$aRb \quad \text{iff} \quad a \bmod 3 > b \bmod 3, \tag{1.25}$$

$$c = a \, o \, b \quad \text{iff} \quad [c \equiv a + b \pmod 3 \text{ and } c \in \{0, 1, 2\}]. \tag{1.26}$$

Thus, $2R1$, $8R10$, etc. Similarly, $2 \, o \, 2 = 1$, $6 \, o \, 3 = 0$. There are three equivalence classes under E, namely:

$$0^* = \{0, 3, 6, 9, 12, 15, 18, 21, 24\},$$

$$1^* = \{1, 4, 7, 10, 13, 16, 19, 22, 25\},$$

and

$$2^* = \{2, 5, 8, 11, 14, 17, 20, 23, 26\}.$$

We have $2^* R^* 1^*$, since $2R1$. Equation (1.24) holds, and hence o^* is well-defined, and $\mathfrak{A} = (A, R, o)$ is shrinkable. We have, for example, $2^* o^* 2^* = 1^*$, since $2 \, o \, 2 = 1$. We shall return to the idea of reduction in Chapter 2.

Exercises

1. If $\mathfrak{A} = (A, E)$ is an equivalence relation, what is the reduction $\mathfrak{A}^*$?
2. Suppose $A = Re \times Re$,

$$(a, b)R(c, d) \Leftrightarrow a > c, \tag{1.27}$$

and

$$(a, b)S(c, d) \Leftrightarrow a = c. \tag{1.28}$$

   (a) Identify the reduction (A^*, R^*).
   (b) Identify the reduction (A^*, S^*).
   (c) Identify the reduction (A^*, R^*, S^*).
3. Suppose $A = Re \times Re$, R is defined on A by Eq. (1.27), and

$$(a, b)S(c, d) \Leftrightarrow b > d.$$

Identify the reduction (A^*, R^*, S^*).

4. Suppose $A = Re$, $R = >$, and $\mathbf{o} = +$. Show that $(A, R, \mathbf{o})$ is shrinkable.

5. Are the following relational systems shrinkable? If so, find their reductions.
   (a) $(A, R, \mathbf{o})$, where $A = Re^+$, $R = >$, $\mathbf{o} = \times$.
   (b) $(A, R, \mathbf{o})$, where $A = Re \times Re$, R is defined by Eq. (1.27), and

$$(a, b)\mathbf{o}(c, d) = (a + c, b + d).$$

   (c) $(A, R, \mathbf{o}, \mathbf{o}')$, where A, R, and $\mathbf{o}$ are as in (b) and

$$(a, b)\,\mathbf{o}'\,(c, d) = (ac, bd).$$

   (d) $(A, R, S, \mathbf{o})$, where A, R, and $\mathbf{o}$ are as in (b) and S is defined by Eq. (1.28).
   (e) $(A, R, S, \mathbf{o}, \mathbf{o}')$, where A, R, S, and $\mathbf{o}$ are as in (d) and $\mathbf{o}'$ is as in (c).
   (f) $(A, R, \mathbf{o})$, where $A = \{0, 1, 2, \ldots, 26\}$, $R = >$, and $\mathbf{o}$ is defined by Eq. (1.26).

6. Show that if (A, R) is a strict weak order, then the perfect substitutes relation E is the same as the tying relation of Eq. (1.19) of Section 1.5.

7. Show that the reduction $\mathfrak{A}^*$ is always irreducible.

References

Baker, K.A., Fishburn, P.C., and Roberts, F.S., "Partial Orders of Dimension 2," *Networks*, 2 (1971), 11–28.

Davidson, R. R., and Bradley, R. A., "Multivariate Paired Comparisons: The Extension of a Univariate Model and Associated Estimation and Test Procedures," *Biometrika*, 56 (1969), 81–95.

Dusnhnik, B., and Miller, E.W., "Partially Ordered Sets," *Amer. J. Math.*, **63** (1941), 600–610.

Farris, D. R., "On the Use of Interpretive Structural Modeling to Obtain Models for Worth Assessment," in M. M. Baldwin (ed.), *Portraits of Complexity: Applications of Systems Methodologies to Social Problems*, Battelle Monograph, Battelle Memorial Institute, Columbus, Ohio, 1975, pp. 153–159.

Folgmann, E. E. E., "An Experimental Study of Composer-Preferences of Four Outstanding Symphony Orchestras, "*J. Exp. Psychol.*, **16** (1933), 709–724.

Hart, W. L., "Goal Setting for a State Environmental Agency," in M. M. Baldwin (ed.), *Portraits of Complexity: Applications of Systems Methodologies to Social Problems*, Battelle Monograph, Battelle Memorial Institute, Columbus, Ohio, pp. 89–94.

Hiraguchi, T., "On the Dimension of Orders," *Sci. Rep. Kanazawa Univ.*, **4** (1955), 1–20.

Komm, H., "On the Dimension of Partially Ordered Sets," *Amer. J. Math.*, **70** (1948), 507–520.

Rothkopf, E. Z., "A Measure of Stimulus Similarity and Errors in Some Paired Associate Learning Tasks," *J. Exp. Psychol.*, **53** (1957), 94–101.

Scott, D., and Suppes, P., "Foundational Aspects of Theories of Measurement," *J. Symbolic Logic*, **23** (1958), 113–128.

Szpilrajn, E., "Sur l'extension de l'ordre partiel," *Fund. Math.*, **16** (1930), 386–389.

Trotter, W.T., "Inequalities in Dimension Theory for Posets," *Proc. Amer. Math. Soc.*, **47** (1975), 311–315.

Trotter, W.T., "Combinatorial Problems in Dimension Theory for Partially Ordered Sets," in *Problèmes Combinatoires et Théorie des Graphes*, Colloques Internationaux C.N.R.S., Paris, 1978, pp. 403–406.

CHAPTER 2

Fundamental Measurement, Derived Measurement, and the Uniqueness Problem

2.1 The Theory of Fundamental Measurement

2.1.1 *Formalization of Measurement*

In this chapter, we introduce the theory of fundamental measurement and the theory of derived measurement, and study the uniqueness of fundamental and derived measures. Fundamental measurement deals with the measurement process that takes place at an early stage of scientific development, when some fundamental measures are first defined. Derived measurement takes place later, when new measures are defined in terms of others previously developed. In this section, we shall begin with fundamental measurement. Derived measurement will be treated in Section 2.5. Our approach to measurement follows those of Scott and Suppes [1958], Suppes and Zinnes [1963], Pfanzagl [1968], and Krantz *et al.* [1971].

Russell [1938, p. 176] defines measurement as follows: "Measurement of magnitudes is, in its most general sense, any method by which a unique and reciprocal correspondence is established between all or some of the magnitudes of a kind and all or some of the numbers, integral, rational, or real as the case may be." Campbell [1938, p. 126] says that measurement is "the assignment of numerals to represent properties of material systems other than number, in virtue of the laws governing these properties." To Stevens [1951, p. 22], "measurement is the assignment of numerals to objects or events according to rules." Torgerson [1958, p. 14] says that "measurement of a property . . . involves the assignment of numbers to systems to represent that property."

ENCYCLOPEDIA OF MATHEMATICS and Its Applications, Gian-Carlo Rota (ed.). Vol. 7: Fred S. Roberts, Measurement Theory

What then is measurement? It seems almost redundant to say that measurement has something to do with assignment of numbers. (In Section 6.1.4, however, we shall argue that measurement without numbers is a perfectly legitimate and useful activity.) All the above definitions suggest that measurement has something to do with assigning numbers that correspond to or represent or "preserve" certain observed relations. This idea fits paradigm examples of physics such as temperature or mass. In the case of temperature, measurement is the assignment of numbers that preserve the observed relation "warmer than." In the case of mass, the relation preserved is the relation "heavier than."

More precisely, suppose A is a set of objects and the binary relation aWb holds if and only if you judge a to be warmer than b. Then we want to assign a real number $f(a)$ to each $a \in A$ such that for all $a, b \in A$,

$$aWb \Leftrightarrow f(a) > f(b). \qquad (2.1)$$

Similarly, if A is a set of objects which you lift and H is the judged relation "a is heavier than b," then we would like to assign a real number $f(a)$ to each $a \in A$ such that for all $a, b \in A$,

$$aHb \Leftrightarrow f(a) > f(b). \qquad (2.2)$$

Measurement in the social sciences can be looked at in a similar manner. Thus, for example, measurement of preference is assignment of numbers preserving the observed binary relation "preferred to." If A is a set of alternatives and aPb holds if and only if you (strictly) prefer* a to b, then we would like to assign a real number $u(a)$ to each $a \in A$ such that for all $a, b \in A$,

$$aPb \Leftrightarrow u(a) > u(b). \qquad (2.3)$$

The function u is often called a *utility function* or an *ordinal utility function* or an *order-preserving utility function*, and the value $u(a)$ is called the *utility* of a. Measurement of loudness is analogous, and calls for an assignment of numbers to sounds preserving the observed relation "louder than." So is measurement of air quality: we are trying to preserve an observed relation like "the air quality on day a was better than the air quality on day b."

In the case of mass, we actually demand more of our "measure." We want it to be "additive" in the sense that the mass of the combination of two objects is the sum of their masses. Formally, we need to speak of a binary operation o on the set A of objects—think of a o b as the object

---

*Strict preference is to be distinguished from weak preference: the former means "better than," the latter "at least as good as." See Section 1.5.

obtained by placing a next to b.[†] We want a real-valued function f on A that not only satisfies condition (2.2) but also "preserves" the binary operation o, in the sense that for all $a, b \in A$,

$$f(a \circ b) = f(a) + f(b). \qquad (2.4)$$

There is no comparable operation in the case of temperature as it is commonly measured.[*] Whether there is a comparable operation in the case of preference depends on the structure of the set of alternatives being considered, and on how demanding we want to be in our measurement. We might want to allow complex alternatives like paper and pencil $(a \circ b)$, and we might want to require utility to be additive, that is, to satisfy

$$u(a \circ b) = u(a) + u(b). \qquad (2.5)$$

A utility function that is also additive is often called a *cardinal utility function*.[††]

2.1.2 *Homomorphisms of Relational Systems*

Abstracting from these examples, let us use the concept of a relational system introduced in Section 1.8. A *relational system* is an ordered $(p + q + 1)$-tuple $\mathfrak{A} = (A, R_1, R_2, \ldots, R_p, o_1, o_2, \ldots, o_q)$, where A is a set, $R_1, R_2, \ldots, R_p$ are (not necessarily binary) relations on A, and $o_1, o_2, \ldots, o_q$ are (binary) operations on A. The *type* of the relational system is a sequence $(r_1, r_2, \ldots, r_p; q)$ of length $p + 1$, where r_i is m if R_i is an m-ary relation. For example, in the case of mass, we are dealing with a relational system (A, H, o) of type $(2; 1)$. In the case of temperature, we are dealing with a relational system (A, W) of type $(2; 0)$. The relational system $\mathfrak{A} = (Re, >, \geq, +)$ has type $(2, 2; 1)$. $\mathfrak{A}$ is an example of what we shall call a *numerical relational system*, that is, one where A is the set of real numbers. A second example of a numerical relational system is $(Re, >, +, \times)$, which has type $(2; 2)$. Although we have been very general in our definition of relational systems, we shall usually deal with only a small number of relations and operations.

In our examples, we have seen that in measurement we start with an observed or empirical relational system $\mathfrak{A}$ and we seek a mapping to a numerical relational system $\mathfrak{B}$ which "preserves" all the relations and

[†]The formal difficulties are the same as those we encountered in combining aircraft engines in Section 1.7.

[*]The exception is if we are dealing with the Kelvin notion of temperature. Later on, we shall want to make other demands in the measurement of temperature.

[††]In the literature, a utility function is often called cardinal if it gives rise to a scale at least as strong as what we shall call an interval scale.

operations in $\mathfrak{A}$. For example, in measurement of mass, we seek a mapping from $\mathfrak{A} = (A, H, \mathbf{o})$ to $\mathfrak{B} = (Re, >, +)$ which "preserves" the relation H and the operation $\mathbf{o}$. In measurement of temperature, we seek a mapping from $\mathfrak{A} = (A, W)$ to $\mathfrak{B} = (Re, >)$ which "preserves" the relation W. A mapping f from one relational system $\mathfrak{A}$ to another $\mathfrak{B}$ which preserves all the relations and operations is called a *homomorphism*. To make this precise, suppose $\mathfrak{B} = (B, R_1', R_2', \ldots, R_p', \mathbf{o}_1', \mathbf{o}_2', \ldots, \mathbf{o}_q')$ is a second relational system of the same type as $\mathfrak{A}$. A function $f{:}A \to B$ is called a *homomorphism* from $\mathfrak{A}$ *into* $\mathfrak{B}$ if, for all $a_1, a_2, \ldots, a_{r_i} \in A$,

$$R_i(a_1, a_2, \ldots, a_{r_i}) \Leftrightarrow R_i'\big[\, f(a_1), f(a_2), \ldots, f(a_{r_i})\,\big], \quad i = 1, 2, \ldots, p, \quad (2.6)$$

and for all $a, b \in A$,

$$f(a\ \mathbf{o}_i\ b) = f(a)\ \mathbf{o}_i'\ f(b), \quad i = 1, 2, \ldots, q. \tag{2.7}$$

(The function f need not be one-to-one or onto.) For example, in the case of mass, a function f that satisfies Eqs. (2.2) and (2.4) gives a homomorphism from the observed relational system $(A, H, \mathbf{o})$ into the numerical relational system $(Re, >, +)$. A one-to-one homomorphism will be called an *isomorphism*. If there is a homomorphism from $\mathfrak{A}$ into $\mathfrak{B}$, we say $\mathfrak{A}$ is *homomorphic* to $\mathfrak{B}$. If there is an *onto* isomorphism, we say $\mathfrak{A}$ is *isomorphic* to $\mathfrak{B}$.

It is important to make one remark about the definition of homomorphism. If $\mathbf{o}_i$ and $\mathbf{o}_i'$ were considered ternary relations on A and B, respectively, then the condition corresponding to Eq. (2.6) would read

$$(a, b, c) \in \mathbf{o}_i \Leftrightarrow \big[\, f(a), f(b), f(c)\,\big] \in \mathbf{o}_i'. \tag{2.8}$$

Equation (2.7) is the implication $\Rightarrow$ of Eq. (2.8), but it does not correspond exactly to (2.8). We shall show this by example below.

To give a concrete example of a homomorphism, suppose $A = \{0, 1, 2, \ldots, 26\}$, and we define R on A by

$$aRb \Leftrightarrow a \bmod 3 > b \bmod 3.$$

(The reader will recall that $a \bmod 3$ is the number among 0, 1, 2 that is congruent to a, modulo 3.) Then the function $f(a) = a \bmod 3$ defines a homomorphism from $\mathfrak{A} = (A, R)$ into $\mathfrak{B} = (Re, >)$. We have $f(11) = 2, f(6) = 0$, etc. To give a second example, suppose $A = \{a, b, c, d\}$ and

$$R = \{(a, b), (b, c), (a, c), (a, d), (b, d), (c, d)\}.$$

Then a homomorphism from $\mathfrak{A} = (A, R)$ into $\mathfrak{B} = (Re, >)$ is given by

$f(a) = 4, f(b) = 3, f(c) = 2, f(d) = 1$. A second homomorphism is given by
$g(a) = 10, g(b) = 4, g(c) = 2, g(d) = 0$. Both f and g are isomorphisms.
Next, suppose $A = \{a, b, c\}$ and

$$R = \{(a, b), (b, c), (c, a)\}.$$

Then there is no homomorphism f from (A, R) into $(Re, >)$, since aRb
implies $f(a) > f(b)$, bRc implies $f(b) > f(c)$, and cRa implies $f(c) > f(a)$.

To give several additional examples, if $A = Re$ and $B = Re$, then $f(a) = - a$ gives a homomorphism from $(A, >)$ into $(B, <)$. If $A = \{0, 1, 2, \ldots\}$ and $B = \{0, 2, 4, \ldots\}$, then $f(a) = 2a$ defines a homomorphism from $\mathfrak{A} = (A, >, +)$ to $\mathfrak{B} = (B, >, +)$. A second homomorphism is
given by $g(a) = 4a$. (There is no requirement that a homomorphism must
be an onto function.) If $\mathfrak{A} = (Re, >, +)$ and $\mathfrak{B} = (Re^+, >, \times)$, where
Re^+ is the positive reals, then $f(a) = e^a$ defines a homomorphism from $\mathfrak{A}$
into $\mathfrak{B}$. For

$$a > b \Leftrightarrow e^a > e^b$$

and

$$e^{a+b} = e^a \times e^b.$$

To give still another example, suppose

$$A = \{a, b, c, d, e\}, \quad R = \{(b, c), (d, a)\}$$

and

$$B = \{\text{plane, train, car, bus, bicycle}\}, \quad S = \{(\text{car, bus}), (\text{plane, bicycle})\}.$$

Then the function $f : A \rightarrow B$ defined by $f(a) = \text{bicycle}$, $f(b) = \text{car}$, $f(c) = \text{bus}$, $f(d) = \text{plane}$, $f(e) = \text{train}$ is a homomorphism, indeed an isomorphism, from (A, R) into (B, S).

To give one final example, let $A = \{x, y\}$ and let $\mathbf{o}$ be the operation
defined as follows:

$$x \mathbf{o} x = x, \quad x \mathbf{o} y = x, \quad y \mathbf{o} x = x, \quad y \mathbf{o} y = y. \tag{2.9}$$

Then the function f defined by $f(x) = f(y) = 0$ is a homomorphism from
$(A, \mathbf{o})$ into $(Re, +)$, that is, it satisfies Eq. (2.7). However, f does not satisfy
Eq. (2.8), since $f(y) = f(x) + f(x)$, yet $y \neq x \mathbf{o} x$. Similarly, if R is the
empty relation on A, then f is a homomorphism from $(A, R, \mathbf{o})$ into
$(Re, >, +)$, even though f violates (2.8).

2.1.3 *The Representation and Uniqueness Problems*

In general, we shall say that *fundamental measurement* has been performed if we can assign a homomorphism from an observed (empirical)
relational system $\mathfrak{A}$ to some (usually specified) numerical relational system
$\mathfrak{B}$. Thus, measurement of temperature is the assignment of a homomorphism from the observed relational system (A, W) to the numerical relational system $(Re, >)$, measurement of mass is the assignment of a
homomorphism from the observed relational system $(A, H, \mathbf{o})$ to the
numerical relational system $(Re, >, +)$, and so on.

The difficult philosophical question—not a mathematical question—is
the specification of the numerical relational system $\mathfrak{B}$. Why try to find a
homomorphism into one relational system rather than another? The
answer will depend on a combination of intuition and theory about what is
being measured, and desired properties of the numerical assignment. If we
have a homomorphism, the homomorphism is said to give a *representation*,
and the triple $(\mathfrak{A}, \mathfrak{B}, f)$ will be called a *scale*, though sometimes we shall be
sloppy and refer to f alone as the scale. The reader may wish to formulate
in these terms other measurement problems, for example, measurement of
length, of area, or of height.*

The first basic problem of measurement theory is the *representation
problem*: Given a particular numerical relational system $\mathfrak{B}$, find conditions
on an observed relational system $\mathfrak{A}$ (necessary and) sufficient for the
existence of a homomorphism from $\mathfrak{A}$ into $\mathfrak{B}$. The emphasis is on finding
sufficient conditions. If all the conditions in a collection of sufficient
conditions are necessary as well, that is all the better. A more important
criterion is that the conditions be "testable" or empirically verifiable in
some sense. (Often it is desired that the conditions be statable in the form
of a law expressible using only universal conditionals.) In any case, the
conditions are usually called *axioms* for the representation, and the theorem stating their sufficiency is usually called a *representation theorem*. If
possible, the proof of a representation theorem should be *constructive*; it
should not only show us that a representation is possible, but it should
show us how actually to construct it. A typical axiom for a representation
theorem is the following. Suppose we seek a homomorphism f from (A, R)
into $(Re, >)$. If such a homomorphism f exists, it follows that (A, R) is
transitive. For if aRb and bRc, then $f(a) > f(b)$ and $f(b) > f(c)$, from which
$f(a) > f(c)$ and aRc follow. Thus, a typical measurement axiom is the
requirement that (A, R) be transitive.

---

*We have chosen to allow all possible real numbers as scale values. In practical measure-
ment, of course, only rational numbers are ever needed. However, it is convenient theoretically to allow all possible real values. Later on, we shall modify our position, and consider
scales that assign mathematical objects other than numbers.

In Sections 3.1, 3.2, and 3.3, we shall present some basic representation theorems. Further representation theorems will be proved in other chapters.

The axioms for a representation give conditions under which measurement can be performed. The axioms can be thought of as giving a foundation on which the process of measurement is based. In a less global sense, the axioms can also be thought of as conditions that must be satisfied in order for us to organize data in a certain way. In any case, it is important to be able to state such foundational axioms, at least for measurement in the social sciences. For we must know under what circumstances certain kinds of scales of measurement can be produced. In the physical sciences, the situation is different. We by now have well-developed scales of measurement, and writing down a representation theorem for these scales is often more a theoretical exercise than a significant practical development.

The second basic problem of measurement theory is the *uniqueness problem*: How unique is the homomorphism f? We shall see later in this chapter that a uniqueness theorem tells us what kind of scale f is, and gives rise to a theory of meaningfulness of statements involving scales. In particular, a uniqueness theorem puts limitations on the mathematical manipulations that can be performed on the numbers arising as scale values. As Hays [1973, p. 87] points out, one can always perform mathematical operations on numbers (add them, average them, take logarithms, etc.). However, the key question is whether, after having performed such operations, one can still deduce true (or better, meaningful) statements about the objects being measured.

Sometimes we shall start with a desired uniqueness result and work backwards. That is, we shall seek a representation that will lead to that result. For example, in measurement of temperature, we shall start with the observation that temperature is measured up to determination of an origin and of a unit, and ask what representations will give rise to a scale of temperature that has these properties.

This chapter will emphasize the uniqueness problem and applications of the theory of uniqueness. In Chapter 3 we shall begin our study of the representation problem.

It should be pointed out in closing this subsection that not all of the theory of measurement, even that in the spirit of this work, fits perfectly into the framework we have described. One sometimes deals with systems formally different from the relational systems we have defined—for example, systems having several underlying sets (see Section 5.6), or having operations defined only on subsets of the underlying set (see, for example, Krantz *et al.* [1971, Section 3.4]), or having sets with structure such as Boolean algebras and vector spaces (see, for example, Narens [1974a] and Domotor [1969]). One sometimes modifies the notion of homo-morphism—for example, using Eq. (2.8) in place of Eq. (2.7), or modifying

(2.6) to read

$$R_i(a_1, a_2, \ldots, a_{r_i}) \Rightarrow R'_i\big[\, f(a_1), f(a_2), \ldots, f(a_{r_i})\,\big].$$

(See, for example, Adams [1965].) One sometimes deals with representations into relational systems where the underlying set is not the set of real numbers. This can involve as simple a change as using the rationals rather than the reals, or it can be as complex as using the nonstandard reals or systems with non-Archimedean properties. (See, for example, Skala [1975] or Narens [1974a,b].) It can also involve maps of objects into sets of numbers such as intervals, or into geometric figures such as rectangles and circles (see Section 6.1.4). Not as much work has been done in any of the directions mentioned as in the specific framework we have outlined for measurement. However, it will probably be fruitful to develop many of these alternative directions in the future. In all these directions, the central idea is unchanged. This is the notion of representation, the translation of "qualitative" concepts such as relations and operations into appropriate "quantitative" (or other concrete) relations and operations enjoying known properties. As we pointed out in the Introduction, it is from a representation that we can learn about empirical phenomena, by applying the concepts and theories that have been developed for the representing relations to the represented ones.

Exercises

1. (a) Show in each of the following that the relational system $\mathfrak{A}$ is homomorphic to the relational system $\mathfrak{B}$ ($2N$ is the set of even positive integers):

	$\mathfrak{A}$	$\mathfrak{B}$
(i)	$(N, >)$	$(2N, >)$
(ii)	$(2N, \geqq)$	$(N, \geqq)$
(iii)	$(N, >, +)$	$(2N, >, +)$
(iv)	$(Re, +)$	$(Re^+, \times)$

(b) Determine which of the following relational systems $\mathfrak{A}$ is homomorphic to the corresponding relational system $\mathfrak{B}$:

	$\mathfrak{A}$	$\mathfrak{B}$
(i)	$(N, =)$	$(2N, =)$
(ii)	$(Re, +)$	$(Re, -)$
(iii)	$(Z, >)$	$(Z, <)$
(iv)	$(N, >)$	$(Z, <)$
(v)	$(Re, =)$	$(Re, \neq)$
(vi)	$(Re^+, >, \times)$	$(Re, >, +)$

2. Show that there is no homomorphism from $(N, >)$ into $(N, <)$.

3. Recall that a homomorphism f is an isomorphism if it is one-to-one.
   (a) Show that there is an isomorphism from $(N, >)$ into $(2N, >)$.
   (b) Show that if $A = \{1, 2, 3\}$ and R is $\{(1, 2), (1, 3)\}$, then (A, R) is homomorphic to $(Re, >)$, but not isomorphic to $(Re, >)$.

4. Suppose $A = B = Re$, $D(a, b, c, d)$ holds on A iff $a + b > c + d$, and $K(a, b, c, d)$ holds on B iff $a + b < c + d$. Show that (A, D) is homomorphic to (B, K).

5. Suppose $A = \{$Tom, Dick, Harry, John$\}$, and

$R = \{$(Tom, Dick, John),

(Tom, Dick, Harry), (Tom, John, Harry), (Dick, John, Harry)$\}$.

Let $B(u, v, w)$ hold on Re if and only if $u < v < w$. Is (A, R) homomorphic to (Re, B)?

6. (a) Suppose $A = \{0, 1, 2, \ldots, 26\}$ and R on A is defined by

aRb iff $a \bmod 3 > b \bmod 3$.

(For the definition of $a \bmod 3$, see Section 1.5.) Suppose

$c = a \mathbf{o} b$ iff $[c \equiv a + b (\bmod 3) \text{ and } c \in \{0, 1, 2\}]$.

Let $f(a) = a \bmod 3$. Show that f is a homomorphism from $(A, R, \mathbf{o})$ into $(N, >, \mathbf{o})$.
   (b) Is f a homomorphism from $(A, R, \mathbf{o})$ into $(Re, >, +)$?

7. Formalize in terms of relational systems the theory of measurement of length, area, or height.

2.2 Regular Scales

2.2.1 Definition of Regularity

In this section, we begin with four statements and consider which seem to make sense. In the remainder of this chapter, we shall develop a theory of scale type and meaningfulness that accounts for our observations, and apply the theory to a wide variety of more complex problems. The theory is aimed at telling us what manipulations of scale values are appropriate in the sense that they lead to results that have unambiguous interpretations as statements about the objects or phenomena being investigated.

The four statements we shall consider are the following:

Statement I: The number of cans of corn in the local supermarket at closing time yesterday was at least 10.

Statement II: One can of corn weighs at least 10.

Statement III: One can of corn weighs twice as much as a second.

Statement IV: The temperature of one can of corn at closing time yesterday was twice as much as that of a second can.

Statement I seems to make sense but Statement II does not, for the number of cans is specified without reference to a particular scale of measurement, whereas the weight of a can is not. Similarly, Statement III seems to make sense but Statement IV does not, for the ratio of weights is the same regardless of the scale of measurement used (if one can has twice as many grams as another, it has twice as many ounces, pounds, kilograms, etc.); whereas the ratio of temperatures is not necessarily the same (if one can is twice as many degrees Fahrenheit as another, it is not twice as many degrees centigrade). To be an adequate description of what we mean by measurement, a theory must account for observations such as these.

In general, to account for such observations, we shall study the uniqueness of the numerical assignment (homomorphism) involved. It is quite possible, given two relational systems $\mathfrak{A}$ and $\mathfrak{B}$ of the same type, for there to be several different functions that map $\mathfrak{A}$ homomorphically into $\mathfrak{B}$. Since this is the case, any statement about measurement should either specify which scale (which homomorphism) is being used or be true independent of scale.

To make this statement more precise, let us recall that a scale is a triple $(\mathfrak{A}, \mathfrak{B}, f)$ where $\mathfrak{A}$ and $\mathfrak{B}$ are relational systems and f is a homomorphism from $\mathfrak{A}$ into $\mathfrak{B}$. Let us call this scale a *numerical scale* if $\mathfrak{B}$ is a numerical relational system, that is, the set underlying $\mathfrak{B}$ is *Re*. We are interested in the uniqueness of the numerical assignment f. Specifically, we shall say that a statement involving numerical scales is *meaningful* if its truth (or falsity) remains unchanged if every scale $(\mathfrak{A}, \mathfrak{B}, f)$ involved is replaced by another (acceptable) scale $(\mathfrak{A}, \mathfrak{B}, g)$. Meaningful statements are unambiguous in their interpretation. Moreover, they say something significant about the fundamental relations among the objects being measured, whereas statements that are dependent on a particular, arbitrary choice of scale do not.

Meaningfulness can be studied by analyzing admissible transformations of scale. Suppose f is one homomorphism from a relational system $\mathfrak{A}$ into a relational system $\mathfrak{B}$, and suppose A is the set underlying $\mathfrak{A}$ and B is the set underlying $\mathfrak{B}$. Suppose ϕ is a function that maps the range of f, the set

$$f(A) = \{ f(a) : a \in A \},$$

into set B. Then the composition $\phi \circ f$ is a function from A into B. If $\phi \circ f$ is a homomorphism from $\mathfrak{A}$ into $\mathfrak{B}$, we call ϕ an *admissible transformation of scale*. For example, suppose $\mathfrak{A} = (N, >)$, $\mathfrak{B} = (Re, >)$, and $f: N \rightarrow Re$

is given by $f(x) = 2x$. Then f is a homomorphism from $\mathfrak{A}$ into $\mathfrak{B}$. If $\phi(x) = x + 5$, then $\phi \circ f$ is also a homomorphism from $\mathfrak{A}$ into $\mathfrak{B}$, for we have $(\phi \circ f)(x) = 2x + 5$, and

$$x > y \quad \text{iff} \quad 2x + 5 > 2y + 5.$$

Thus, $\phi{:}f(A) \to B$ is an admissible transformation of scale. However, if $\phi(x) = -x$ for all $x \in f(A)$, then ϕ is not an admissible transformation, for $\phi \circ f$ is not a homomorphism from $\mathfrak{A}$ into $\mathfrak{B}$.

If $(\mathfrak{A}, \mathfrak{B}, f)$ is any scale and $(\mathfrak{A}, \mathfrak{B}, g)$ is any other scale, then it is sometimes possible to find a function $\phi{:}f(A) \to B$ so that $g = \phi \circ f$. For example, if $\mathfrak{A}$, $\mathfrak{B}$, and f are as above and $g(x) = 7x$, then $\phi(x) = \frac{7}{2}x$ will suffice. If for every scale $(\mathfrak{A}, \mathfrak{B}, g)$, there is a transformation $\phi{:}f(A) \to B$ such that $g = \phi \circ f$, then we shall call the *scale* $(\mathfrak{A}, \mathfrak{B}, f)$ *regular*. If every homomorphism f from $\mathfrak{A}$ into $\mathfrak{B}$ is regular, we shall call the *representation* $\mathfrak{A} \to \mathfrak{B}$ *regular*. A representation $\mathfrak{A} \to \mathfrak{B}$ is regular if, given any two scales f and g, we can map each into the other by an admissible transformation. Almost all the representations we shall encounter in this volume are regular. For the regular representation, there will be a very nice theory of uniqueness, of meaningfulness, and of scale type. In particular, if every scale in a statement is regular, we may use the following simpler definition of meaningfulness: A statement involving (numerical) scales is *meaningful* if and only if its truth or falsity is unchanged under admissible transformations of all the scales in question. If not every scale is regular, we may salvage this definition of meaningfulness by generalizing our notion of admissible transformation. See Roberts and Franké [1976]. This modified definition of meaningfulness is originally due to Suppes [1959] and Suppes and Zinnes [1963]. For variants of it, see Robinson [1963], Adams, Fagot, and Robinson [1965], Pfanzagl [1968], and Luce [1978]. We shall test the definition against the examples given at the beginning of this section and then apply it to other, more complex examples.

Before doing this, however, we present two examples of irregular scales. Define a binary relation $>_1$ on Re by

$$x >_1 y \quad \text{iff} \quad x > y + 1. \tag{2.10}$$

Let

$$A = \{r, s, t\} \quad \text{and} \quad R = \{(r, s), (r, t)\}. \tag{2.11}$$

Let

$$f(r) = 2, \ f(s) = 0, \ f(t) = 0, \ g(r) = 2, \ g(s) = 0.1, \ g(t) = 0. \tag{2.12}$$

Then f and g are homomorphisms from $\mathfrak{A} = (A, R)$ into $\mathfrak{B} = (Re, >_1)$, as is easy to verify. However, there is no function $\phi:f(A) \to Re$ such that $g = \phi \circ f$. For $(\phi \circ f)(s) = (\phi \circ f)(t)$, while $g(s) \neq g(t)$. (Homomorphisms into $(Re, >_1)$ will play a crucial role in Section 6.1 in our study of utility functions if indifference is not transitive.)

To give a second example of an irregular scale, define a binary relation M on Re by

$$xMy \quad \text{iff} \quad [x = y - 1 \text{ or } y = x - 1 \text{ or } x = y]. \tag{2.13}$$

Let

$$A = \{r, s\} \text{ and } R = \{(r, r), (s, s), (r, s), (s, r)\}. \tag{2.14}$$

Let

$$f(r) = 0, \ f(s) = 0, \ g(r) = 0, \ g(s) = 1. \tag{2.15}$$

Then f and g are homomorphisms from $\mathfrak{A} = (A, R)$ into $\mathfrak{B} = (Re, M)$, as the reader can readily verify. However, there is no function $\phi:f(A) \to Re$ such that $g = \phi \circ f$. For $(\phi \circ f)(r) = (\phi \circ f)(s)$, while $g(r) \neq g(s)$.

These two examples suggest the following characterization of regular scales. See Exer. 10 for a related result.

THEOREM 2.1 (Roberts and Franke [1976]). $(\mathfrak{A}, \mathfrak{B}, f)$ *is regular if and only if for every other homomorphism g from $\mathfrak{A}$ into $\mathfrak{B}$, and for all a, b in A, $f(a) = f(b)$ implies $g(a) = g(b)$.*

Proof. If f is regular, then there is a function $\phi:f(A) \to B$ such that $g = \phi \circ f$. Thus, $f(a) = f(b)$ implies $g(a) = \phi[f(a)] = \phi[f(b)] = g(b)$. Conversely, given g, define $\phi[f(a)] = g(a)$. Then, since $f(a) = f(b)$ implies $g(a) = g(b)$, ϕ is well-defined. Moreover, $g = \phi \circ f$, so f is regular. ∎

COROLLARY. *Every isomorphism is regular.*

2.2.2 Reduction and Regularity

The Corollary to Theorem 2.1 provides us with a means for avoiding the difficulty posed by irregular scales. The idea is to use the process of reduction described in Section 1.8 to guarantee that all homomorphisms are isomorphisms. We briefly sketch the idea.

Given a relational system $\mathfrak{A} = (A, R_1, R_2, \ldots, R_p, \mathbf{o}_1, \mathbf{o}_2, \ldots, \mathbf{o}_q)$, we define a binary relation E, the "perfect substitutes" relation, on A. Then if

$$(aEa' \ \& \ bEb') \Rightarrow (a \ \mathbf{o}_i \ b)E(a' \ \mathbf{o}_i \ b'), \quad \text{all } i, \tag{2.16}$$

we say $\mathfrak{A}$ is *shrinkable* and define a *reduction* $\mathfrak{A}^*$ of $\mathfrak{A}$. $\mathfrak{A}^*$ is a relational system $(A^*, R_1^*, R_2^*, \ldots, R_p^*, \mathbf{o}_1^*, \mathbf{o}_2^*, \ldots, \mathbf{o}_q^*)$ with relations R_i^* and operations $\mathbf{o}_i^*$ defined on the set A^* of equivalence classes under E. See Section 1.8 for details.

It is interesting to observe that if $\mathfrak{A}$ is irreducible—that is, if every equivalence class with respect to E has exactly one element—then it is shrinkable and $\mathfrak{A}$ is isomorphic to $\mathfrak{A}^*$ using the function $f(a) = a^*$. Moreover, it is interesting to note that if f is a homomorphism from a relational system $\mathfrak{A}$ into a relational system $\mathfrak{B}$, then $f(a) = f(b)$ implies aEb. Thus, if $\mathfrak{A}$ is irreducible, then every homomorphism f from $\mathfrak{A}$ into $\mathfrak{B}$ is an isomorphism, for $f(a) = f(b)$ implies aEb, which implies $a = b$. Finally, suppose $\mathfrak{A}$ is homomorphic to $\mathfrak{B}$ via a homomorphism f, and $\mathfrak{A}$ is shrinkable. Then we can find a homomorphism F from $\mathfrak{A}^*$ to $\mathfrak{B}$ by letting $F(a^*)$ be $f(a)$ for some representative a of a^*.

These ideas provide us with a way of avoiding irregular scales. If $\mathfrak{A}$ is irreducible, we know that every homomorphism from $\mathfrak{A}$ to a relational system $\mathfrak{B}$ is an isomorphism, and so must be regular by the Corollary to Theorem 2.1. If $\mathfrak{A}$ is not irreducible but is shrinkable, and $\mathfrak{A}$ is homomorphic to $\mathfrak{B}$, then $\mathfrak{A}^*$ is homomorphic to $\mathfrak{B}$. Since $\mathfrak{A}^*$ is always irreducible (Exer. 7, Section 1.8), every homomorphism from $\mathfrak{A}^*$ to $\mathfrak{B}$ is an isomorphism, and so the representation $\mathfrak{A}^* \to \mathfrak{B}$ is regular. Thus, to guarantee that all scales in question are regular, it is sufficient first to reduce the relational systems in question by canceling out the perfect substitutes relation E.

Let us illustrate these ideas by taking $A = \{0, 1, 2, \ldots, 26\}$ and defining R and $\mathbf{o}$ on A by

$$aRb \quad \text{iff} \quad a \bmod 3 > b \bmod 3,$$

$$c = a \, \mathbf{o} \, b \quad \text{iff} \quad [\, c \equiv a + b \,(\bmod 3) \text{ and } c \in \{0, 1, 2\} \,],$$

where $a \bmod 3$ is defined as that number in $\{0, 1, 2\}$ which is congruent to a modulo 3. Then, as we pointed out in Section 1.8, $\mathfrak{A} = (A, R, \mathbf{o})$ is shrinkable. If $\mathbf{o}'$ is addition mod 3 on $\{0, 1, 2\}$, then it is easy to see that $\mathfrak{A}$ is homomorphic to

$$\mathfrak{B} = (\{0, 1, 2\}, >, \mathbf{o}')$$

by the homomorphism $f(a) = a \bmod 3$. Moreover, $\mathfrak{A}^*$ is isomorphic to $\mathfrak{B}$ by the function $F(0^*) = f(0) = 0$, $F(1^*) = f(1) = 1$, $F(2^*) = f(2) = 2$. The scale $(\mathfrak{A}^*, \mathfrak{B}, F)$ is regular.

Exercises

1. (a) Suppose $A = N$ and $B = Q$. Then $(A, >)$ is homomorphic to $(B, >)$ by the homomorphism $f(a) = 2a$.

(i) Show that the function $\phi(x) = 4x + 3$ is an admissible transformation of scale.

(ii) Show that the function $\phi(x) = 2^x$ is also admissible.

(iii) Show that the function $\phi(x) = -x + 10$ is not admissible.

(b) Suppose $A = N$, $B = Q$, and $f(a) = -a$ is a homomorphism from $(A, >)$ to $(B, <)$. Which of the following transformations are admissible?

(i) $\phi(x) = 4x + 3$.

(ii) $\phi(x) = 2^x$.

(iii) $\phi(x) = -x + 10$.

2. The function $f(x) = e^x$ is a homomorphism from $(Re, +)$ to $(Re, \times)$. Which of the following ϕ are admissible transformations of scale?

(a) $\phi(x) = 2x$.

(b) $\phi(x) = x + 5$.

(c) $\phi(x) = 2^x$.

3. Suppose $A, R, \mathbf{o}$, and f are defined as in Exer. 6, Section 2.1, and f is considered a homomorphism from $(A, R, \mathbf{o})$ into $(N, >, \mathbf{o})$.

(a) Show that one admissible transformation of f is the function ϕ defined by $\phi(0) = 0$, $\phi(1) = 1$, $\phi(2) = 5$.

(b) Which of the following transformations are admissible?

(i) $\phi(x) = x + 2$.

(ii) $\phi(x) = 2x$.

(iii) $\phi(x) = 3x$.

(iv) $\phi(x) = x^3$.

(c) If f is instead considered a homomorphism from $(A, R, \mathbf{o})$ into $(B, >, \mathbf{o})$, where $B = \{0, 1, 2\}$, what are all admissible transformations ϕ?

4. (a) Suppose $A = \{x, y, z\}$ and $R = \{(x, y), (z, y)\}$. Let S be defined on Re by

$$uSv \Leftrightarrow u > v + 2.$$

Show that the following functions are homomorphisms from (A, R) into (Re, S) and hence that $((A, R), (Re, S), f)$ is irregular:

$$f(x) = 9, \ f(y) = 0, \ f(z) = 9,$$
$$g(x) = 9, \ g(y) = 0, \ g(z) = 10.$$

(b) Identify the reduction (A^*, R^*) and find a regular scale $((A^*, R^*), (Re, S), F)$.

5. (a) Define S as in Exer. 4. Let $A = \{x, y, z, w\}$ and

$$T = \{(x, z), (x, w), (y, z), (y, w)\}.$$

Show that (A, T) is homomorphic to (Re, S) and that there is a homomorphism f which defines an irregular scale.

(b) Identify the reduction (A^*, T^*) and find a homomorphism from (A^*, T^*) into (Re, S).

6. Define Δ on Re by

$$x \Delta y \quad \text{iff} \quad |x - y| < 1.$$

Let

$$A = \{r, s, t\}, \quad R = \{(r, r), (s, s), (t, t), (r, s), (s, r)\},$$

and let

$$f(r) = 0, \, f(s) = 0, \, f(t) = 10, \, g(r) = 0, \, g(s) = 1/2, g(t) = 10.$$

Conclude that $((A, R), (Re, \Delta), f)$ is irregular.

7. Suppose $\mathbf{o}$ is defined on Re by $x \mathbf{o} y = \min\{x, y\}$. Let $f : Re \to Re$ be defined by

$$f(x) = \begin{cases} -1 & \text{if } x < 0, \\ 0 & \text{if } x = 0, \\ 1 & \text{if } x > 0. \end{cases}$$

(a) Show that f is a homomorphism from $\mathfrak{A} = (Re, \mathbf{o})$ into $\mathfrak{B} = (Re, \mathbf{o})$.

(b) Show that $(\mathfrak{A}, \mathfrak{B}, f)$ is an irregular scale.

(c) Is $\mathfrak{A}$ shrinkable? If so, find $\mathfrak{A}^*$ and a homomorphism F from $\mathfrak{A}^*$ into $\mathfrak{B}$.

8. Suppose $A = \{x, y\}$ and $\mathbf{o}$ is defined as in Eq. (2.9). Show the following:

(a) $f(x) = 0, f(y) = 1$ defines a homomorphism from $\mathfrak{A} = (A, \mathbf{o})$ into $\mathfrak{B} = (Re, \times)$.

(b) $(\mathfrak{A}, \mathfrak{B}, f)$ is regular.

(c) However, the representation $\mathfrak{A} \to \mathfrak{B}$ is irregular.

9. Given a system $\mathfrak{A} = (A, R_1, R_2, \ldots, R_p, \mathbf{o}_1, \mathbf{o}_2, \ldots, \mathbf{o}_q)$, suppose we define a binary relation E' on A by $aE'b$ if and only if a and b are perfect substitutes for each other with respect to all the relations R_i and with respect to all the binary operations $\mathbf{o}_i$ considered as ternary relations. Show that if f is a homomorphism from $\mathfrak{A}$ to $\mathfrak{B}$, then $f(a) = f(b)$ does not imply that $aE'b$. [The example of Eq. (2.9) illustrates this point.]

10. (Roberts and Franke [1976]) Let E be the perfect substitutes relation. Prove that a representation $\mathfrak{A} \to \mathfrak{B}$ is regular if and only if, for every homomorphism f from $\mathfrak{A}$ into $\mathfrak{B}$, aEb implies $f(a) = f(b)$.

2.3 Scale Type

If a representation $\mathfrak{A} \to \mathfrak{B}$ is regular—that is, if all scales $(\mathfrak{A}, \mathfrak{B}, f)$ are regular—then the class of admissible transformations defines how unique each such scale is, and can be used to define *scale type*. We shall assume throughout this section (unless mentioned otherwise) that all scales in question come from regular representations. The idea of defining scale type from the class of admissible transformations is due to S. S. Stevens [1946, 1951, 1959].

Table 2.1 gives several examples of scale types. The simplest example of a scale is where the only admissible transformation is $\phi(x) = x$. There is only one way to measure things in this situation. Such a scale is called *absolute*. Counting is an example of an absolute scale.

To give a second example, let us suppose the admissible transformations are all the functions $\phi: f(A) \to B$ of the form $\phi(x) = \alpha x, \alpha > 0$. Such a

Table 2.1. Some Common Scale Types*

Admissible Transformations	Scale Type	Example
$\phi(x) = x$ (identity)	Absolute	Counting
$\phi(x) = \alpha x, \quad \alpha > 0$ Similarity transformation	Ratio	Mass Temperature on the Kelvin scale Time (intervals) Loudness (sones)[†] Brightness (brils)[†]

$\phi(x) = \alpha x + \beta, \quad \alpha > 0$
Positive linear
    transformation
	Interval	Temperature
		    (Fahrenheit,
		    centigrade, etc.)
		Time (calendar)
		Intelligence tests,
		    "standard scores" ?

$x \geqq y \quad \text{iff} \quad \phi(x) \geqq \phi(y)$
(Strictly) monotone
    increasing transformation
	Ordinal	Preference?
		Hardness
		Air quality
		Grades of leather,
		    lumber, wool, etc.
		Intelligence tests,
		    raw scores

Any one-to-one ϕ	Nominal	Number uniforms Label alternative plans Curricular codes

*See Table 2.2 for some other scale types.
[†]According to the work of S. S. Stevens—see Chapter 4.

function ϕ is called a *similarity transformation*, and a scale with the similarity transformations as its class of admissible transformations is called a *ratio scale*. Mass defines a ratio scale, as we can fix a zero point and then change the unit of mass by multiplying by a positive constant. Thus, for example, we change from grams to kilograms by multiplying by 1000. The term ratio scale arose because ratios of quantities on a ratio scale—for example, mass—make sense. Temperature also defines a ratio scale if we allow absolute zero, as in the Kelvin scale. Intervals of time (in minutes, hours, etc.) define a ratio scale. According to Stevens (see Chapter 4), various sensations such as loudness and brightness can also be measured in ratio scales.

To give a third example, suppose we let the class of admissible transformations be all functions $\phi: f(A) \to B$ of the form $\phi(x) = \alpha x + \beta, \alpha > 0$. Such a function is called a *positive linear transformation*, and a corresponding scale is called an *interval scale*. Temperature (as it is commonly measured) is an example of an interval scale. We vary the 0 point (this amounts to changing β) and also the unit (this amounts to changing α). In this way, we can change, for example, from Fahrenheit to centigrade. (We take $\alpha = 5/9$ and $\beta = -160/9$.) Time on the calendar (for example, the year 1980) defines an interval scale. It is often argued that the "standard scores" from an intelligence test define an interval scale (Stevens [1959, p. 25]).

Some scales are unique only up to order. For example, the scale of air quality being used in a number of cities is such a scale. It assigns a number 1 to unhealthy air, 2 to unsatisfactory air, 3 to acceptable air, 4 to good air, and 5 to excellent air. We could just as well use the numbers 1, 7, 8, 15, 23, or the numbers 1.2, 6.5, 8.7, 205.6, 750, or any numbers that preserve the order. If a scale is unique only up to order, the admissible transformations are monotone increasing functions $\phi(x)$, that is, functions $\phi: f(A) \to B$ satisfying the condition that

$$x \geqq y \Leftrightarrow \phi(x) \geqq \phi(y),$$

or equivalently the condition

$$x > y \Leftrightarrow \phi(x) > \phi(y).$$

Such scales are called *ordinal scales*. Another example of an ordinal scale is the Mohs scale of hardness. Numbers are assigned to minerals, reflecting their relative hardness subject to the restriction that mineral a gets a larger number than mineral b if and only if mineral a is harder than mineral b. (In practice, a is judged to be harder than b if a scratches b.) Raw scores on an intelligence test probably define only an ordinal scale (Stevens [1959, p. 25]). Later we shall see that preference may be no more than an ordinal scale.

Finally, in some scales, all one-to-one functions ϕ define admissible transformations. Such scales are called *nominal*. Examples of nominal scales are numbers on the uniforms of baseball players or the numbering of alternative plans as plan 1, plan 2, etc. Many coding systems such as curricular codes used in college catalogues to identify the department define nominal scales. The actual number has no significance, and any change of numbers will contain the same information: identification of the elements of the set A.

In general, the scale types listed in Table 2.1 go from "strongest" to "weakest," in the sense that absolute scales and ratio scales contain much more information than ordinal scales or nominal scales. It is often a goal of measurement to obtain as strong a scale as possible.

It is interesting to note that measurement can progress from lower to higher scale types. Stevens [1959, p. 24] gives a very nice discussion of this point. Early men probably distinguished only between cold and warm, thus using a nominal scale. Later, degrees of warmer and colder might have been introduced, corresponding to various natural events. This would give an ordinal scale. Later, introduction of thermometers led to interval scales of temperature. Finally, the development of thermodynamics led to a ratio scale of temperature, the Kelvin scale.

There are several less common scale types which are important in the social sciences, and we mention them briefly. (See Table 2.2.) A scale is called a *log-interval scale* if the admissible transformations are functions of the form αx^{β}, $\alpha, \beta > 0$. Log-interval scales are important in psychophysics, where they are considered as scale types for the psychophysical functions relating a physical quantity (for example, intensity of a sound) to a psychological quantity (for example, loudness of a sound). We shall encounter these psychophysical functions and log-interval scales in Chapter 4.

Another less common scale type is the *difference scale*. Here, the admissible transformations are functions of the form $\phi(x) = x + \beta$. We shall not encounter many difference scales in this volume. Suppes and Zinnes [1963, Section 4.2] give an example from the psychological literature, the so-called Thurstone Case V scale, which is a measure of response strength. Difference scales also arise when we make logarithmic transformations of ratio scales. For example, if $f(x)$ measures the mass of x, then f is unique

Table 2.2. Some Other Scale Types

Admissible Transformations	Scale Type	Example
$\phi(x) = \alpha x^{\beta}$, $\alpha, \beta > 0$	Log-interval	Psychophysical function
$\phi(x) = x + \beta$	Difference	Thurstone Case V

up to multiplication by a positive constant α. But $\log f$ is unique to addition of a constant β, for multiplication of $f(x)$ by $\alpha > 0$ corresponds to addition to $\log f(x)$ of $\beta = \log \alpha$. (Similarly, log-interval scales correspond to exponential transformations of interval scales.)

To illustrate the definition of scale type, let $A = \{r, s, t\}$ and let $R = \{(r, s), (s, t), (r, t)\}$. Then (A, R) is homomorphic to $(Re, >)$. One homomorphism is given by $f(r) = 2$, $f(s) = 1$, $f(t) = 0$. By the Corollary to Theorem 2.1, $((A, R), (Re, >), f)$ is regular. Moreover, $\phi: f(A) \to Re$ is admissible if and only if $(\phi \circ f)(r) > (\phi \circ f)(s)$, $(\phi \circ f)(s) > (\phi \circ f)(t)$, and $(\phi \circ f)(r) > (\phi \circ f)(t)$ — that is, if and only if $\phi(2) > \phi(1)$, $\phi(1) > \phi(0)$, and $\phi(2) > \phi(0)$. Thus, ϕ is admissible if and only if ϕ is a monotone increasing function on $f(A)$. Thus, f is an ordinal scale.

Before closing this section, let us observe that if $\mathfrak{A} \to \mathfrak{B}$ is not a regular representation, then the notion of scale type runs into trouble. For example, there can be two homomorphisms f and g from $\mathfrak{A}$ to $\mathfrak{B}$ such that one is one type of scale and the other is another. However, if $\mathfrak{A} \to \mathfrak{B}$ is a regular representation, this problem does not arise.[*]

THEOREM 2.2 (Roberts and Franke [1976]). *If the representation $\mathfrak{A} \to \mathfrak{B}$ is regular and f and g are homomorphisms from $\mathfrak{A}$ to $\mathfrak{B}$, then f is an absolute, ratio, interval, ordinal, or nominal scale if and only if g is, respectively, an absolute, ratio, interval, ordinal, or nominal scale.*

Proof. We shall give the proof for the ordinal case. Suppose f is an ordinal scale. By regularity, we know that $g = \phi \circ f$, for some $\phi: f(A) \to B$. Since g is a homomorphism, ϕ is an admissible transformation of f. Moreover, since f is an ordinal scale, ϕ must be monotone increasing. To show that g is an ordinal scale, suppose first that $\phi': g(A) \to B$ is monotone increasing. Then ϕ' is an admissible transformation of g. For $\phi' \circ g = \phi' \circ (\phi \circ f) = (\phi' \circ \phi) \circ f$. Since $\phi' \circ \phi$ is monotone increasing, it is an admissible transformation of f, and we conclude that $\phi' \circ g = (\phi' \circ \phi) \circ f$ is a homomorphism from $\mathfrak{A}$ into $\mathfrak{B}$. Thus, ϕ' is an admissible transformation of g. Finally, suppose that $\phi': g(A) \to B$ is an admissible transformation of g. We show that ϕ' is monotone increasing. For $\phi' \circ g = \phi' \circ (\phi \circ f) = (\phi' \circ \phi) \circ f$ is a homomorphism from $\mathfrak{A}$ into $\mathfrak{B}$, and so $\phi' \circ \phi$ is an admissible transformation of f. Therefore, $\phi' \circ \phi$ is a monotone increasing function on $f(A)$. Now if x, y are in $g(A)$ and $x > y$, we shall show that $\phi'(x) > \phi'(y)$. We know that $x = \phi(u)$ and $y = \phi(v)$ for some u, v in $f(A)$. Moreover, since ϕ is monotone increasing, $x > y$ implies that $u > v$. Finally, since $\phi' \circ \phi$ is monotone increasing, it follows that $(\phi' \circ \phi)(u) > (\phi' \circ \phi)(v)$, that is, $\phi'(x) > \phi'(y)$. ∎

---

[*]After reading the statement of the next theorem, the reader may wish to skip immediately to the remarks at the end of this section, which he may do with no loss of continuity.

To show that the definition of scale type can lead to dilemmas for irregular scales, let us recall the homomorphisms f and g from (A, R) into (Re, M), where f, g, A, R, and M are as defined in Eqs. (2.13) through (2.15). The representation $(A, R) \rightarrow (Re, M)$ is irregular, since g is not $\phi \circ f$, any $\phi : f(A) \rightarrow Re$. The scale f is ordinal, for every transformation of $f(A)$ is monotone increasing, and every monotone increasing transformation $\phi : f(A) \rightarrow Re$ is an admissible transformation.* However, g is not an ordinal scale, since the monotone increasing function $\phi : g(A) \rightarrow Re$ defined by $\phi(x) = 2x$ is not admissible. For $(\phi \circ g)(r) = 0$, $(\phi \circ g)(s) = 2$, and not $0 M 2$, even though rRs holds. Thus, $\phi \circ g$ is not a homomorphism from (A, R) into (Re, M), and ϕ is not admissible.

This example is a bit unsatisfying because the range of f, $f(A)$, has just one element. The following example, due to Roberts and Franke [1976], should be more convincing. Define R on Re by

$$ xRy \quad \text{iff} \quad [(x < 0 \text{ and } y \geqq 0) \text{ or } (x = 0 \text{ and } y > 0)]. \quad (2.17) $$

That is, all negative numbers are in the relation R to all nonnegative numbers, and 0 is in the relation R to all positive numbers. Let $\mathfrak{A} = (Re, R, \times)$, where $\times$ is ordinary multiplication on Re. Define $g: Re \rightarrow Re$ and $f: Re \rightarrow Re$ by

$$ g(x) = \begin{cases} -1 & \text{if } x < 0, \\ 0 & \text{if } x = 0, \\ 1 & \text{if } x > 0, \end{cases} \quad (2.18) $$

and $f(x) = x$, for all $x \in Re$. Then f and g are clearly homomorphisms from $\mathfrak{A}$ into $\mathfrak{B} = (Re, R, \times)$. Moreover, we shall show that $(\mathfrak{A}, \mathfrak{B}, g)$ is an absolute scale, while $(\mathfrak{A}, \mathfrak{B}, f)$ is not. The latter follows, since $\phi(x) = g(x)$ is an admissible transformation. To show the former, suppose $\phi : g(Re) \rightarrow Re$ is an admissible transformation. Then $-1 R 0$ and $0 R 1$, so $\phi(-1)R\phi(0)$ and $\phi(0)R\phi(1)$. Hence, by definition of R, $\phi(-1)$ must be negative, $\phi(0)$ must be 0, and $\phi(1)$ must be positive. Now $1 \times 1 = 1$, so $\phi(1) \times \phi(1) = \phi(1)$. Hence, $\phi(1) = 1$. Moreover, $(-1) \times (-1) = 1$, so $\phi(-1) \times \phi(-1) = \phi(1) = 1$. Thus, $\phi(-1) = -1$, since $\phi(-1)$ is negative. Thus, we have shown that ϕ on $g(Re) = \{-1, 0, 1\}$ is the identity.

It would be interesting to find other examples of irregular representations where one homomorphism with a nontrivial range gives rise to an ordinal, interval, or ratio scale, while another homomorphism does not. Such examples have not been given in the literature. It would also be helpful to determine whether broader conditions than regularity of a representation $\mathfrak{A} \rightarrow \mathfrak{B}$ are sufficient for all homomorphisms from $\mathfrak{A}$ into $\mathfrak{B}$ to have the same scale type.

---

*Indeed, f is even an interval scale. (A scale can have several types.)

Remarks

1. Given empirically obtained measurements, the problem of determining what kind of scale they define cannot always be solved by using a formal approach of the type we have described. We may, for example, not have a specific representation in mind. In these cases, we must resort to an alternative definition of admissible transformation, as one that keeps "intact the empirical information depicted by the scale," to follow Stevens [1968, p. 850]. The scale type, which is based on the definition of admissible transformation, can then be difficult to determine in practice, because it involves capturing the vague notion of "empirical information" depicted by a scale. We shall see this in Section 2.6. As Adams, Fagot, and Robinson [1965, p. 122] point out, much of the criticism of the applications of Stevens' theory of scale type has centered around measurements where the class of admissible transformations is not clearly defined. It seems likely that such criticism will continue until the scales used to measure loudness, brightness, IQ, etc., are put on a firmer measurement-theoretic foundation.

2. Other writers have classified scale type a little differently than we do in this section. Coombs [1952] considered the four scales—nominal, ordinal, interval, and ratio—which are the four types of scales considered by Stevens [1951]. However, Coombs then added a partially ordered scale, falling between the nominal and the ordinal. He also obtained a more detailed classification of scales by asking first whether objects are just classified, partially ordered, or completely ordered, and then whether distances between objects are classified (large, small, etc.), partially ordered, or completely ordered. In all, this two-way classification led to eleven different scale types. Coombs, Raiffa, and Thrall [1954] made a similar distinction. Torgerson [1958] argued that a nominal scale is not really an example of measurement, since the numbers assigned do not reflect any real properties of the systems or objects being measured. He also distinguished between ordinal scales with natural origins and ordinal scales without natural origins.

Exercises

1. (a) If $(\mathfrak{A}, \mathfrak{B}, f)$ is an ordinal scale with $f(A) = Re$, which of the following functions $\phi: Re \to Re$ are admissible transformations?
- (i) $\phi(x) = e^x$.
- (ii) $\phi(x) = x + 7$.
- (iii) $\phi(x) = 801x$.
- (iv) $\phi(x) = x^2$.
- (v) $\phi(x) = 8x^4$.
- (vi) $\phi(x) = x^2 + 10$.

(b) Repeat for f an interval scale.
(c) A ratio scale.
(d) A nominal scale.
(e) An absolute scale.
(f) A difference scale.
(g) A log-interval scale.

2. In football, the numbering of uniforms is not totally arbitrary. In some numbering schemes, offensive backs receive numbers lower than 50, ends receive numbers in the 80's and 90's, etc. Discuss when the numbering of uniforms, plans, alternatives, etc., defines a nominal scale.

3. Suppose $A = \{a, b, c\}$, $R = \{(a, b), (a, c), (b, c)\}$, and $f(a) = 3$, $f(b) = 2$, $f(c) = 1$. Show that $((A, R), (Re, >), f)$ is a (regular) ordinal scale, but not an interval scale or a ratio scale.

4. Let $A = \{a, b, c\}$, $R = \{(a, a), (b, b), (c, c), (a, b), (a, c), (b, c)\}$. Show that (A, R) is homomorphic to $(Re, \geqq)$ and that every homomorphism f defines a (regular) ordinal scale.

5. Suppose $A = \{a, b, c\}$, $R = \{(a, a), (b, b), (c, c)\}$. Show that (A, R) is homomorphic to $(Re, =)$ and every homomorphism f defines a (regular) nominal scale.

6. Suppose $A = \{0, 1\}$, $R = >$, and o is defined by

$$0 \circ 0 = 0 \circ 1 = 1 \circ 0 = 0, \quad 1 \circ 1 = 1.$$

Show that (A, R, o) is homomorphic to $(Re, >, \times)$ and every homomorphism f defines a (regular) absolute scale.

7. Suppose $A = \{r, s\}$, $R = \{(r, s)\}$, $f(r) = 1$, and $f(s) = 0$. Show that f is a (regular) homomorphism from (A, R) into $(Re, >)$ which defines both an ordinal and an interval scale.

8. Suppose o and f are defined as in Exer. 7 of Section 2.2.

(a) Show that every admissible transformation ϕ of f is monotone nondecreasing, that is, it satisfies

$$x \geqq y \Rightarrow \phi(x) \geqq \phi(y),$$

for all x, y in $f(A)$.

(b) Show that every monotone nondecreasing ϕ on $f(A)$ is admissible.

(c) Show that statements (a) and (b) hold for any homomorphism from (Re, o) into (Re, o).

9. Suppose $(\mathfrak{A}, \mathfrak{B}, f)$ defines a (regular) interval scale and g is another homomorphism from $\mathfrak{A}$ into $\mathfrak{B}$. Show that it follows that ratios of intervals are the same under f and g, that is, that

$$\frac{f(a) - f(b)}{f(c) - f(d)} = \frac{g(a) - g(b)}{g(c) - g(d)}.$$

10. Suppose R is defined on Re by (2.17), g on Re by (2.18), and f on Re by $f(x) = x$.

(a) Observe that g and f are homomorphisms from (Re, R) into (Re, R). [In the text, we considered these as homomorphisms from $(Re, R, \times)$ into $(Re, R, \times)$.]

(b) Observe that every admissible transformation of g is monotone increasing, but there are admissible transformations of f that are not.

11. Prove Theorem 2.2 for
    (a) absolute scales;
    (b) ratio scales;
    (c) interval scales;
    (d) nominal scales.

12. Is Theorem 2.2 true for
    (a) log-interval scales?
    (b) difference scales?

13. (Hays [1973, pp. 133,134], etc.) Consider what scale types (if any) are most appropriate for the following data:
    (a) The nationality of an individual's male parent.
    (b) Hand pressure as applied to a bulb or dynamometer.
    (c) Memory ability as measured by the number of words recalled from an initially memorized list.
    (d) Distance by air between New York and other cities.
    (e) U.S. Department of Agriculture classification of cuts of meat ("choice," etc.).
    (f) The tensile strength or force required to break a wire.
    (g) The excellence of a baseball team as measured by the number of games won during a season.
    (h) Zip codes.
    (i) Commerce Department three-digit industry classification codes.

2.4 Examples of Meaningful and Meaningless Statements

Given a theory of scale type, let us return to our definition of meaningfulness and test it on the examples we stated at the beginning of Section 2.2. *We shall assume that all the scales in question come from regular representations*, and so we may use our second definition of meaningfulness: A statement involving numerical scales is meaningful if and only if its truth (or falsity) remains unchanged under all admissible transformations of all the scales involved.

Let us first consider the statement

$$f(a) = 2f(b), \tag{2.19}$$

where $f(a)$ is some quantity assigned to a, for example, its mass or its temperature. We ask under what circumstances this statement is meaningful. According to the definition, it is meaningful if and only if its truth value is preserved under all admissible transformations ϕ, that is, if and only if, under all such ϕ,

$$f(a) = 2f(b) \Leftrightarrow (\phi \mathbin{o} f)(a) = 2[(\phi \mathbin{o} f)(b)].$$

If ϕ is a similarity transformation, that is, if $\phi(x) = \alpha x$, some $\alpha > 0$, then we do indeed have

$$f(a) = 2f(b) \Leftrightarrow \alpha f(a) = 2\alpha f(b).$$

We conclude that the statement (2.19) is meaningful if the scale f is a ratio scale, as is the case in the measurement of mass. On the other hand, suppose it is only an interval scale, as is the case (usually) in the measurement of temperature. Then a typical admissible transformation has the form $\phi(x) = \alpha x + \beta$, $\alpha > 0$. Certainly we can find examples of interval scales f such that $f(a) = 2f(b)$, but $\alpha f(a) + \beta \neq 2[\alpha f(b) + \beta]$. In particular, if we choose $f(a) = 2\,f(b) = 1$, $\alpha = 1$, and $\beta = 1$, this is the case. Thus, in general, the statement (2.19) is meaningless if f is an interval scale. We use the term "in general" because we have only given an example of an interval scale for which (2.19) is meaningless. In the future, we shall drop this term, and call a statement given in terms of an abstract scale of a given type meaningless if it is meaningless for some example of a scale of the given type. In our case, it is easy to see that the statement (2.19) is meaningless for *every* interval scale. For whenever $f(a) = 2f(b)$, taking $\alpha = \beta = 1$ gives us $\alpha f(a) + \beta \neq 2[\alpha f(b) + \beta]$. Conversely, whenever $f(a) \neq 2f(b)$, taking $\alpha = 1$ and $\beta = f(a) - 2f(b)$ gives us $\alpha f(a) + \beta = 2[\alpha f(b) + \beta]$.

The above discussion explains why Statement III of Section 2.2 about one can of corn weighing twice as much as another makes sense, whereas Statement IV about one can having twice the temperature of a second does not.

In the same way, one can explain why Statement I about the number of cans makes sense, whereas Statement II about the weight of a can does not. To see this, consider the statement

$$f(a) \geq 10. \tag{2.20}$$

If f is an absolute scale, as in the case of counting, we have for every admissible transformation ϕ,

$$f(a) \geq 10 \Leftrightarrow (\phi \circ f)(a) \geq 10,$$

for the only admissible ϕ is the identity transformation. Notice that $f(a)$ need not be greater than or equal to 10 for the statement $f(a) \geq 10$ to be meaningful. Meaningfulness is different from truth; we simply want to know whether or not it makes sense to make the assertion. If f is a ratio scale, as in the case of weight, then the statement (2.20) is meaningless; for example, if $f(a) \geq 10$ is true for some f, then taking α sufficiently small but positive makes $\alpha f(a) \geq 10$ false.

We continue with several other examples. Suppose $(\mathfrak{A}, \mathfrak{B}, f)$ is a scale and a, b are in A, the underlying set of $\mathfrak{A}$. Let us consider the statement

$$f(a) + f(b) = 20. \tag{2.21}$$

Thus, (2.21) might be the statement that the sum of the weight of a and the weight of b is a constant, 20. Is this meaningful? The answer is no if f is a ratio scale. For if $f(a) + f(b) = 20$, then $\alpha f(a) + \alpha f(b) = 20\alpha \neq 20$ for $\alpha \neq 1$. However, the statement that $f(a) + f(b)$ is constant for all a, b in A is meaningful if f is a ratio scale.

To give yet another example, consider the statement

$$f(a) > f(b). \tag{2.22}$$

If $\phi(x) = \alpha x + \beta$, $\alpha > 0$, then

$$\alpha f(a) + \beta > \alpha f(b) + \beta \Leftrightarrow \alpha f(a) > \alpha f(b)$$
$$\Leftrightarrow f(a) > f(b).$$

Thus, the statement (2.22) is meaningful if f is an interval scale. Indeed, it is meaningful if f is an ordinal scale. (Why?) Thus, for example, to say that the hardness of a is greater than the hardness of b is meaningful.

Next, let us consider the statement

$$f(a) - f(b) > f(c) - f(d). \tag{2.23}$$

If $\phi(x) = \alpha x + \beta$, $\alpha > 0$, then

$$[\alpha f(a) + \beta] - [\alpha f(b) + \beta] > [\alpha f(c) + \beta] - [\alpha f(d) + \beta]$$
$$\Leftrightarrow$$
$$f(a) - f(b) > f(c) - f(d).$$

Thus, the statement (2.23) is meaningful if f is an interval scale and of course if f is a ratio scale. It is not meaningful if f is an ordinal scale. To give an example, let $A = \{a, b, c, d\}$, let $f(a) = 10$, $f(b) = 6$, $f(c) = 4$, $f(d) = 2$, and let $\phi(10) = 11$, $\phi(6) = 9$, $\phi(4) = 7$, $\phi(2) = 3$. Then

$$f(a) - f(b) > f(c) - f(d).$$

Moreover, ϕ is monotone increasing on $f(A) = \{10, 6, 4, 2\}$. But

$$(\phi \circ f)(a) - (\phi \circ f)(b) = 11 - 9 = 2,$$

which is less than

$$(\phi \circ f)(c) - (\phi \circ f)(d) = 7 - 3 = 4.$$

The reader might wish to consider whether (2.23) could ever be meaningful for f an ordinal scale. We conclude from our analysis that it is meaningful, for example, to compare temperature differences, that is, to say that the difference in temperature between a and b is greater than the difference in temperature between c and d. However, a similar comparison of differences in hardness is not meaningful.

In general, ordinal scales can be used to make comparisons of size, like

$$f(a) > f(b),$$

interval scales to make comparisons of difference, like

$$f(a) - f(b) > f(c) - f(d),$$

and ratio scales to make more quantitative comparisons such as

$$f(a) = 2f(b)$$

and

$$f(a)/f(b) = \lambda.$$

Continuing with examples, let us consider two scales, $(\mathfrak{A}, \mathfrak{B}, f)$ and $(\mathfrak{A}', \mathfrak{B}', g)$, where $\mathfrak{A}$ and $\mathfrak{A}'$ have the same underlying set, and let us consider the statement

$$f(a) + g(a) \text{ is constant.} \tag{2.24}$$

This might be the statement that a certain gas's temperature plus its pressure is constant. If f and g are both ratio scales, then to be meaningful, the truth or falsity of (2.24) should be unchanged under (possibly different) admissible transformations of each scale. That is, if $\phi(x) = \alpha x, \alpha > 0$, and $\phi'(x) = \beta x, \beta > 0$, then (2.24) should hold if and only if

$$\alpha f(a) + \beta g(a) \text{ is constant.} \tag{2.25}$$

But (2.25) might very well not be true even if (2.24) is. For if we have $f(a) = -g(a)$ for all a, then $f(a) + g(a) = 0$, all a; but if $\alpha \neq \beta$ and $f(a)$ is not constant, then $\alpha f(a) + \beta g(a) = (\alpha - \beta)f(a)$ is not constant. Thus, (2.24) is not meaningful if f and g are ratio scales. On the other hand, certainly (2.24) is meaningful if f and g are both absolute scales.

We shall mention more complex examples of meaningful and meaningless statements in the next two sections. For now, let us point out one complication.

It is possible that in statements like (2.24), the scale g is defined in terms of the scale f. Then, we might not want to allow all possible admissible transformations of both f and g independently, but only admissible transformations of f and the "induced" admissible transformations of g. This situation will arise in the next section, when we discuss derived measurement, where one scale is defined in terms of another. In derived measurement, there will be several versions of meaningfulness, narrow and wide, depending on whether or not we pick admissible tranformations of all scales independently.

Exercises

1. (a) Suppose f is a ratio scale. Which of the following statements are meaningful?

      (i) $f(a) + f(b) > f(c)$.
      (ii) $f(a) = f(b)$.
      (iii) $f(a) = 1.8f(b)$.
      (iv) $f(a) + f(b)$ is constant for all a, b in A.
      (v) $f(a)f(b) > f(c)^2$.
      (vi) $f(a) - f(b) > 2[f(c) - f(d)]$.
      (vii) $f(a) > 16[f(b) - f(c)]$.
      (viii) $f(a) + f(b) > f(c)^2$.

(b) Repeat for f an ordinal scale.
(c) An interval scale.
(d) A nominal scale.
(e) An absolute scale.
(f) A difference scale.
(g) A log-interval scale.

2. (a) Show that if f and g are (independent) ratio scales, the statement

$$f(a)g(a) \text{ is constant for all } a \text{ in } A$$

is meaningful.

(b) Is this statement meaningful if f and g are interval scales?
(c) What if f is an interval scale and g is a ratio scale?
(d) What if f is an interval scale and g is an absolute scale?

3. (a) Suppose f is an interval scale and g is an absolute scale. Show that neither of the following statements is meaningful:

      (i) $\log_{10}|f(a)| = 2 \log_{10}|f(b)|$.
      (ii) $f(a) + g(a)$ is constant for all a in A.

(b) Consider the meaningfulness of the statements in part (a) if f is a ratio scale and g is an absolute scale.

(c) Repeat for f and g both ratio scales.

4. Is there any ordinal scale f for which the statement (2.23) is meaningful?

5. Are there any ratio scales f and g for which the statement (2.24) is meaningful? [What if (2.25) is false?]

6. If the representation $\mathfrak{A} \to \mathfrak{B}$ is irregular, then there is always a meaningless statement which would be judged meaningful under our second definition, namely, the one in terms of admissible transformations. For let $(\mathfrak{A}, \mathfrak{B}, f)$ be an irregular scale. Then by Theorem 2.1, there are a, b in A such that $f(a) = f(b)$ and $g(a) \neq g(b)$. Show that the statement $f(a) = f(b)$ is meaningful under the second definition, but it is meaningless.

7. Consider the relational system (A, R) of Eq. (2.11), and the homomorphism g of Eq. (2.12) from (A, R) into $(Re, >_1)$, where $>_1$ is defined in Eq. (2.10).
   (a) Show that $[(A, R), (Re, >_1), g]$ defines an irregular scale.
   (b) Comment on the meaningfulness of the assertion $g(s) > g(t)$.
   (c) Comment on the meaningfulness of the assertion $g(r) > g(t)$.

8. Consider the relational system (A, R) of Exer. 6, Section 2.2, and the homomorphism g from (A, R) into (Re, Δ) given there. Comment on the meaningfulness of the following assertions:
   (a) $g(s) > g(r)$.
   (b) $g(t) > g(r)$.
   (c) $g(t) \neq g(r)$.

9. Comment on the meaningfulness of the following statements:
   (a) This shelf is three times as long as that one.
   (b) A patient's height is twice his weight.
   (c) Glass is twice as hard as paper.
   (d) The school day in England is one and one-half times as long as that in the United States.
   (e) A can's weight is greater than its height.
   (f) A circle's area is three times that of a second circle.
   (g) The wind yesterday was calmer than it is today (the Beaufort wind scale classifies winds as calm, light air, light breeze, . . .).
   (h) This rat weighs more than any of the others.
   (i) This rat weighs more than those two combined.

2.5 Derived Measurement

Very often we are given certain numerical scales or assignments and we want to introduce new scales defined in terms of the old ones. Most scales in the physical sciences have this property. A typical one is density d, which can be defined in terms of mass m and volume V as $d = m/V$. If density is simply defined from mass and volume, then density is not measured fundamentally in the sense we have been describing, but rather it is derived from other scales, which may or may not be fundamental. In this section, we shall discuss the process of obtaining derived scales.

We do not present an elaborate formal representation theory of derived measurement. Indeed, there is no generally accepted theory. The approach to derived measurement which we shall present is based on that of Suppes and Zinnes [1963]. This is to be contrasted with the approaches of Campbell [1920, 1928] and Ellis [1963], who emphasize "dimensional" parameters (see below for an example), and of Causey [1967, 1969], who deals with "classes of similar systems." Some writers, for example Pfanzagl [1968, p. 31], argue that derived measurement is not measurement at all. Pfanzagl argues that if a property measured by a numerical scale had any empirical meaning of its own, then there would also be a fundamental scale. The defining relation then simply becomes an empirical law between fundamental scales.

Certainly it is true that derived measurement is often a relative matter. The same scale can be developed as either fundamental or derived. Usually, some basic scales are chosen and others are derived from them. In this sense, it is possible to define density using fundamental measurement; see the discussion in Section 5.4, Exer. 26. However, it is not always so easy or so natural to treat derived scales as fundamental. Besides, derived measurement corresponds to a process which is frequently used in practice. Hence, it is important to have a theory for this kind of measurement.

To have in mind a firm idea of what we mean by derived measurement, let us suppose that A is a set and $f_1, f_2, \ldots, f_n$ are given real-valued functions on A. We call these functions *primitive scales*, and define a new real-valued function g on A in terms of these primitive scales. The function g is called the *derived scale*. This definition is very broad. Indeed, Causey [1969] argues that it is deficient because the derived scale is not required to reflect in any direct manner the characteristics of empirical relational systems. As Adams [1966] points out, weight times volume and weight plus volume are equally good derived measures according to the definition, regardless of empirical significance. In spite of these objections, we feel that this broad notion of derived scale leads to some empirically useful results, and we shall adopt it.

The definition of the derived scale g in terms of the f_i need not be in terms of an equation. A simple example to illustrate this last point is the following contrived example. Suppose $u:A \to Re$ is an ordinal utility function, that is, a function satisfying Eq. (2.3). Let v be a function on A with the property that

$$u(a) > u(b) \quad \text{iff} \quad v(a) < v(b).$$

Then v is a derived scale. A *representation theorem* will state (necessary and) sufficient conditions for the existence of a function satisfying the definition. (In this case, the function v always exists. However, the function v is not unique.) We present a less contrived example in Section 6.2.2. In general, the derived scale g and the primitive scales $f_1, f_2, \ldots, f_n$ will satisfy a certain condition $C(f_1, f_2, \ldots f_n, g)$, and any function g satisfying this condition will be acceptable. Condition C may of course be an equation relating g to $f_1, f_2, \ldots, f_n$. In the case of density, $C(m, V, d)$ holds if and only if $d = m/V$.

A more important problem for derived measurement is the *uniqueness problem*. There are two different senses of uniqueness, depending on whether or not we allow the primitive scales $f_1, f_2, \ldots, f_n$ to vary. For example, in the case of density, if m and V are not allowed to vary, then d is defined uniquely in terms of m and V. However, both m and V are ratio scales. If we allow m and V to be replaced by other allowable scales, then d can vary. If m' and V' are other allowable scales, then there are positive numbers α and β such that $m'(a) = \alpha m(a)$ and $V'(a) = \beta V(a)$, for all a in A, the set of objects being measured. The corresponding derived scale of density is given by

$$d'(a) = \frac{m'(a)}{V'(a)} = \frac{\alpha m(a)}{\beta V(a)} = \frac{\alpha}{\beta} d(a).$$

Thus, d' is related to d by a similarity transformation.

If a derived scale g is defined from primitive scales $f_1, f_2, \ldots, f_n$ by condition C, we say that a function $\phi : g(A) \to Re$ is *admissible in the narrow sense* if $g' = \phi \circ g$ satisfies

$$C(f_1, f_2, \ldots, f_n, g').$$

We say ϕ is *admissible in the wide sense* if there are acceptable replacement scales $f_1', f_2', \ldots, f_n'$ for $f_1, f_2, \ldots, f_n$, respectively, so that

$$C(f_1', f_2', \ldots, f_n', g').$$

(In particular, if each f_i is a regular scale, the f_i' would be defined by taking appropriate admissible transformations of the f_i.) In the case of density, the identity is the only admissible transformation in the narrow sense. We have shown that every admissible transformation in the wide sense is a similarity transformation. Conversely, every similarity transformation is admissible in the wide sense. For suppose $d' = \alpha d, \alpha > 0$. Then $C(\alpha m, V, d')$ holds, so d' is obtained from d by an admissible transformation in the wide sense.

We say that the derived scale g is *regular in the narrow sense* if, whenever $C(f_1, f_2, \ldots, f_n, g')$ holds, then there is a transformation $\phi : g(A) \to Re$ such that $g' = \phi \circ g$. The transformation ϕ is an admissible transformation in the narrow sense. We say g is *regular in the wide sense* if, whenever $f_1', f_2', \ldots, f_n'$ are acceptable replacement scales for scales $f_1, f_2, \ldots, f_n$, and whenever $C(f_1', f_2', \ldots, f_n', g')$ holds, then there is a transformation $\phi : g(A) \to Re$ such that $g' = \phi \circ g$. Here, ϕ is an admissible transformation in the wide sense. Thus, density is a regular scale in both the narrow and wide senses.

For regular scales, scale type can be defined analogously to the definition in fundamental measurement, except that there are narrow and wide senses of scale type. For example, if g is a regular scale in the narrow (wide) sense, then we say it is a *ratio scale in the narrow (wide) sense* if the class of admissible transformations in the narrow (wide) sense is *exactly* the class of similarity transformations. Thus, density is an absolute scale in the narrow sense, since the only admissible transformation in the narrow sense is the identity. However, density is a ratio scale in the wide sense.*

We shall adopt the same theory of meaningfulness as for fundamental measurement. Thus, for example, even in the wide sense, it makes sense to assert that one medium is twice as dense as another. We shall usually want statements to be meaningful in this wide sense.

In Chapter 4, we shall apply the theory of uniqueness in derived measurement to psychophysical scaling. In the next section, we shall apply this theory to energy use, air pollution, and the consumer price index.

Remark: Luce [1959, 1962] and Rozeboom [1962a, b] point out that certain parameters can enter into relationships $C(f_1, f_2, \ldots, f_n, g)$ in a manner different from that in either the narrow or wide senses as defined above. For example, sometimes $f_1, f_2, \ldots, f_n$ represent different measurements using the same scale. Then it may only be reasonable to use the same admissible transformation of each of the f_i in determining the wide-scale type of g. We shall encounter an example of this in the next section. More complicated situations arise. For example, in the law of decay for a radioactive material, we have

$$q = q_0 e^{-0.14t},$$

where q is the quantity of material at time t measured in seconds, and q_0 is the initial quantity. If time is measured in hours, the law changes to

$$q = q_0 e^{-50.4t},$$

since $50.4 = (360) \times (0.14)$. Thus, the law of radioactive decay could be restated as

$$q = q_0 e^{-kt},$$

where k is a parameter. We have

$$C(t, k, q) \quad \text{iff} \quad q = q_0 e^{-kt}.$$

However, transformations of k are determined by transformations of t, and not by any measurement theory for k. Luce calls k a "dimensional parameter." We may not take independent transformations of k as well as

---

*Note that the proof of this required two steps: proof that every admissible transformation in the wide sense is a similarity, and proof that every similarity is an admissible transformation in the wide sense.

independent transformations of t. Thus, the discussion above must be modified. What is true is that transformations of t lead to transformations of k such that kt is a constant. This relationship is implicit in the statement of the condition $C(t, k, q)$. Thus, in a wide sense, q/q_0 is an absolute scale. In applying the theory of uniqueness to derived measurement, the reader should be careful to modify the theory when needed to consider these sorts of complications.

Exercises

1. The inefficiency I of a particular fossil fuel power plant might be measured as $I =$ tons of emissions per ton of fuel burned. Show the following:

(a) I is regular in both the narrow and wide senses.

(b) I is an absolute scale in the narrow sense and a ratio scale in the wide sense.

(c) Even in the wide sense, it makes sense to assert that plant a is twice as inefficient as plant b or that the inefficiency of plant a plus the inefficiency of plant b is greater than the inefficiency of plant c.

2. Suppose $f(a)$ is the height of a and $g(a)$ the weight of a. Let

$$h(a) = \frac{f(a) + g(a)}{2}$$

and

$$k(a) = \sqrt{f(a)g(a)} .$$

Then h and k are derived scales. Show the following:

(a) Both h and k are (regular) absolute scales in the narrow sense.

(b) The scale k is a (regular) ratio scale in the wide sense, but h is not necessarily a (regular) ratio scale in the wide sense.

3. Suppose $u(a)$ is the utility of a, measured on an interval scale, and $m(a)$ is the dollar value of a, measured on a ratio scale. Then the utility per dollar is measured as $D(a) = u(a)/m(a)$. Show that there are admissible transformations of $D(a)$ in the wide sense that are not positive linear transformations.

4. The physical scales work, momentum, etc., are derived scales. Consider what types of scales they are.

5. (a) Suppose f_1 and f_2 are both (regular) ratio scales. Show that $g = f_1 f_2$ is a (regular) ratio scale in the wide sense.

(b) If f_1 and f_2 are both (regular) interval scales, is $g = f_1 f_2$ necessarily an interval scale in the wide sense?

6. Repeat Exer. 5 for $g = f_1 + f_2$.

2.6 Some Applications of the Theory of Meaningfulness: Energy Use, Air Pollution, and the Consumer Price Index

2.6.1 *Energy Use; Arithmetic and Geometric Means*

In this section, we give several more complicated applications of the theory of meaningfulness, to both fundamental and derived scales. We shall assume that all scales in question come from regular representations or are regular in both the narrow and wide senses.

Let us first imagine that we study n animals under one kind of experimental treatment (say a special diet) and m animals under a second kind of treatment. We want to say that the average weight of the animals under the first kind of treatment is larger than the average weight of the animals under the second treatment. We treat weight as a fundamental scale. Specifically, if f is the scale of weight in question, we want to consider the statement

$$\frac{1}{n} \sum_{i=1}^{n} f(a_i) > \frac{1}{m} \sum_{i=1}^{m} f(b_i). \tag{2.26}$$

Here, we calculate arithmetic means over two different sets, and compare them. We consider the statement (2.26) meaningful if for all admissible transformations ϕ, (2.26) holds if and only if

$$\frac{1}{n} \sum_{i=1}^{n} (\phi \circ f)(a_i) > \frac{1}{m} \sum_{i=1}^{m} (\phi \circ f)(b_i). \tag{2.27}$$

If ϕ is a similarity transformation, say $\phi(x) = \alpha x$, $\alpha > 0$, then certainly (2.27) holds if and only if (2.26) holds. This is even the case when ϕ is a positive linear transformation, that is, $\phi(x) = \alpha x + \beta$, $\alpha > 0$. For then (2.27) becomes

$$\frac{1}{n} \sum_{i=1}^{n} \left[\alpha f(a_i) + \beta \right] > \frac{1}{m} \sum_{i=1}^{m} \left[\alpha f(b_i) + \beta \right],$$

which reduces to (2.26). Thus, (2.26) is meaningful if f is a ratio scale or an interval scale. If f is an ordinal scale, then (2.26) is meaningless. Proof is left to the reader. This result can be applied to say that the statement "group A has higher average IQ than group B" is meaningless if IQ is only an ordinal scale. (Raw scores on intelligence tests define ordinal scales, according to Stevens [1959].) However, the statement is meaningful if IQ is an interval scale or a ratio scale. (It has been argued that "standard scores" on an intelligence test define an interval scale—see Stevens [1959].) All too often in the social sciences, comparisons of arithmetic means (as well as other comparisons) are made with little attention paid to whether or not these comparisons are meaningful.

Suppose next that we have several experts or individuals and each rates two alternatives, a and b. Let $f_i(a)$ be the rating of a by expert i, and $f_i(b)$ be the rating of b by expert i. We might want to consider the statement

$$\frac{1}{n} \sum_{i=1}^{n} f_i(a) > \frac{1}{n} \sum_{i=1}^{n} f_i(b). \tag{2.28}$$

Here, we say that the average rating of alternative a is higher than the average rating of alternative b. If we think of each f_i as a fundamental scale, then the statement (2.28) may be thought of as a statement involving several fundamental scales or a statement involving one derived scale, the average $\frac{1}{n}\sum f_i$. We use the former interpretation, though the latter interpretation gives the same results (using the wide sense of scale). Even if each f_i is a ratio scale, statement (2.28) is now meaningless. For, we must consider simultaneously transformations $\phi_i(x) = \alpha_i x$ of each f_i, and certainly there are α_i such that (2.28) holds, while

$$\frac{1}{n} \sum_{i=1}^{n} \alpha_i f_i(a) > \frac{1}{n} \sum_{i=1}^{n} \alpha_i f_i(b) \tag{2.29}$$

does not. On the other hand, comparison of geometric means* over individuals is meaningful, for

$$\sqrt[n]{\prod_i f_i(a)} > \sqrt[n]{\prod_i f_i(b)} \Leftrightarrow \sqrt[n]{\prod_i \alpha_i f_i(a)} > \sqrt[n]{\prod_i \alpha_i f_i(b)} .$$

Another way to phrase this conclusion is that if $g = \sqrt[n]{\prod_i f_i}$ is thought of as a derived scale, then it forms a ratio scale in the wide sense, and so $g(a) > g(b)$ is a meaningful statement (in the wide sense).

We shall briefly mention an application of this last example. As part of a larger study (Roberts [1972, 1973]), a set of variables relevant to the growing demand for energy was presented to a panel of experts, who were asked to judge their relative importance using the method of magnitude estimation. In this method, the expert first selects that variable which seems most important and assigns it the rating 100. Then he rates the other variables in terms of the most important one, so that a variable receiving a rating of 50 is considered "half as important" as one receiving a rating of 100, etc. A typical set of variables and ratings of one of the experts is shown in Table 2.3. It seems plausible that the magnitude estimation

---

*The *geometric mean* of a collection of numbers $x_1, x_2, \ldots, x_n$ is $\sqrt[n]{\prod_i x_i}$, where $\prod_i x_i$ means the product $x_1 \cdot x_2 \cdots x_n$.

Table 2.3. Magnitude Estimation by One Expert of Relative Importance for Energy Demand of Variables Related to Commuter Bus Transportation in a Given Region *

Variable	Relative Importance Rating
1. Number of passenger miles (annually, by bus)	80
2. Number of trips (annually)	100
3. Number of miles of bus routes	50
4. Number of miles of special bus lanes	50
5. Average time home to office (or office to home, or sum)	70
6. Average distance home to office	65
7. Average speed	10
8. Average number passengers per bus	20
9. Distance to bus stop from home (or office, or sum)	50
10. Number of buses in the region	20
11. Number of stops (home to office, or vice versa, or sum)	20

*Adapted from Roberts [1972] with permission of the RAND Corporation.

procedure leads to a ratio scale—this is presumed by Stevens [1957, 1968].* Thus, comparisons of geometric mean relative importance ratings (over experts) are probably meaningful, while comparisons of arithmetic means are probably not. This observation led to the use of geometric means. (See Roberts [1979] for a related application of the theory of meaningfulness.)

Remark: The analysis of this example must be done with some care. Since the scale value of the most important element was fixed to be the same for all the experts, the scales are not really independent, and so it is probably not reasonable to demand that the truth value of a statement be preserved under *different* transformations of all the scales, but rather only under *the same* transformation of each scale. But now it probably becomes meaningful to speak of comparisons of arithmetic means. For if we take $\alpha_i = \alpha$, all i, then (2.28) holds if and only if (2.29) holds. This reasoning points up the fact that the theory of meaningfulness has to be applied with care.

Remark: A more detailed discussion of measurement theory and the use of statistical summaries such as means and of statistical tests can be found in Pfanzagl [1968]. For a discussion of many particular statistics, see Adams, Fagot, and Robinson [1965]. There is a need for a more systematic development, and such a development would be of widespread practical significance, if its results were made widely known.

There have been views expressed in the measurement literature which differ from the point of view taken here, and assert that the choice of

*We have remarked earlier that where there is no obvious representation, admissible transformations must be defined as functions preserving empirical information depicted by a scale, and hence scale type is not formally defined, but depends on our interpretation of what the empirical information content of the scale is.

statistic (arithmetic mean, geometric mean, etc.) or of statistical test (t-test, etc.) does not depend on the type of scale involved. For example, Anderson [1961] argues that statistical tests concern numbers, and it does not matter what kind of scale these numbers come from. For a survey of such arguments, and a criticism of them, see Adams, Fagot, and Robinson [1965] or Stevens [1968]. Hays [1973, Section 3.2] also discusses this issue. Luce [1967] gives references to some of the literature on this issue, and argues that the use of a statistical test is limited by the class of transformations under which a null hypothesis is unchanged. Adams, Fagot, and Robinson [1965] argue that applying any sort of statistic to numbers is always *appropriate*. However, statements made using this statistic might be *inappropriate* if they are meaningless. Finally, some authors have formalized the risk involved in applying a statistical test (for example, comparison of arithmetic means) in a situation (for example, an ordinal scale) when such a test is inappropriate. See Abelson and Tukey [1959, 1963] for an example of such an attempt.

2.6.2 Consumer Price Index

Turning to another example, let us consider the computation of the consumer price index.* This index relates current prices of certain basic commodities, including food, clothing, fuel, etc., to prices at some reference time. Suppose the index will be based on n fixed commodities. Suppose $p_i(0)$ is the price of commodity i at the reference or base time, and $p_i(t)$ is the price of commodity i at time t. Then one consumer price index, due to Bradstreet and Dutot (see Fisher [1923, p. 40]), is given by

$$I(p_1(t), p_2(t), \ldots, p_n(t)) = \sum_{i=1}^{n} p_i(t) \Big/ \sum_{i=1}^{n} p_i(0). \qquad (2.30)$$

This is an example of a derived scale: I is defined in terms of $p_1, p_2, \ldots, p_n$. Now each price is measured on a ratio scale—the admissible transformations (in the wide sense) are conversions from dollars to cents, cents to francs, etc. But the scales are independent, so an admissible transformation of I in the wide sense results from independent choice of positive numbers $\alpha_1, \alpha_2, \ldots, \alpha_n$. But now even the statement

$$I(p_1(t), p_2(t), \ldots, p_n(t)) > I(p_1(s), p_2(s), \ldots, p_n(s)) \qquad (2.31)$$

is meaningless in the wide sense. For it is quite possible to choose the α_i so that

$$\sum_{i=1}^{n} p_i(t) \Big/ \sum_{i=1}^{n} p_i(0) > \sum_{i=1}^{n} p_i(s) \Big/ \sum_{i=1}^{n} p_i(0), \qquad (2.32)$$

*Our discussion in part follows Pfanzagl [1968, p. 49] who references Fisher [1923, p. 40]. The discussion easily generalizes to other indices, such as of productivity and consumer confidence. See Eichhorn [1978] for some recent discussion, and a variety of references, especially on page 158. See also Allen [1975] and Samuelson and Swamy [1974].

while

$$\sum_{i=1}^{n} \alpha_i p_i(t) \bigg/ \sum_{i=1}^{n} \alpha_i p_i(0) < \sum_{i=1}^{n} \alpha_i p_i(s) \bigg/ \sum_{i=1}^{n} \alpha_i p_i(0). \qquad (2.33)$$

For example, suppose $n = 2$, $p_1(0) = p_2(0) = 1$, $p_1(t) = 5$, $p_2(t) = 10$, $p_1(s) = 6$, $p_2(s) = 8$, $\alpha_1 = 2$, $\alpha_2 = \frac{1}{2}$. Then (2.32) and (2.33) hold, since

$$\frac{5 + 10}{1 + 1} > \frac{6 + 8}{1 + 1}$$

and

$$\frac{10 + 5}{2 + \frac{1}{2}} < \frac{12 + 4}{2 + \frac{1}{2}}.$$

If we insist that all prices be measured in the same units, which is reasonable, then various comparisons of the index I are meaningful, even in the wide sense. For an admissible transformation of I in the wide sense results from multiplication of each $p_i(t)$ and $p_i(0)$ by the same positive number α. Then

$$\sum_i \alpha p_i(t) \bigg/ \sum_i \alpha p_i(0) = \sum_i p_i(t) \bigg/ \sum_i p_i(0),$$

so it is meaningful to assert that

$$I(p_1(t), p_2(t), \ldots, p_n(t)) > I(p_1(s), p_2(s), \ldots, p_n(s)).$$

It is now even meaningful to assert that the index has doubled or increased by 20% between time s and time t, for the following statements are now easily seen to be meaningful:

$$I(p_1(t), p_2(t), \ldots, p_n(t)) = 2I(p_1(s), p_2(s), \ldots, p_n(s)),$$

$$I(p_1(t), p_2(t), \ldots, p_n(t)) = 1.2I(p_1(s), p_2(s), \ldots, p_n(s)).$$

A consumer price index for which comparisons of size are meaningful even if different prices may be measured in different units is

$$J(p_1(t), p_2(t), \ldots, p_n(t)) = \left(\prod_{i=1}^{n} p_i(t) \bigg/ \prod_{i=1}^{n} p_i(0) \right)^{1/n}.* \qquad (2.34)$$

*Index J is discussed in Boot and Cox [1970, p. 499].

For

$$\left(\prod_{i=1}^{n} \alpha_i p_i(t) \Big/ \prod_{i=1}^{n} \alpha_i p_i(0) \right)^{1/n} = \left(\prod_{i=1}^{n} p_i(t) \Big/ \prod_{i=1}^{n} p_i(0) \right)^{1/n}.$$

Indeed, with the index J of Eq. (2.34), it again is meaningful to say that the index doubled between time s and time t or that it increased by 20%. For the statements

$$J(p_1(t), p_2(t), \ldots, p_n(t)) = 2J(p_1(s), p_2(s), \ldots, p_n(s))$$

and

$$J(p_1(t), p_2(t), \ldots, p_n(t)) = 1.2J(p_1(s), p_2(s), \ldots, p_n(s))$$

are meaningful in the wide sense, even allowing independent changes of scale for different prices. The reader should note that simply to say that these comparisons or statements are meaningful in our technical sense does not say they are meaningful in an economic sense. It is not clear exactly what is the economic content of a consumer price index as measured by index J. Meaningfulness in our technical sense is necessary but not sufficient for meaningfulness in practical terms.

Some economic arguments can be raised against the index I of Eq. (2.30) as well. In actual practice, the consumer price index is measured by the Bureau of Labor Statistics by taking a ratio of weighted sums,

$$K(p_1(t), p_2(t), \ldots, p_n(t)) = \sum_i \lambda_i p_i(t) \Big/ \sum_i \lambda_i p_i(0). \qquad (2.35)$$

(See Samuelson [1961, p. 135], Lapin [1973], Boot and Cox [1970], Rothwell [1964], or Mudgett [1951].) The weight λ_i measures the quantity of item i in an "average market basket," a weighting factor disregarded in index I. The weighting factors are obtained from surveys of spending patterns. These weighting factors differ over time, of course. (See Table 2.4 for an example of different spending patterns.) But one set of weighting factors is fixed over all time in calculating the index K. (See Exers. 5 and 6.)

In terms of the consumer price index K, if weights λ_i are assumed fixed, the following statements are meaningless in the wide sense if the prices are allowed to be measured in different units, but meaningful otherwise:

$$K(p_1(t), p_2(t), \ldots, p_n(t)) > K(p_1(s), p_2(s), \ldots, p_n(s)),$$

$$K(p_1(t), p_2(t), \ldots, p_n(t)) = 2K(p_1(s), p_2(s), \ldots, p_n(s)),$$

$$K(p_1(t), p_2(t), \ldots, p_n(t)) = 1.2K(p_1(s), p_2(s), \ldots, p_n(s)).$$

We return to the consumer price index, and index numbers in general, in the exercises below and in Exer. 15, Sec. 4.2.

Table 2.4. Percentage of Consumer Spending for Various Years of Urban Wage Earners and Clerical Worker Families, by Broad Categories*

	1917–1919	1934–1936	1952	1963
Food	40	33	30	22
Housing	27	32	33	33
Apparel	18	11	9	11
Transportation	3	8	11	14
Other goods and services	12	16	17	20

*Source: Boot and Cox [1970, p. 493].

2.6.3 *Measurement of Air Pollution*

For our last application, we turn to measurement of air pollution. There are various pollutants present in the air, which have quite different effects. As far as the damaging health effects of pollution are concerned, the most important pollutants seem to be carbon monoxide (CO), hydrocarbons (HC), nitrogen oxides (NOX), sulfur oxides (SOX), and particulate matter (dust, etc.) (PM). Also damaging are the products of chemical reactions among these pollutants, the most serious ones being the oxidants (such as ozone) produced by hydrocarbons and nitrogen oxides reacting in the presence of sunlight. Finally, some of these pollutants are more serious in the presence of others; for example, sulfur oxides are more harmful in the presence of particulate matter (National Air Pollution Control Administration [1969]). We say that these are *synergistic effects*. We shall disregard chemical reactions among pollutants and synergistic effects.

To be able to compare alternative pollution control policies, one should be able to compare the effects of different pollutants. Indeed, some policies might result in net increases in the emissions of some (it is hoped less harmful) pollutants while achieving cutbacks in emissions of other pollutants. How can two such strategies be compared? One proposal given in the literature is that some form of combined pollution index be used, which would give one number indicating how bad the air pollution is, based on the levels of emissions of the different pollutants. There are other advantages for having such a single index. For one, daily or yearly pollution forecasts and reports could be more easily given, and progress could be easily measured.

A simple way of producing a single, combined pollution index is to measure the total weight of emissions of each pollutant i over a fixed period of time (one hour, one day, one year, or whatever the time period in question is), and then to sum up these numbers. Let $e(i, t, k)$ be the total weight of emissions of pollutant i (per cubic meter) over the tth time period and due to the kth source or measured in the kth location. Then the simple pollution index we have just described is given by

$$A(t, k) = \sum_i e(i, t, k). \tag{2.36}$$

This is again an example of derived measurement. Use of the derived index $A(t, k)$ leads to the conclusion that transportation is the largest source of air pollution, accounting for over 50% of all pollution, and that stationary fuel combustion (especially by electric power plants) is second largest; use of the numbers $e(i, t, k)$ leads to the conclusion that carbon monoxide accounts for over half of all emitted air pollution (Walther, 1972).

These conclusions are meaningful (in the wide sense). For statements like the following are meaningful if we measure all $e(r, t, k)$ in the same units of mass (a unit often used in the air pollution literature is $\mu g/m^3$):

$$A(t, k) > A(t, k'),$$

$$A(t, k_r) > \sum_{k \neq k_r} A(t, k),$$

$$\sum_{t, k} e(i, t, k) > \sum_{t, k} \sum_{j \neq i} e(j, t, k).$$

These statements are meaningful because an admissible transformation in the wide sense amounts to multiplying each $e(i, t, k)$ by the same positive number α, and the comparison

$$\sum_i e(i, t, k) > \sum_i e(i, t, k')$$

holds if and only if

$$\sum_i \alpha e(i, t, k) > \sum_i \alpha e(i, t, k'),$$

the comparison

$$\sum_i e(i, t, k_r) > \sum_{k \neq k_r} \sum_i e(i, t, k)$$

holds if and only if

$$\sum_i \alpha e(i, t, k_r) > \sum_{k \neq k_r} \sum_i \alpha e(i, t, k),$$

and the comparison

$$\sum_{t, k} e(i, t, k) > \sum_{t, k} \sum_{j \neq i} e(j, t, k)$$

holds if and only if

$$\sum_{t, k} \alpha e(i, t, k) > \sum_{t, k} \sum_{j \neq i} \alpha e(j, t, k).$$

Although such comparisons using the numbers $e(i, t, k)$ and $A(t, k)$ are meaningful in the technical sense, there is some question about whether they are meaningful comparisons of the pollution level in a practical sense. A unit of mass of carbon monoxide is far less harmful than a unit of mass of nitrogen oxide. For example, 1971 U.S. Environmental Protection Agency (EPA) ambient air quality standards, based on health effects and corrected for a 24-hour period, allowed 7800 units of carbon monoxide as compared to 330 units of nitrogen oxides, 788 of hydrocarbons, 266 of sulfur oxides, and 150 of particulate matter. (For these EPA standards, see Environmental Protection Agency, 1971; for corrections to 24 hours, see Babcock and Nagda, 1973). Babcock and Nagda call these numbers *tolerance factors*. They are also called Minimum Acute Toxicity Effluent (MATE) criteria. (See, for example, Hangebrauck [1977], Industrial Environmental Research Laboratory [1976] or Schalit and Wolfe [1978].) The tolerance factors are levels above which adverse effects are known or thought to occur. Let $t(i)$ be the tolerance factor for the ith pollutant. The *severity factor* or *effect factor* is $1/t(i)$ or $t(CO)/t(i)$, the ratio of tolerance factor for CO to that for i.* Babcock and Nagda and others (Babcock [1970], Walther [1972], Caretto and Sawyer [1972]) suggest weighting the emission levels (in mass) by the severity factor and obtaining a combined pollution index by using a weighted sum. This amounts to using the indices

$$\frac{1}{t(i)} e(i, t, k) \tag{2.37}$$

and

$$B(t, k) = \sum_i \frac{1}{t(i)} e(i, t, k) \tag{2.38}$$

or

$$\frac{t(CO)}{t(i)} e(i, t, k)$$

and

$$B'(t, k) = \sum_i \frac{t(CO)}{t(i)} e(i, t, k). \tag{2.39}$$

*The severity factors appearing in the literature differ from reference to reference, for various reasons. First, federal air quality standards are not all laid out for the same time period. Rather, some are for one hour, some for eight hours, etc. There is a difference of opinion about how to extrapolate these standards to the same time period, for example 24 hours. There are also differing approaches to bringing in chemical reactions and synergistic effects.

The index (2.37) is sometimes called *degree of hazard* (see, for example, Industrial Environmental Research Laboratory [1976] and Schalit and Wolfe [1978]). The combined pollution index (2.38) (called *pindex* by Babcock [1970]) is designed to measure the harmful effect of pollution. Under this index, transportation still is the largest source of pollutants by effect, but now accounting for less than 50%. Stationary sources fall to fourth place on the list. Use of the weighted factors $[1/t(i)]e(i, t, k)$ drops carbon monoxide to the bottom of the list of pollutants by effect, with $[1/t(CO)]e(CO, t, k)$ just over 2% of the total— it was over 50% of the total by mass. (See Walther [1972] for a discussion of these results, and Babcock and Nagda [1973] for a discussion of implications of the results for air pollution control strategies.) A similar analysis could be applied to a particular factory or power plant which puts out a variety of different pollutants in its effluent stream. For the procedure see, for example, Hangebrauck [1977], Industrial Environmental Research Laboratory [1976] or Schalit and Wolfe [1978].

These results are meaningful in our technical sense. For, again assuming that all emission weights are measured in the same units, an admissible transformation in the wide sense now amounts to multiplication of each $e(i, t, k)$ and each $t(i)$ by the same number α. Since

$$\frac{1}{\alpha t(i)} \alpha e(i, t, k) = \frac{1}{t(i)} e(i, t, k),$$

the statements

$$B(t, k) > B(t, k'),$$
$$B(t, k_r) > \sum_{k \neq k_r} B(t, k),$$

and

$$\sum_{t, k} e(i, t, k) = (.02) \sum_{t, k} \sum_{j} e(j, t, k)$$

are meaningful. Similar statements are also meaningful if we use $[t(CO)/t(i)]e(i, t, k)$ and $B'(t, k)$ of Eq. (2.39).

The measure $B(t, k)$ of Eq. (2.38) amounts to the following. For a given pollutant, take the percentage of a given harmful level of emissions that is reached in a given period of time, and add up these percentages over all pollutants. The resulting number is a measure of total air pollution. (It can, of course, come out larger than 100%.) This is the procedure that was introduced for use in the San Francisco Bay Area in the 1960's (see Bay Area Pollution Control District [1968] or Sauter and Chilton [1970]). There are some serious problems with this measure. First, if 100% of the carbon monoxide tolerance level is attained, this is known to have some damaging effects. The measure implies that the effects are equally severe if the levels of all five major pollutants are relatively low, say 20% of their known harmful levels. Similarly, the measure assumes that reaching 95% of the tolerance level for carbon monoxide and 5% of the level for one other pollutant is as bad as reaching 100% of the tolerance level for carbon monoxide. Is this really the case? It seems unlikely. Thus, once again, the comparisons made using the combined pollution index are meaningful in our technical sense, but there is some doubt as to whether or not they have real meaning.

Exercises

1. If geometric means are used in place of arithmetic means in Eq. (2.26), and f is weight, is the comparison still meaningful?

2. (a) In the situation where there are different experts, and each f_i is an ordinal scale, show that comparison of arithmetic means is meaningless.
   (b) What about comparison of geometric means?
   (c) Show that comparison of medians is meaningful if there is an odd number of experts and we only allow the same admissible transformation of each f_i.
   (d) What if we allow different transformations?

3. In Section 8.3, we shall consider direct estimates of subjective probability by various experts. If $p_i(A)$ is the estimate by expert i of the subjective probability of event A, consider the meaningfulness of the statement

$$\frac{1}{n} \sum_{i=1}^{n} p_i(A) > \frac{1}{n} \sum_{i=1}^{n} p_i(B).$$

(It is not clear what kind of scale direct estimates of subjective probability define. In Section 8.3, we shall argue that they might be absolute scales, or they might even be ratio scales, with each expert choosing a unit independently.)

4. If each $p_i(t)$ is measured on the same interval scale, and if K is defined by Eq. (2.35), are either of the following statements meaningful?
   (a) $K(p_1(t), p_2(t), \ldots, p_n(t)) > K(p_1(s), p_2(s), \ldots, p_n(s))$.
   (b) $K(p_1(t), p_2(t), \ldots, p_n(t)) = 2K(p_1(s), p_2(s), \ldots, p_n(s))$.

5. The *Laspeyres price index* (Laspeyres [1871]) is the consumer price index K_L obtained from K of Eq. (2.35) by setting the weight λ_i equal to the quantity $q_i(0)$ of good i in the "average market basket" in year 0, and multiplying the value of K by 100. The *Paasche price index* (Paasche [1874]) is the consumer price index K_P obtained from K by setting the weight λ_i equal to the quantity $q_i(t)$ of good i in the "average market basket" in year t and multiplying the value of K by 100. (The Laspeyres

Table 2.5. Hypothetical Prices and Quantities of Oil, Grain, and Wine in Italy in 1500 and 1750

Item i	$p_i(0)^*$	$p_i(t)^*$	$q_i(0)$	$q_i(t)$
Oil (qt)	\$.10	\$.15	10	12
Grain (bushel)	\$.30	\$.25	10	16
Wine (gal)	\$.10	\$.20	10	10

*Year 0 = 1500; year t = 1750.

index is the one usually used.)* The first price index was computed in 1764 by Carli, to compare Italian prices in 1750 with prices in 1500.[†] Carli limited consideration to three items—oil, grain, and wine. Let us suppose for the sake of discussion that Carli had obtained the basic price and quantity data of Table 2.5.

(a) Calculate the Laspeyres price index for the year 1750 relative to the year 1500.

(b) Calculate the Paasche index for the year 1750 relative to the year 1500.

(c) Calculate both indices for the year 1500 relative to the year 1500.

6. If all prices are measured in the same units, and quantities $q_i(0)$ and $q_i(t)$ are assumed fixed, consider whether it is meaningful (in the wide sense) to compare the Laspeyres and the Paasche price indices, that is, whether the statement

$$K_L(p_1(t), p_2(t), \ldots, p_n(t)) > K_P(p_1(t), p_2(t), \ldots, p_n(t))$$

is meaningful (in the wide sense).

7. Comparing the consumer price index for two different cities, say New York and Los Angeles, one would use different base prices and different weighting factors, but the same base year. Consider the meaningfulness of the following statements (using the index K).

(a) In a given year, the consumer price index in New York was greater than the consumer price index in Los Angeles.

(b) In a given year, the consumer price index in New York rose by a higher percentage over the previous year than did the consumer price index in Los Angeles.

8. In a detailed study of consumer confidence, Pickering et al. [1973] identified twenty-three variables related to consumer confidence. These included financial position compared with the previous year, personal expectations for economic development for the next three years, whether it is viewed as a good time to buy consumer durables, whether one has a desire to buy durables, and what are the employment expectations for next

---

*We return to these indices in Exer. 19, Sec. 4.2.
†Boot and Cox [1970].

Table 2.6. Typical Semantic Differential*

	Agree Strongly	Agree	Agree Slightly	Agree with Neither	Agree Slightly	Agree	Agree Strongly	
Financially, we, as a family, are better off than we were a year ago	☐	☐	☐	☐	☐	☐	☐	Financially, we, as a family, are less well off than we were a year ago

*From Pickering *et al.* [1973].

year. Questions about each variable were posed using the so-called seven-point semantic differential scale. A typical question looked like that shown in Table 2.6. Answers to these questions were scaled 1 to 7, where 7 was strongly agreeing with the most optimistic answer, 4 was agreeing with neither, etc. Arithmetic means of scale values were calculated. The survey was repeated later.

(a) The mean answer to the financial position question was 4.325 in the first survey (February 1971) and 4.722 in the second survey (May 1971). Consider whether or not it is meaningful to assert that the general financial position of families relative to their year-ago position improved from February 1971 to May 1971.

(b) The twenty-three variables identified do not contribute equally to consumer confidence, and many of these variables overlap. A statistical analysis called principle components analysis was performed to find relative weights of importance for each variable. (The analysis depended on the original data, not the arithmetic means.) If $q_i(t)$ is the relative weight of variable i at the tth survey, and $p_i(t)$ is the mean value of the answer to the question about variable i in the tth survey, Pickering *et al.* point out that one can use the standard Laspeyres index (Exer. 5) to calculate an index of consumer confidence:

$$\hat{I}(t) = \frac{\Sigma q_i(0)p_i(t)}{\Sigma q_i(0)p_i(0)}.$$

Consider the meaningfulness of the statement $\hat{I}(t) > \hat{I}(0)$.

(c) Pickering *et al.* suggest using a cross between a Laspeyres and a Paasche index as a measure of consumer confidence. The measure proposed is sometimes called an *ideal Fisher index*, and is calculated as

$$\hat{J}(t) = \sqrt{\frac{\Sigma q_i(0)p_i(t)}{\Sigma q_i(0)p_i(0)} \times \frac{\Sigma q_i(t)p_i(t)}{\Sigma q_i(t)p_i(0)}}.$$

This index, the authors claim, has the advantage of combining both current and base weights, and so allows for change in relative importance of different variables and their pattern of relation to purchasing behavior. Assuming that weights $q_i(t)$ and $q_i(0)$ are fixed, consider the meaningfulness of the following statements:

(i) $\hat{J}(t) > \hat{J}(0)$.

(ii) $\hat{J}(t) = 2.7\,\hat{J}(0)$.

(If the weights are not thought of as fixed, but are thought of as derived scales based on the raw data, the situation becomes more difficult to analyze.)

9. If mass were just an interval scale and all $e(r, t, k)$ used the same unit of mass, would the statement

$$\sum_{t,\,k} e(i, t, k) > \sum_{t,\,k}\sum_{j\neq i} e(j, t, k)$$

be meaningful?

10. The *severity tonnage* of a given pollutant i due to a given source is the actual tonnage times the severity factor. Table 2.7 shows various

Table 2.7.   Annual Severity Tonnage due to Emissions by Various Pollutants from Various Sources*

Pollutant	Source	Annual Quantity (10^6 tons)	Severity Factor	Severity Tonnage
Hydrocarbons	Transportation	19.8	125	2480.0
	Miscellaneous	9.2	125	1150.0
	Industry	5.5	125	688.0
	Solid waste disposal	2.0	125	250.0
	Stationary fuel combustion	0.9	125	112.5
	TOTAL	37.4		4680.5
Nitrogen oxides	Transportation	11.2	22.4	251.0
	Stationary fuel combustion	10.0	22.4	224.0
	Miscellaneous	2.0	22.4	44.8
	Solid waste disposal	0.4	22.4	9.0
	Industry	0.2	22.4	4.5
	TOTAL	23.8		533.3
Sulfur oxides	Stationary fuel combustion	24.4	15.3	373.2
	Industry	7.5	15.3	114.5
	Transportation	1.1	15.3	16.8
	Miscellaneous	0.2	15.3	3.1
	Solid waste disposal	0.2	15.3	3.1
	TOTAL	33.4		510.7
Carbon monoxide	Transportation	111.5	1	111.5
	Miscellaneous	18.2	1	18.2
	Industry	12.0	1	12.0
	Solid waste disposal	7.9	1	7.9
	Stationary fuel combustion	1.8	1	1.8
	TOTAL	151.4		151.4

*Data from Walther [1972].

pollutants, sources, and severity tonnages. From the table we can make the following statements. Consider which of them is meaningful.

(a) Hydrocarbon emissions are more severe (have greater severity tonnage) than nitrogen oxide emissions.

(b) The effects of hydrocarbon and nitrogen oxide emissions from transportation are more severe than those of hydrocarbon and nitrogen oxide emissions from industry.

(c) The effects of hydrocarbon and nitrogen oxide emissions from transportation are more severe than those of carbon monoxide emissions from industry.

(d) The effects of hydrocarbon emissions from transportation are more than twenty times as severe as the effects of carbon monoxide emissions from transportation.

(e) The total effect of hydrocarbon emissions due to all sources is more than eight times as severe as the total effect of nitrogen oxide emissions due to all sources.

11. (Pfanzagl [1968, p. 47] Show that if f is an interval scale, then the following comparison is meaningful:

$$\left[\frac{1}{n-1}\sum_{i=1}^{n}\left(f(a_i) - \bar{f}_a\right)^2\right]^{1/2} > \left[\frac{1}{n-1}\sum_{i=1}^{n}\left(f(b_i) - \bar{f}_b\right)^2\right]^{1/2},$$

where

$$\bar{f}_a = \frac{1}{n}\sum_{i=1}^{n} f(a_i), \qquad \bar{f}_b = \frac{1}{n}\sum_{i=1}^{n} f(b_i).$$

12. Table 2.8a shows the results of an experiment conducted in England to compare the performance of different stereo speakers. Relevant listening parameters were identified. Each expert rated each speaker on each parameter (rating up to 10), and the sum of the experts' scores multiplied by a weighting factor is shown in each column under the brand of speaker. The total *subjective score* of a particular speaker is obtained by adding the numbers in its column. The subjective scores are used to rank-order the speakers in Table 2.8b, first column. That is, suppose w_i is the weighting factor of the ith parameter, and $f_{ij}(a)$ is the rating of speaker a by the jth expert on the ith parameter. Speaker a is ranked over speaker b if and only if

$$\sum_i\left[w_i\sum_j f_{ij}(a)\right] > \sum_i\left[w_i\sum_j f_{ij}(b)\right]. \tag{2.40}$$

(a) Suppose f_{ij} defines an absolute scale and $w = w(i) = w_i$ defines a ratio scale.

(i) Show that the comparison (2.40) is meaningful.

(ii) Consider what happens if w_i defines an interval scale.

(b) Suppose w is an absolute scale. Show that the situation is similar to the importance ratings comparisons: if for all i, j, f_{ij} defines the same

Table 2.8. Comparisons of Stereo Speakers*

(a)
Subjective Scores
(Five assessors, each awarding up to full mark of ten for each parameter [5 × 10 = 50]. Wide range of music, known voice, white and pink noise)

Listening Parameters	Weighting Factor (W)	Maximum Possible Score (W × 50)	Marsden Hall	Omal	SMC	Goodmans	Quasar	Dahlquist	Sansui
Smoothness	1.0	50	37	$36\frac{1}{2}$	$28\frac{1}{2}$	33	$33\frac{1}{2}$	$35\frac{1}{2}$	37
Mid-frequency coloration	1.0	50	34	34	31	$34\frac{1}{2}$	$35\frac{1}{2}$	$32\frac{1}{2}$	33
Overall tonal balance	0.95	$47\frac{1}{2}$	35	32	$33\frac{1}{2}$	37	$37\frac{1}{2}$	$36\frac{1}{2}$	$36\frac{1}{2}$
Transients	0.87	$43\frac{1}{2}$	32	31	31	$32\frac{1}{2}$	32	31	32
High-frequency performance	0.71	$35\frac{1}{2}$	26	25	$22\frac{1}{2}$	25	$24\frac{1}{2}$	$22\frac{1}{2}$	$24\frac{1}{2}$
Low-frequency performance	0.66	33	24	$26\frac{1}{2}$	$22\frac{1}{2}$	24	$23\frac{1}{2}$	$24\frac{1}{2}$	21
Total Score	—	$259\frac{1}{2}$	188	185	169	186	$186\frac{1}{2}$	$182\frac{1}{2}$	184
Rank Order (on above scores)			1	4	7	3	2	6	5

(b)

Order of Preference (from Subjective Score)	Order on Basis of Performance for the Money (Subjective Score Divided by Price)
Marsden-Hall	SMC
Quasar	Sansui
Goodmans	Quasar
Omal	Marsden-Hall
Sansui	Goodmans
Dahlquist	Omal
SMC	Dahlquist

*Data from *HiFi News & Record Review*, 1975 page 107. The author thanks Issie Rabinovitch for showing him this data.

ratio scale, then the comparison (2.40) is meaningful. (It is not clear that this is a reasonable assumption for this example; what instructions were given the experts is a crucial missing piece of information.)

(c) If the weighting factor w defines a ratio scale and the f_{ij} all define a common ratio scale, is the comparison (2.40) still meaningful?

(d) Table 2.8b rank-orders the speakers on the basis of subjective score per dollar of price, a measure of "performance for the money." Consider whether this ranking is meaningful. We return to this example, and present some critical remarks, in the exercises of Sections 5.4 and 6.2.

13. Table 2.9 presents the results of a survey of seven experts by the President's Council on Physical Fitness and Sports. The entry in column j, row i is the sum of the ratings by the experts of the value of exercise j under criterion i. (Ratings by each expert were based on a scale of 0 to 3.) Consider what conclusions attainable from this table are meaningful.

14. Apply the theory developed in this section to decide when it is meaningful to say that

(a) one school district's average reading score or average intelligence test score is higher than another's;

Table 2.9. Ratings of Different Forms of Exercise*

	Jogging	Bicycling	Swimming	Skating (Ice or Roller)	Handball/Squash	Skiing (Nordic)	Basketball	Skiing (Alpine)	Tennis	Calisthenics	Walking	Golf	Softball	Bowling
PHYSICAL FITNESS														
Cardiorespiratory endurance (stamina)	21	19	21	18	19	19	19	16	16	10	13	8	6	5
Muscular endurance	20	18	20	17	18	19	17	18	16	13	14	8	8	5
Muscular strength	17	16	14	15	15	15	15	15	14	16	11	9	7	5
Flexibility	9	9	15	13	16	14	13	14	14	19	7	8	9	7
Balance	17	18	12	20	17	16	16	21	16	15	8	8	7	6
GENERAL WELL-BEING														
Weight control	21	20	15	17	19	17	19	15	16	12	13	6	7	5
Muscle definition	14	15	14	14	11	12	13	14	13	18	11	6	5	5
Digestion	13	12	13	11	13	12	10	9	12	11	11	7	8	7
Sleep	16	15	16	15	12	15	12	12	11	12	14	6	7	6
TOTAL	148	142	140	140	140	139	134	134	128	126	102	66	64	51

*Data from *Medical Times*, May 1976; data obtained from a survey of seven experts by the President's Council on Physical Fitness and Sports.

†Ratings for golf are based on the fact that many people ride a golf cart. Physical-fitness values improve if one walks in golf.

(b) one student's grade point average is higher than a second student's;

(c) one President's average popularity rating over his term was higher than another President's.

References

Abelson, R. P., and Tukey, J. W., "Efficient Conversion of Non-metric Information into Metric Information," *Proc. Stat. Assoc., Social Statistics Sec.*, (1959), 226–230.

Abelson, R. P., and Tukey, J. W., "Efficient Utilization of Non-numerical Information in Quantitative Analysis: General Theory and the Case of Simple Order," *Ann. Math. Statist.*, **34** (1963), 1347–1369.

Adams, E. W., "Elements of a Theory of Inexact Measurement," *Phil. Sci.*, **32** (1965), 205–228.

Adams, E. W., "On the Nature and Purpose of Measurement," *Synthese*, **16** (1966), 125–169.

Adams, E. W., Fagot, R. F., and Robinson, R. E., "A Theory of Appropriate Statistics," *Psychometrika*, **30** (1965), 99–127.

Allen, R. G. D., *Index Numbers in Theory and Practice*, Macmillan, London, 1975.

Anderson, N. H., "Scales and Statistics: Parametric and Non-Parametric," *Psychol. Bull.*, **58** (1961), 305–316.

Babcock, L. R., "A Combined Pollution Index for Measurement of Total Air Pollution," *J. Air Poll. Control Assoc.*, **20** (1970), 653–659.

Babcock, L. R., and Nagda, N., "Cost Effectiveness of Emission Control," *J. Air Poll. Control Assoc.*, **23** (1973), 173–179.

Bay Area Pollution Control District, "Combined Pollutant Indexes for the San Francisco Bay Area," Information Bulletin 10-68, San Francisco, California, 1968.

Boot, J. C. G., and Cox, E. B., *Statistical Analysis for Managerial Decisions*, McGraw-Hill, New York, 1970.

Campbell, N. R., *Physics: The Elements*, Cambridge University Press, Cambridge, Massachussetts, 1920. Reprinted as *Foundations of Science: The Philosophy of Theory and Experiment*, Dover, New York, 1957.

Campbell, N. R., *An Account of the Principles of Measurement and Calculation*, Longmans, Green, London, 1928.

Campbell, N. R., "Symposium: Measurement and Its Importance for Philosophy," *Proc. Arist. Soc. Suppl.*, **17** (1938), 121–142, Harrison, London.

Caretto, L. S., and Sawyer, R. F., "The Assignment of Responsibility for Air Pollution," presented at the Annual Meeting of the Society of Automotive Engineers, Detroit, Michigan, January 10–14, 1972.

Causey, R. L., "Derived Measurement and the Foundations of Dimensional Analysis," Tech. Rept. 5, Measurement Theory and Mathematical Models Reports, University of Oregon, Eugene, Oregon, April 1967.

Causey, R. L., "Derived Measurement, Dimensions, and Dimensional Analysis," *Phil. Sci.*, **36** (1969), 252–270.

Coombs, C. H., "A Theory of Psychological Scaling," *Eng. Res. Bull.*, No. 34, University of Michigan Press, Ann Arbor, Michigan, 1952.

Coombs, C. H., Raiffa, H., and Thrall, R. M., "Some Views on Mathematical Models and Measurement Theory," *Psychol. Rev.*, **61** (1954), 132–144.

Domotor, Z., "Probabilistic Relational Structures and Their Applications, Tech. Rept. 144, Institute for Mathematical Studies in the Social Sciences, Stanford University, Stanford, California, 1969.

Eichhorn, W., *Functional Equations in Economics*, Addison-Wesley, Reading, Massachusetts, 1978.

Ellis, B., "Derived Measurement, Universal Constants, and the Expression of Numerical Laws," in B. Baumrin (ed.), *Philosophy of Science: The Delaware Seminar*, Vol. 2, Interscience, New York, 1963, pp. 371–392.

Environmental Protection Agency, "National Primary and Secondary Ambient Air Quality Standards," *Federal Register*, 36:84, Part II, 8186, April 30, 1971.

Fisher, I., *The Making of Index Numbers*, Houghton Mifflin, Boston, Massachusetts, 1923.

Hangebrauck, R. P., "Environmental Assessment Methodology for Fossil Energy Processes," Industrial Environmental Research Laboratory, Office of Energy, Minerals, and Industry, Office of Research and Development, Environmental Protection Agency, Research Triangle Park, North Carolina, September 1977.

Hays, W. L., *Statistics for the Social Sciences*, Holt, Rinehart, Winston, New York, 1973.

Industrial Environmental Research Laboratory, "IERL-RTP Environmental Assessment Guideline Document, First Edition," Industrial Environmental Research Laboratory, Office of Energy, Minerals and Industry, Office of Research and Development, Environmental Protection Agency, Research Triangle Park, North Carolina, March 1976.

Krantz, D. H., Luce, R. D., Suppes, P., and Tversky, A., *Foundations of Measurement*, Vol. I, Academic Press, New York, 1971.

Lapin, L. L., *Statistics for Modern Business Decisions*, Harcourt, Brace, Jovanovich, New York, 1973.

Laspeyres, E., "Die Berechnung einer Mittleren Warrenpreissteigerung," *Jahr. Nationalökonomie und Statistik*, 16 (1871), 296–314.

Luce, R. D., "On the Possible Psychophysical Laws," *Psychol. Rev.*, 66 (1959), 81–95.

Luce, R. D., "Comments on Rozeboom's Criticisms of 'On the Possible Psychophysical Laws'," *Psychol. Rev.*, 69 (1962), 548–551.

Luce, R. D., "Remarks on the Theory of Measurement and Its Relation to Psychology," in *Les Modèles et la Formalisation du Comportement*, Editions du Centre National de la Recherche Scientifique, Paris, 1967.

Luce, R. D., "Dimensionally Invariant Numerical Laws Correspond to Meaningful Qualitative Relations," *Phil. Sci.*, 45(1978), 1–16.

Mudgett, B. D., *Index Numbers*, Wiley, New York, 1951.

Narens, L., "Measurement without Archimedean Axioms," *Phil. Sci.*, 41 (1974a), 374–393.

Narens, L., "Minimal Conditions for Additive Conjoint Measurement and Qualitative Probability," *J. Math. Psychol.*, 11 (1974b), 404–430.

National Air Pollution Control Administration, "Air Quality Criteria for Particulate Matter," NAPCA No. AP-49, Durham, North Carolina, January 1969.

Paasche, A., "Über die Preisentwicklung der Letzten Jahre, Nach den Hamburger Börsennotierungen," *Jahr. Nationalökonomie und Statistik*, 23 (1874), 168–178.

Pfanzagl, J., *Theory of Measurement*, Wiley, New York, 1968.

Pickering, J. F., Harrison, J. A., and Cohen, C. D., "Identification and Measurement of Consumer Confidence: Methodology and Some Preliminary Results," *J. Roy. Stat. Soc.*, A (1973), 43–63.

Roberts, F. S., "Building an Energy Demand Signed Digraph I: Choosing the Nodes," Rept. 927/1-NSF, The RAND Corporation, Santa Monica, California, April 1972.

Roberts, F. S., "Building and Analyzing an Energy Demand Signed Digraph," *Environment and Planning*, 5 (1973), 199–221.

Roberts, F. S., "Structural Modeling and Measurement Theory," *Tech. Forecasting and Social Change*, 1979, to appear.

Roberts, F. S., and Franke, C. H., "On the Theory of Uniqueness in Measurement," *J. Math. Psychol.*, 14 (1976), 211–218.

Robinson, R. E., "A Set-Theoretical Approach to Empirical Meaningfulness of Empirical Statements," Tech. Rept. 55, Institute for Mathematical Studies in the Social Sciences, Stanford University, Stanford, California, 1963.

Rothwell, D. P., *The Consumer Price Index Pricing and Calculation Procedures*, U.S. Department of Labor Statistics, Washington, D.C., March 1964.

Rozeboom, W. W., "The Untenability of Luce's Principle," *Psychol. Rev.*, **69** (1962a), 542–547.

Rozeboom, W. W., "Comment," *Psychol. Rev.*, **69** (1962b), 552.

Russell, B., *Principles of Mathematics*, Norton, New York, 1938.

Samuelson, P. A., *Economics*, 5th ed., McGraw-Hill, New York, 1961.

Samuelson, P. A., and Swamy, S., "Invariant Economic Index Numbers and Canonical Duality: Survey and Synthesis," *Amer. Economic Rev.*, **64** (1974), 566–593.

Sauter, G. D., and Chilton, E. G. (eds.), "Air Improvement Recommendations for the San Francisco Bay Area," The Stanford-Ames NASA/ASEE Summer Faculty Systems Design Workshop, Final Report, published by School of Engineering, Stanford University, Stanford, California, under NASA Contract NGR-05-020-409, October 1970.

Schalit, L. M., and Wolfe, K. L., "SAM/1A: A Rapid Screening Method for Environmental Assessment of Fossil Energy Process Effluents," Rep. EPA-600/7-78-015, Acurex Corporation/Aerotherm Division, Mountain View, California, February 1978.

Scott, D., and Suppes, P., "Foundational Aspects of Theories of Measurement," *J. Symbolic Logic*, **23** (1958), 113–128.

Skala, H. J., *Non-Archimedean Utility Theory*, Reidel, Boston, 1975.

Stevens, S. S., "On the Theory of Scales of Measurement," *Science*, **103** (1946), 677–680.

Stevens, S. S., "Mathematics, Measurement and Psychophysics," in S. S. Stevens (ed.), *Handbook of Experimental Psychology*, Wiley, New York, 1951, pp. 1–49.

Stevens, S. S., "On the Psychophysical Law," *Psychol. Rev.*, **64** (1957), 153–181.

Stevens, S. S., "Measurement, Psychophysics, and Utility," in C. W. Churchman and P. Ratoosh (eds.), *Measurement: Definitions and Theories*, Wiley, New York, 1959, pp. 18–63.

Stevens, S. S., "Measurement, Statistics, and the Schemapiric View," *Science*, **161** (1968), 849–856.

Suppes, P., "Measurement, Empirical Meaningfulness and Three-Valued Logic," in C. W. Churchman and P. Ratoosh (eds.), *Measurement: Definitions and Theories*, Wiley, New York, 1959, pp. 129–143.

Suppes, P. and Zinnes, J., "Basic Measurement Theory," in R. D. Luce, R. R. Bush, and E. Galanter (eds.), *Handbook of Mathematical Psychology*, Vol. I, Wiley, New York, 1963, pp. 1–76.

Torgerson, W. S., *Theory and Methods of Scaling*, Wiley, New York, 1958.

Walther, E. G., "A Rating of the Major Air Pollutants and their Sources by Effect," *J. Air Poll. Control Assoc.*, **22** (1972), 352–355.

Three Representation Problems: Ordinal, Extensive, and Difference Measurement

3.1 Ordinal Measurement

3.1.1 *Representation Theorem in the Finite Case*

In this chapter, we study the representation problem of fundamental measurement. We study two representations that arose in our discussion of temperature, preference, mass, and the like. These are the representations $(A, R) \to (Re, >)$, and $(A, R, \mathbf{o}) \to (Re, >, +)$, where R is a binary relation on A, and $\mathbf{o}$ is an operation on A. We present axioms on (A, R) and $(A, R, \mathbf{o})$ necessary and sufficient for the existence of the desired homomorphisms, and we present a uniqueness theorem for each of the representations. We then study a third representation, which also arises in the measurement of temperature and of preference.

In this section we illustrate the simplest case of fundamental measurement, that dealing with the relational system (A, R). We seek a real-valued function f on A such that for all $a, b \in A$,

$$aRb \Leftrightarrow f(a) > f(b). \tag{3.1}$$

This representation arose in our discussion of temperature and our discussion of preference, and it arises in many measurement situations. We begin with a representation theorem for the case where A is finite.

THEOREM 3.1. *Suppose A is a finite set and R is a binary relation on A. Then there is a real-valued function f on A satisfying*

$$aRb \Leftrightarrow f(a) > f(b) \tag{3.1}$$

ENCYCLOPEDIA OF MATHEMATICS and Its Applications, Gian-Carlo Rota (ed.). Vol. 7: Fred S. Roberts, Measurement Theory

if and only if (A, R) is a strict weak order.

Before beginning the proof, let us recall that (A, R) is strict weak if and only if it is

(i) asymmetric: $aRb \Rightarrow \sim bRa$

and

(ii) negatively transitive: $\sim aRb \ \& \ \sim bRc \Rightarrow \sim aRc$.

If R is (strict) preference, a function f satisfying Eq. (3.1) is called an (ordinal) utility function. Thus, to see if we can measure a person's preferences to the extent of producing an ordinal utility function, we simply check whether or not these preferences satisfy the conditions of asymmetry and negative transitivity. In general, we could do this by doing a *pair comparison experiment*. For every pair of alternatives a and b in A, we present a and b and ask the individual to tell us which, if any, he prefers. We present these pairs in a random order, and use his judgments to define preference. Then we check if asymmetry and negative transitivity are satisfied. (The results might be presented in a table like Table 3.1, which shows an individual's preferences among several composers.)

As we remarked in the Introduction, there are two interpretations for axioms such as asymmetry and negative transitivity. One is that these are testable conditions which describe what a person's preferences must be for measurement to take place. We then take the *descriptive approach*, and simply ask whether or not a person's preferences satisfy these conditions. Alternatively, we could use these axioms to define *rationality*. We could say that an individual who violates these axioms is acting irrationally. Indeed, many would say that an individual presented with a violation of, say, negative transitivity would say: "Oh, I've made a mistake." This approach is the *prescriptive* or *normative approach*, and it is usually the approach taken in economic theory. The representation theorem is used to define the class of individuals to whom the theory applies, the so-called rational individuals.* It is this second approach or interpretation that we apply to this representation theorem if it is used to study measurement of temperature. If R is the relation "warmer than," we think of the conditions of asymmetry and negative transitivity as conditions of rationality, which must be satisfied before measurement can take place.

Of course, in the case of warmer than, we can also think of these conditions as testable, and we would be surprised if an individual violated them, or at least we might be tempted to think of violations more as experimental errors than "real" violations. In the case of preference, the situation is different. If we subject the asymmetry and negative transitivity conditions to an experimental test, with R taken to be strict preference, then we often find they are violated. For example, an individual may think

---

*See footnote on page 4 for a distinction between prescriptive and normative, between the idealized "superrational being" and the "normally intelligent" individual. This is a distinction made in Keeney and Raiffa [1976].

Table 3.1. Preferences among Composers
(Item i is preferred to item j iff the i, j entry is 1.)

	Mozart	Haydn	Brahms	Beethoven	Wagner	Bach	Mahler	Strauss	Row Sum
Mozart	0	1	1	0	0	0	1	1	4
Haydn	0	0	1	0	0	0	1	0	2
Brahms	0	0	0	0	0	0	0	0	0
Beethoven	1	1	1	0	1	0	1	1	6
Wagner	0	1	1	0	0	0	1	1	4
Bach	1	1	1	0	1	0	1	1	6
Mahler	0	0	1	0	0	0	0	0	1
Strauss	0	1	1	0	0	0	1	0	3

price is more important than quality, but choose on the basis of quality if prices are close. Thus, if a and b are close in price and b and c are close in price, he may prefer a to b because a is of higher quality than b, and he may prefer b to c because b is of higher quality than c. But he may prefer c to a because c is sufficiently lower in price than a to make a difference. Then transitivity of preference is violated. Moreover, so is negative transitivity, for he does not prefer c to b and he does not prefer b to a, but he prefers c to a.[†] If one of the axioms such as negative transitivity is violated, then the representation (3.1) cannot be achieved.

If the violation of a measurement axiom is systematic—that is, if it has some sort of pattern—then often a different measurement representation must be sought. In the case of preference where some form of transitivity is violated, we shall describe in detail one such alternative representation (using the notion of semiorder) in Section 6.1, and we shall mention another (the additive difference model) in Section 5.5.

Alternatively, seeing a systematic violation of a measurement axiom, one can find some "statistical" pattern to it. Indeed, as Falmagne [1976a] argues, statistical regularities are the only regularities one is likely to find, at least in the behavioral sciences. Hence, a statistical or random analogue of the deterministic fundamental measurement theories must be developed. Without this, Falmagne argues, the measurement theory models we describe will not pass many experimental tests. Falmagne [1976a, 1978, 1979] and Falmagne, Iverson, and Marcovici [1978] begin to make progress on such random analogues of deterministic measurement theories. A related approach based on the notion of probabilistic consistency is described in Sec. 6.2 below.

[†]This argument is due to Krantz et al. [1971, p. 17]. We shall present other arguments against these simple axioms for preference later in this volume.

Sometimes, the violation of a measurement axiom could be due to a matter of minor errors or "noise." In that case, a theory of error or noise is called for. See Adams [1965] or Krantz *et al.* [to appear] for a discussion of error theories and Adams and Carlstrom [1979] for a new approach to error. In a situation of error or noise, some statistical discussion of goodness of fit of a measurement representation is called for. The literature of measurement theory has not been very helpful on the development of such statistical tests. For some recent foundational work on this subject, see Falmagne [1976b]. In general, much foundational work still needs to be done along these lines. Much practical work also needs to be done, to subject the measurement axioms we shall present in different places in this volume to a systematic experimental test in a variety of contexts.

To prove Theorem 3.1, let us suppose first that there is a homomorphism f satisfying Eq. (3.1). We show that (A, R) is strict weak. First, (A, R) is asymmetric. For if aRb, then $f(a) > f(b)$, whence not $f(b) > f(a)$, and $\sim bRa$. Second, (A, R) is negatively transitive. For if $\sim aRb$ and $\sim bRc$, then $\sim [f(a) > f(b)]$ and $\sim [f(b) > f(c)]$, so $f(a) \leqq f(b)$ and $f(b) \leqq f(c)$. It follows that $f(a) \leqq f(c)$, so $\sim [f(a) > f(c)]$, so $\sim aRc$.

Conversely, suppose (A, R) is a strict weak order. The proof that a homomorphism f exists is constructive. We define $f(x)$ as follows:

$$f(x) = \text{the number of } y \text{ in } A \text{ such that } xRy. \qquad (3.2)$$

Let us begin by illustrating this construction. Suppose $A = \{a, b, c, d\}$ and

$$R = \{(a, c), (a, d), (b, c), (b, d), (c, d)\}.$$

Then it is easy to show that (A, R) is a strict weak order. The function f defined by (3.2) is given by

$$f(a) = 2 \qquad (aRc, aRd),$$
$$f(b) = 2,$$
$$f(c) = 1,$$
$$f(d) = 0.$$

It is easy to check that f is a homomorphism.

To prove formally that a function f defined by Eq. (3.2) satisfies Eq. (3.1), we recall that by Theorem 1.3, every strict weak order is transitive. If aRb, then by transitivity of R, bRy implies aRy for every y. Thus, the number of y such that aRy is at least as big as the number of y such that bRy. It follows that $f(a) \geqq f(b)$. Moreover, aRb but not bRb, since a strict weak order is irreflexive. Thus, $f(a) > f(b)$. Conversely, if $\sim aRb$, then $\sim bRy$ implies $\sim aRy$, by negative transitivity. Hence, aRy implies bRy, so $f(b) \geqq f(a)$, whence $\sim [f(a) > f(b)]$. This proves (3.1) and completes the proof of Theorem 3.1.

It should be remarked that if A is finite and f defined by (3.2) does not give a homomorphism, then (A, R) is not strict weak, and so there is no homomorphism. Thus, this gives us another test for existence of a homomorphism—simply try to build one by a fixed procedure and verify whether or not you have succeeded. We state the result as a corollary.

COROLLARY 1. *Suppose A is a finite set and R is a binary relation on A. Then there is a real-valued function f on A satisfying Eq. (3.1) if and only if the following function f satisfies Eq. (3.1):*

$$f(x) = \text{the number of } y \text{ such that } xRy. \tag{3.2}$$

Let us apply these ideas to measurement of preference. Suppose an individual is asked his preferences among composers in a pair comparison format, and gives the data shown in Table 3.1. Then $f(x)$ as defined by Eq. (3.2) is given by the row sum of row x. Specifically, we obtain

$$f(\text{Beethoven}) = f(\text{Bach}) = 6,$$

$$f(\text{Wagner}) = f(\text{Mozart}) = 4,$$

$$f(\text{Strauss}) = 3,$$

$$f(\text{Haydn}) = 2,$$

$$f(\text{Mahler}) = 1,$$

$$f(\text{Brahms}) = 0.$$

If the matrix is rearranged so that alternatives are listed in descending order of row sums (with arbitrary ordering in case of ties), then we can easily test whether f is a homomorphism by checking that there are 1's in row x for all those y with $f(y) < f(x)$. This should give a block of 1's in row x from some point to the end. In our example, f is a homomorphism. (The rearranged matrix is shown in Table 3.2.) Thus, f is an ordinal utility function for this individual.

On the other hand, if preferences are expressed as in Table 3.3, then $f(x)$ as defined by Eq. (3.2) is

$$f(\text{Mozart}) = f(\text{Haydn}) = 6,$$

$$f(\text{Beethoven}) = 5,$$

$$f(\text{Brahms}) = 4,$$

$$f(\text{Wagner}) = 3,$$

$$f(\text{Mahler}) = 2,$$

$$f(\text{Bach}) = 1,$$

$$f(\text{Strauss}) = 0.$$

Table 3.2.   Table 3.1 Rearranged, with the Blocks of 1's Shown

	Beethoven	Bach	Wagner	Mozart	Strauss	Haydn	Mahler	Brahms	Row Sum
Beethoven	0	0	1	1	1	1	1	1	6
Bach	0	0	1	1	1	1	1	1	6
Wagner	0	0	0	0	1	1	1	1	4
Mozart	0	0	0	0	1	1	1	1	4
Strauss	0	0	0	0	0	1	1	1	3
Haydn	0	0	0	0	0	0	1	1	2
Mahler	0	0	0	0	0	0	0	1	1
Brahms	0	0	0	0	0	0	0	0	0

Table 3.3.   Preferences among Composers

(Item i is preferred to item j iff the i,j entry is 1.)

	Mozart	Haydn	Brahms	Beethoven	Wagner	Bach	Mahler	Strauss	Row Sum
Mozart	0	1	1	0	1	1	1	1	6
Haydn	0	0	1	1	1	1	1	1	6
Brahms	0	0	0	0	1	1	1	1	4
Beethoven	1	0	0	0	1	1	1	1	5
Wagner	0	0	0	0	0	1	1	1	3
Bach	0	0	0	0	0	0	0	1	1
Mahler	0	0	0	0	0	1	0	1	2
Strauss	0	0	0	0	0	0	0	0	0

Rearranging the matrix, we see in Table 3.4 that the row corresponding to Mozart has its block of 1's broken by a 0 in the Mozart, Beethoven entry. We discover that Beethoven is preferred to Mozart, even though f(Beethoven) $<$ f(Mozart). Thus, f does not define a homomorphism and so, by Corollary 1 to Theorem 3.1, there can be none. There is no ordinal utility function. This implies that one of the strict weak order axioms will have to be broken, and it is not hard to discover that negative transitivity is violated: Beethoven is preferred to Mozart, but Haydn is not preferred to Mozart, and Beethoven is not preferred to Haydn.

Table 3.4.   Table 3.3 Rearranged

	Mozart	Haydn	Beethoven	Brahms	Wagner	Mahler	Bach	Strauss	Row Sum
Mozart	0	1	⓪	1	1	1	1	1	6
Haydn	0	0	1	1	1	1	1	1	6
Beethoven	1	0	0	0	1	1	1	1	5
Brahms	0	0	0	0	1	1	1	1	4
Wagner	0	0	0	0	0	1	1	1	3
Mahler	0	0	0	0	0	0	1	1	2
Bach	0	0	0	0	0	0	0	1	1
Strauss	0	0	0	0	0	0	0	0	0

It can be a rather time-consuming process to perform a pair comparison experiment and make an individual compare every pair of elements. The set of pairs can be very large. In Chapter 7 we describe an alternative practical procedure for calculating an ordinal utility function, if one exists, which avoids this difficulty. The procedure is based on the expected utility hypothesis. An extensive discussion of practical techniques for calculating utility functions which are quite different from the techniques we have described can be found in Keeney and Raiffa [1976]. Their idea is to assess the general shape of an individual's utility function by judging his attitude toward risk. Then several data points are obtained and a curve of the determined shape is fitted.

Before leaving Theorem 3.1, we state an additional corollary.

COROLLARY 2. *Suppose A is a finite set and R is a binary relation on A. Then there is a real-valued function f on A satisfying*

$$aRb \Leftrightarrow f(a) \geq f(b) \tag{3.3}$$

if and only if (A, R) is a weak order.

Proof. If (A, R) is a weak order, define S on A by

$$aSb \Leftrightarrow \sim bRa.$$

Then (A, S) is a strict weak order. (Proof is left to the reader.) If f on A satisfies Eq. (3.1) with S in place of R, then it satisfies Eq. (3.3) for (A, R). The proof that if f satisfies Eq. (3.3) then (A, R) is weak is left to the reader. ∎

3.1.2 *The Uniqueness Theorem*

Turning next to the uniqueness problem, we state a uniqueness theorem for the representation (3.1).

THEOREM 3.2. *Suppose A is a finite set, R is a binary relation on A, and f is a real-valued function on A satisfying*

$$aRb \Leftrightarrow f(a) > f(b). \tag{3.1}$$

Then $\mathfrak{A} = (A, R) \to \mathfrak{B} = (Re, >)$ is a regular representation and $(\mathfrak{A}, \mathfrak{B}, f)$ is an ordinal scale.

Proof. We first show that every function f satisfying Eq. (3.1) is a regular scale, so we are dealing with a regular representation. Suppose g is another function satisfying Eq. (3.1). Then $f(a) = f(b)$ implies $\sim aRb$ and $\sim bRa$, which implies $g(a) = g(b)$. Regularity of f follows by Theorem 2.1.

Next we show that $(\mathfrak{A}, \mathfrak{B}, f)$ is an ordinal scale. If $\phi: f(A) \to Re$ is any monotone increasing function, then $\phi \circ f$ satisfies (3.1) whenever f does. For

$$(\phi \circ f)(a) > (\phi \circ f)(b) \Leftrightarrow f(a) > f(b) \Leftrightarrow aRb.$$

Conversely, suppose f satisfies Eq. (3.1) and we are given a function $\phi: f(A) \to Re$ such that $\phi \circ f$ also satisfies (3.1). We shall show that ϕ is monotone increasing. Suppose α and β are in $f(A)$, with $\alpha = f(a)$ and $\beta = f(b)$. Then

$$\alpha > \beta \Leftrightarrow aRb \Leftrightarrow (\phi \circ f)(a) > (\phi \circ f)(b) \Leftrightarrow \phi(\alpha) > \phi(\beta).$$

We conclude that the class of admissible transformations is the class of monotone increasing functions, and so $(\mathfrak{A}, \mathfrak{B}, f)$ is an ordinal scale. ∎

Let us apply the uniqueness theorem (Theorem 3.2) to the case of temperature. The representation (3.1) arises in the measurement of temperature if R is interpreted as "warmer than." Applying Theorem 3.2, we conclude that temperature is an ordinal scale, whereas in Section 2.3 we suggested that temperature is an interval scale. Is there something wrong with our whole formalism? The answer is that we have not made use of all the properties that a scale of temperature preserves. We can obtain temperature as an interval scale if we observe that it is possible to make judgments of comparative temperature difference. To make this precise in the theory, one introduces a quaternary relation D on a set A of objects whose temperatures are being compared. The relation $D(a, b, s, t)$, or $abDst$, is interpreted to mean that the difference between the temperature of a and the temperature of b is judged to be more than the difference between the temperature of s and the temperature of t. One seeks a real-valued function f on A such that for all $a, b, s, t \in A$,

$$abDst \Leftrightarrow f(a) - f(b) > f(s) - f(t). \tag{3.4}$$

Under some reasonable assumptions, this numerical assignment is regular and unique up to a positive linear transformation and hence defines an interval scale. We return to the representation (3.4) in Section 3.3.

It should be emphasized here how the desired properties of a scale can influence our choice of a representation. We were able to measure temperature in the sense of Eq. (3.1), but obtained a scale that did not have strong enough properties. Thus, we were led to seek a more stringent representation, but also one that can be based on sensible empirical relations.

3.1.3 *The Countable Case*

Most practical applications of measurement or scaling deal with the case of a finite set of objects A. However, if measurement representations are used to put measurement on a firm theoretical foundation, it is important to consider theoretically infinite populations. Thus, we shall frequently ask whether a result has an analogue for infinite sets of objects. Theorems 3.1 and 3.2 actually hold in the more general case where A is *countable*, that is, may be put in one-to-one correspondence with a set of positive integers.* The representation part of this result is due to Cantor [1895]. We proceed to prove it.

THEOREM 3.3 (Cantor). *Suppose A is a countable set and R is a binary relation on A. Then there is a real-valued function f on A satisfying*

$$aRb \Leftrightarrow f(a) > f(b) \tag{3.1}$$

if and only if (A, R) is a strict weak order. Moreover, if there is such an f, then $\mathfrak{A} = (A, R) \to \mathfrak{B} = (Re, >)$ is a regular representation and $(\mathfrak{A}, \mathfrak{B}, f)$ is an ordinal scale.

Proof. Suppose (A, R) is strict weak. Since A is countable, we may list the elements of A as $x_1, x_2, x_3, \ldots$. Then, we define

$$r_{ij} = \begin{cases} 1 & \text{if } x_i R x_j, \\ 0 & \text{otherwise.} \end{cases}$$

*Our use of the word countable includes finite sets, and we use the word *denumerable* for sets which are in one-to-one correspondence with the whole set of positive integers.

The definition of a function f that satisfies Eq. (3.1) is analogous to the explicit definition in Eq. (3.2). Namely, we take*

$$f(x_i) = \sum_{j=1}^{\infty} \frac{1}{2^j} r_{ij}. \tag{3.5}$$

The sum in (3.5) converges, since

$$\sum_{j=1}^{\infty} \frac{1}{2^j}$$

converges. Now, given a and b in A, we have $a = x_i$ and $b = x_k$, some i and k. If aRb, then, as in the proof of Theorem 3.1, we observe that $bRy \Rightarrow aRy$ and moreover that aRb but not bRb. Hence, by construction,

$$f(x_i) \geq f(x_k) + \frac{1}{2^k} > f(x_k);$$

that is, $f(a) > f(b)$. Conversely, if $\sim aRb$, then again as in the proof of Theorem 3.1, $aRy \Rightarrow bRy$, so $f(b) \geq f(a)$, and $\sim [f(a) > f(b)]$. This proves (3.1).

The converse is proved just as it was in Theorem 3.1 and the uniqueness is proved just as it was in Theorem 3.2.   ■

We next state several corollaries of Theorem 3.3. The first uses the notion of reduction defined in Section 1.5.

COROLLARY 1. *Suppose* (A, R) *is a strict weak order*, (A^*, R^*) *is its reduction, and* A^* *is countable. Then there is a real-valued function f on A satisfying*

$$aRb \Leftrightarrow f(a) > f(b). \tag{3.1}$$

Proof. Recall that by Theorem 1.4, if (A, R) is a strict weak order, then (A^*, R^*) is a strict simple order and hence a strict weak order. By Theorem 3.3, there is a real-valued function F on A^* satisfying Eq. (3.1) with R^* in place of R. Then let $f(a) = F(a^*)$, where a^* is the equivalence class in A^* containing a. The function f satisfies (3.1) for R.   ■

COROLLARY 2. *Suppose A is a countable set and R is a binary relation on A. Then there is a real-valued function f on A satisfying*

$$aRb \Leftrightarrow f(a) \geq f(b) \tag{3.3}$$

*This idea is due to David Radford (personal communication).

if and only if (A, R) is a weak order. Moreover, if there is such an f, then the representation $\mathfrak{A} = (A, R) \rightarrow \mathfrak{B} = (Re, \geq)$ is regular and $(\mathfrak{A}, \mathfrak{B}, f)$ is an ordinal scale.

Proof. The proof is analogous to that of Corollary 2 of Theorem 3.1. ∎

3.1.4 *The Birkhoff–Milgram Theorem*

It is easy to show that Theorem 3.3 is false without the assumption that A be countable. To give a counterexample, suppose we let $A = Re \times Re$, and we define R on A by

$$(a, b)R(s, t) \Leftrightarrow [a > s \text{ or } (a = s \,\&\, b > t)]. \tag{3.6}$$

This relation (A, R) is called the *lexicographic ordering of the plane*. The lexicographic ordering of the plane corresponds to the ordering of words in a dictionary. We order first by first letter, in case of the same first letter, by second letter, and so on. It is easy to see that (A, R) is a strict weak order, indeed that it is strict simple. We shall show that there is no real-valued function f on A satisfying (3.1). For suppose that such an f exists. Now we know that $(a, 1)R(a, 0)$. Thus, by (3.1), $f(a, 1) > f(a, 0)$. We know that between any two real numbers there is a rational. Thus, there is a rational number $g(a)$ so that

$$f(a, 1) > g(a) > f(a, 0).$$

Now the function g is defined on the set Re and maps it into the set of rationals. Moreover, it maps Re into the rationals in a one-to-one fashion. For, if $a \neq b$, then either $a > b$ or $b > a$, say $a > b$. Then we have

$$g(a) > f(a, 0) > f(b, 1) > g(b),$$

from which we conclude that $g(a) > g(b)$. It is well known that there can be no one-to-one mapping from the reals into the rationals. Thus, we have reached a contradiction. We conclude that the strict simple (weak) order (A, R) cannot be represented in the form (3.1).

A theorem stating conditions on a binary relation (A, R) both necessary and sufficient for the existence of a representation satisfying (3.1) can be stated, even if A is uncountable. Any (A, R) that can be mapped homomorphically into the real numbers must reflect the properties of the reals. In particular, the reals have a countable subset, the rationals, which is order-dense in the sense that whenever $a > b$ for *nonrationals* a and b, there is a rational c such that $a > c > b$. In general, if (A, R) is a binary relation and $B \subseteq A$, let us say that B is *order-dense* in (A, R) if, whenever a and b are in $A - B$, aRb and $\sim bRa$,* then there is a c in B such that

---

*For asymmetric relations, we do not have to assume $\sim bRa$, for this follows from aRb.

aRcRb. The concept of order-denseness is different from the more well-known concept of denseness: B is said to be *dense* in (A, R) if whenever a and b are in A, aRb and $\sim bRa$, then there is c in B such that $aRcRb$. Every dense subset B of (A, R) is order-dense. But the converse is not true. Take A to be the integers and B to be the even integers. Then between any two odd integers (elements of $A - B$) there is an even integer, but it is not true that between every two integers there is an even integer.

Suppose we concentrate momentarily on strict simple orders. It turns out that what characterizes those strict simple orders representable in the form (3.1) is that they have a countable order-dense subset. (It is not necessarily true that they have a countable dense subset: the integers under the ordering $>$ do not.) From the result about strict simple orders it will follow that a strict weak order is representable in the form (3.1) if and only if its reduction (A^*, R^*) has a countable order-dense subset.

THEOREM 3.4 (Birkhoff–Milgram). *Suppose (A, R) is a strict simple order. Then there is a real-valued function f on A satisfying*

$$aRb \Leftrightarrow f(a) > f(b) \tag{3.1}$$

if and only if (A, R) has a countable order-dense subset. Moreover, if there is such an f, then the representation $\mathfrak{A} = (A, R) \to \mathfrak{B} = (Re, >)$ is a regular representation and $(\mathfrak{A}, \mathfrak{B}, f)$ is an ordinal scale.

The representation part of this theorem was apparently first proved in Milgram [1939]. It is proved in Birkhoff [1948], though his proof is incomplete. We shall present a proof at the end of this section.*

We now state several corollaries of Theorem 3.4.

COROLLARY 1 (Birkhoff–Milgram Theorem).[†] *Suppose (A, R) is a binary relation. Then there is a real-valued function f on A satisfying*

$$aRb \Leftrightarrow f(a) > f(b) \tag{3.1}$$

if and only if (A, R) is a strict weak order and its reduction (A^*, R^*) has a countable order-dense subset. Moreover, if there is such an f, then the representation $\mathfrak{A} = (A, R) \to \mathfrak{B} = (Re, >)$ is regular and $(\mathfrak{A}, \mathfrak{B}, f)$ is an ordinal scale.

---

*It should be remarked that, while the condition of having a countable order-dense subset seems reasonable, it is not empirically testable, for it would be impossible to explicitly verify this condition for real data. Any experiment would only give finite data. Thus, this axiom is one we either accept or reject as a reasonable idealization.

[†] Both Theorem 3.4 and this Corollary are called the Birkhoff–Milgram Theorem.

Proof. Sufficiency follows from Theorem 3.4, for by Theorem 1.4, (A^*, R^*) is strict simple, and if F gives a representation for (A^*, R^*), then $f(a) = F(a^*)$ gives a representation for (A, R). To prove the converse, one proceeds exactly as in the proof of Theorem 3.1 to show that (A, R) is strict weak. To prove that (A^*, R^*) has a countable order-dense subset, let $F : A^* \to Re$ be defined by $F(a^*) = f(a)$. It is simple to show that F is well-defined, for if a and b are in the same equivalence class, then $\sim aRb$ and $\sim bRa$, so $f(a) = f(b)$. Now F satisfies Eq. (3.1) for (A^*, R^*) and so, by Theorem 3.4, (A^*, R^*) has a countable order-dense subset. Uniqueness of f is proved just as in the proof of Theorem 3.2. ∎

COROLLARY 2. *Suppose (A, R) is a binary relation. Then there is a real-valued function f on A satisfying*

$$aRb \Leftrightarrow f(a) \geqq f(b) \qquad (3.3)$$

if and only if (A, R) is a weak order and its reduction (A^*, R^*) has a countable order-dense subset. Moreover, if there is such an f, then the representation $\mathfrak{A} = (A, R) \to \mathfrak{B} = (Re, \geqq)$ is regular and $(\mathfrak{A}, \mathfrak{B}, f)$ is an ordinal scale.

We now present several examples to illustrate the Birkhoff–Milgram Theorem. Suppose first that (A, R) is the lexicographic ordering of the plane, defined in Eq. (3.6). Then (A, R) is strict simple, and $(A^*, R^*) = (A, R)$. Since we already know that (A, R) is not homomorphic to $(Re, >)$, Theorem 3.4 tells us that (A, R) can have no countable order-dense subset.

To give a second example, let $A = [0, 1] \cup [2, 3]$ and let R be $>$ on A. Then (A, R) is homomorphic to $(Re, >)$, by the homomorphism $f(a) = a$. Now (A, R) is strict simple, so it follows by Theorem 3.4 that (A, R) has a countable order-dense subset. Such a subset is the set of all rationals in $[0, 1] \cup [2, 3]$.

To give a third example, let $A = [0, \pi] \cup [2\pi, 3\pi]$, and let R be $>$ on A. Then (A, R) is strict simple, and it is homomorphic to $(Re, >)$ by the map $f(a) = a$. However, the set B of rationals in $[0, \pi] \cup [2\pi, 3\pi]$ does not form a countable order-dense subset of A. For π and 2π are in $A - B$, $2\pi R\pi$ (and hence $\sim \pi R 2\pi$), but there is no element c of B such that $2\pi R c R\pi$. To obtain a countable order-dense set, we must add π and 2π to B.

To give a fourth example, let $A = \{0, 1\}$ and let R be $>$ on A. Then (A, R) is homomorphic to $(Re, >)$ by the map $f(a) = a$. The set $\{0, 1\}$ is a countable order-dense subset of A.

To give one final example, suppose that $A = Re \times Re$ and that $(a, b)R(s, t)$ holds if and only if $a > s$. Then (A, R) is strict weak and (A^*, R^*) has a countable order-dense subset, consisting of all equivalence classes that contain an element (s, t) with both s and t rational. It follows from Corollary 1 to Theorem 3.4 that (A, R) is homomorphic to $(Re, >)$. The homomorphism is easy to describe explicitly; it is $f[(a, b)] = a$.

We close this section by proving Theorem 3.4. The reader may skip the proof without loss of continuity.*

To present the proof, we use the following notion: If (A, R) is an asymmetric binary relation and $a, b \in A$ are such that aRb holds but $aRcRb$ fails for all c, then (a, b) is called a *gap*, and the points a and b are called, respectively, the *lower* and *upper end points* of the gap. In our third example above, $(2\pi, \pi)$ is a gap. Our preliminary result, which follows one of Krantz *et al.* [1971], is about the set G of all a which are end points of some gap.

LEMMA. *Suppose (A, R) is an asymmetric, complete binary relation and G is the set of all a which are end points of some gap. Then G is countable if either (A, R) has a countable order-dense subset or there is a real-valued function f on A satisfying (3.1).*

Proof. Let G_1 be the set of lower end points of gaps and G_2 be the set of upper end points. Suppose first that there is a countable order-dense subset B. Whenever (a, b) is a gap, order-denseness of B implies that either $a \in B$ or $b \in B$. Thus we may map $G_1 - B$ into B in a one-to-one manner: if $a \notin B$, map it into b. Similarly, we may map $G_2 - B$ into B in a one-to-one manner. (Proof that the maps are one-to-one requires completeness of (A, R), and is left to the reader.) It follows that $G_1 - B$ and $G_2 - B$ are countable, since they are in a one-to-one correspondence with a subset of a countable set. Thus, G is countable, since

$$G = (G_1 - B) \cup (G_2 - B) \cup (B \cap G)$$

and the union of three countable sets is countable.

Next, suppose that f satisfies (3.1). If (a, b) is a gap, then there is a rational r such that $f(a) > r > f(b)$. Thus, we may define a one-to-one correspondence between G_1 and a subset of the set of rationals, and similarly for G_2. We shall show that the correspondences are one-to-one. Since the rationals are countable, so are G_1 and G_2, and hence so is G, which completes the proof.

We show that the correspondence between G_1 and a subset of the rationals is one-to-one. The proof for G_2 is similar. Suppose (a, b) and

---

*An alternative proof, which for the most part avoids the unpleasant notion of "gap" that arises in the following, is outlined in Exers. 19 through 21. This proof is due to David Radford (personal communication).

(a', b') are two different gaps, and r and r' are rationals so that

$$f(a) > r > f(b),$$
$$f(a') > r' > f(b').$$

It is easy to show that $a \neq a'$ and $b \neq b'$. Since (a, b) is a gap, $\sim aRa'$ or $\sim a'Rb$. Hence, by completeness, $a'Ra$, $b = a'$, or bRa'. In the latter two cases,

$$r > f(b) \geqq f(a') > r',$$

so $r \neq r'$. In the former case, since (a', b') is a gap and $a'Ra$, we have $\sim aRb'$. Thus, we have $b' = a$ or $b'Ra$. In either case,

$$r' > f(b') \geqq f(a) > r,$$

so $r \neq r'$.   ■

   Suppose (A, R) is strict simple. Suppose first that (A, R) has a countable order-dense subset B and let $B' = B \cup G$. We construct a function f satisfying (3.1). By the lemma, B' is countable. Thus we may list the elements of B' as $x_1, x_2, x_3, \ldots$. If x and y are any elements of A, define

$$r(x, y) = \begin{cases} 1 & \text{if } xRy, \\ 0 & \text{otherwise.} \end{cases}$$

Analogously to Eq. (3.5), define*

$$f(x) = \sum_{j=1}^{\infty} \frac{1}{2^j} r(x, x_j). \tag{3.7}$$

Then, the sum in (3.7) converges, since

$$\sum_{j=1}^{\infty} \frac{1}{2^j}$$

does. Moreover, f satisfies Eq. (3.1). For, suppose that aRb. Then bRx_j implies aRx_j. Moreover, there is some j so that aRx_j but not bRx_j. This is immediate if b is in B'. If b is not in B', it is not in G, so (a, b) is not a gap and there is $c \in A$ with $aRcRb$. If c is in B', then we may take $x_j = c$. If c is not in B', then by order-denseness of B, there is d in B so that $cRdRb$. We may take $x_j = d$. In any case, the conclusion is that

$$f(a) \geqq f(b) + \frac{1}{2^j} > f(b),$$

*Once again, this idea is due to David Radford (personal communication).

so aRb implies $f(a) > f(b)$. Conversely, if $\sim aRb$, then as in the proof of Theorem 3.1, aRy implies bRy, so $f(b) \geq f(a)$, and $\sim [f(a) > f(b)]$.

To prove the converse, suppose that there is a function f on A satisfying Eq. (3.1). We show that there is a countable order-dense subset B of A.*
Let J be the set of ordered pairs (r, r') such that r and r' are rational numbers and $r' > f(a) > r$ for some $a \in A$. For all $(r, r') \in J$, let $a_{rr'}$ be one such element a, and let $B_1 = \{a_{rr'}\}$.† The set B_1 is countable because the Cartesian product of the set of rationals with itself can be mapped *onto* B_1, and it is well-known that the Cartesian product of a countable set with itself is countable. Let G be the collection of end points of gaps in (A, R). By the lemma, G is countable, and hence so is $B = G \cup B_1$. We show that B is order-dense. Suppose aRb holds. If (a, b) is a gap, then $a, b \in B$. If (a, b) is not a gap, then there is a $c \in A$ such that $aRcRb$. Choose rationals r and r' such that $f(a) > r' > f(c) > r > f(b)$. Then $(r, r') \in J$, and $a_{rr'}$ is such that $f(a) > f(a_{rr'}) > f(b)$ and $a_{rr'}$ is in B.

The uniqueness statement of Theorem 3.4 is proved just as that of Theorem 3.2. This completes the proof.

Exercises

1. Let $A = \{0, 1, 2, 3, 4\}$, and let R be $<$ on A. Show the following:
   (a) The function f of Eq. (3.2) is given by $f(x) = 4 - x$.
   (b) (A, R) is homomorphic to $(Re, >)$ via f.

2. A strict weak ordering R can be defined by ranking alternatives in vertical order, with aRb if and only if a is higher than b in the order. Show that the order is not necessarily strict simple, as long as two elements can be equally high in the list.

3. (a) The following preference ranking of cities as places to visit defines a strict weak ordering as in Exer. 2:

> Rome–London–Copenhagen
> Paris
> Athens
> Vienna
> Moscow–Stockholm
> Brussels

---

*Our proof again follows that of Krantz *et al.* [1971].

†The choice of one representative from each of an infinite class of sets involves the assumption known as the Axiom of Choice. We do not dwell on this point here, but refer the reader to works on set theory like that of Bernays [1968] for a discussion of the Axiom of Choice and its numerous equivalent versions. The referees have pointed out that the Birkhoff–Milgram Theorem can be proved without the use of the Axiom of Choice.

Show that an ordinal utility function is given by

$$f(\text{Rome}) = f(\text{London}) = f(\text{Copenhagen}) = 6,$$
$$f(\text{Paris}) = 5,$$
$$f(\text{Athens}) = 4,$$
$$f(\text{Vienna}) = 3,$$
$$f(\text{Moscow}) = f(\text{Stockholm}) = 1,$$
$$f(\text{Brussels}) = 0.$$

(b) Find ordinal utility functions for the following preference rankings:

    (i) Preference for cars
        Buick
        Cadillac–Volkswagen
        Datsun–Toyota
        Chevrolet
    (ii) Preference for foods
        Steak
        Lobster
        Roast beef–Chicken
        Sole–Flounder–Cod
        Hamburger

4. (a) Show that an ordinal utility function for the preference data of Table 3.5 is given by

$$f(\text{tennis}) = f(\text{football}) = 10,$$
$$f(\text{track}) = 7,$$
$$f(\text{swimming}) = f(\text{basketball}) = 1,$$
$$f(\text{baseball}) = 0.$$

(b) Determine whether or not an ordinal utility function exists for the data of Table 3.6.

(c) Repeat for Table 3.7.

(d) Consider the judgments of relative importance among objectives for a library system in Dallas shown in Table 1.8 of Chapter 1. If R is

Table 3.5. Preferences among Sports Activities
(Sport i is preferred to sport j iff the i,j entry is 1.)

	Tennis	Baseball	Football	Basketball	Track	Swimming
Tennis	0	1	0	1	1	1
Baseball	0	0	0	0	0	0
Football	0	1	0	1	1	1
Basketball	0	1	0	0	0	0
Track	0	1	0	1	0	1
Swimming	0	1	0	0	0	0

Table 3.6. Preferences among Cars
(Car i is preferred to car j iff the i,j entry is 1.)

	Cadillac	Buick	Oldsmobile	Toyota	Volkswagen	Chevrolet	Datsun
Cadillac	0	0	0	0	0	0	0
Buick	1	0	1	1	1	1	1
Oldsmobile	1	0	0	0	0	1	1
Toyota	1	0	0	0	0	1	1
Volkswagen	1	0	0	0	0	1	1
Chevrolet	0	0	0	0	0	0	0
Datsun	0	0	0	0	0	0	0

Table 3.7. Preferences among Cars
(Car i is preferred to car j iff the i,j entry is 1.)

	Cadillac	Buick	Oldsmobile	Toyota	Volkswagen	Chevrolet	Datsun
Cadillac	0	1	0	0	0	0	0
Buick	0	0	1	1	1	1	1
Oldsmobile	1	0	0	0	0	1	1
Toyota	1	0	0	0	0	1	1
Volkswagen	1	0	0	0	0	1	1
Chevrolet	0	0	0	0	0	0	0
Datsun	0	0	0	0	0	0	0

Table 3.8. Taste Preference for
Vanilla Puddings* (Entry i,j is 1 iff
pudding i is preferred to pudding j by
the group of judges.)

	1	2	3	4	5
1	0	0	1	1	0
2	1	0	0	1	0
3	0	1	0	0	0
4	0	0	1	0	0
5	1	1	0	1	0

*Data obtained from an experiment of
Davidson and Bradley [1969].

"more important than," is there a function f on $\{a, b, c, d, e, f\}$ satisfying Eq. (3.1)?

(e) Repeat part (d) for the judgments of relative importance among goals for a state environmental agency in Ohio shown in Table 1.9 of Chapter 1.

5. (a) The data of Table 3.8 is obtained from an experiment of Davidson and Bradley [1969] in which various brands of vanilla pudding were compared as to taste or flavor. It represents the consensus preferences of a group (the group is said to prefer i to j if a majority of its members do). Does an ordinal utility function exist?

(b) Group preferences can be nontransitive even when each individual is transitive. To give an example, suppose (A, P_i) is the preference relation of the ith member of a group for elements in a set of alternatives A. Suppose we define group preference P on A by the *simple majority rule*: aPb if and only if for a majority of i, aP_ib. Show that even if all P_i are transitive, even if they are all strict simple orders, P does not have to be transitive. (This result is known as the *voter's paradox* or *Condorcet's paradox*, after the philosopher and social scientist Marie Jean Antoine Nicolas Caritat, Marquis de Condorcet, who discovered it in the eighteenth century. For a discussion, see, for example, Riker and Ordeshook [1973] or Roberts [1976].)

6. Suppose A is a set of sounds. In psychophysics, one often runs an experiment in which sounds a and b from A are presented and a subject is asked which is louder. Subjects can be inconsistent. Hence, pairs a, b from A are presented together a number of times, and the experimenter records p_{ab}, the proportion of times that a is judged louder than b. It is commonly assumed that a is judged definitely louder than b, denoted aRb, if and only if $p_{ab} \geq .75$. Suppose $A = \{x, y, u, v, w\}$ and

$$
(p_{ab}) = \begin{array}{c} x \\ y \\ u \\ v \\ w \end{array}
\begin{array}{ccccc}
x & y & u & v & w \\
\left[\begin{array}{ccccc}
.50 & .81 & .91 & .81 & .85 \\
.19 & .50 & .61 & .55 & .78 \\
.09 & .39 & .50 & .63 & .82 \\
.19 & .45 & .37 & .50 & .79 \\
.15 & .22 & .18 & .21 & .50
\end{array}\right]
\end{array}.
$$

Find R and determine if there is a real-valued function f on A satisfying Eq. (3.1). (See Section 6.2 for a number of measurement representations arising from data p_{ab}.)

7. (a) If $A = \{1, 2, \ldots, 10\}$, $R = >$, and $B = \{1, 2, \ldots, 10\}$, show that B is a countable order-dense subset of A.

(b) Is it countable dense?

8. If A = the irrationals and $R = >$, show from the Birkhoff–Milgram Theorem that (A, R) is homomorphic to $(Re, >)$.

9. If $A = Re$ and $R = <$, show from the Birkhoff–Milgram Theorem that (A, R) is homomorphic to $(Re, >)$.

10. Show that the lexicographic ordering on $Q \times Q$ is homomorphic to $(Re, >)$.

11. Which of the following relational systems (A, R) are homomorphic to $(Re, >)$?

(a) $A = \{0, 1\} \times \{0, 1\}$, R = lexicographic ordering on A.*

---

*The lexicographic ordering R on a Cartesian product $A_1 \times A_2 \times \cdots \times A_n$ of sets of real numbers is defined as follows:

$$(a_1, a_2, \ldots, a_n)R(b_1, b_2, \ldots, b_n) \Leftrightarrow (a_1 > b_1) \quad \text{or} \quad (a_1 = b_1 \ \& \ a_2 > b_2) \quad \text{or} \ldots$$

$$\text{or} \quad (a_1 = b_1 \ \& \ a_2 = b_2 \ \& \ldots \& \ a_{n-1} = b_{n-1} \ \& \ a_n > b_n).$$

More generally, if R_i is a strict weak order on the set A_i, $i = 1, 2, \ldots, n$, and if, on A_i, aE_ib holds if and only if $\sim aR_ib$ and $\sim bR_ia$, then we can define a lexicographic ordering R on $A_1 \times A_2 \times \cdots \times A_n$ relative to $R_1, R_2, \ldots, R_n$ as follows:

$$(a_1, a_2, \ldots, a_n)R(b_1, b_2, \ldots, b_n) \Leftrightarrow (a_1R_1b_1) \quad \text{or} \quad (a_1E_1b_1 \ \& \ a_2R_2b_2) \quad \text{or} \ldots$$

$$\text{or} \quad (a_1E_1b_1 \ \& \ a_2E_2b_2 \ \& \ldots \& \ a_{n-1}E_{n-1}b_{n-1} \ \& \ a_nR_nb_n).$$

(b) $A = Re \times \{0, 1\}$, R = lexicographic ordering on A.

(c) $A = \{0, 1\} \times \{0, 1\} \times Re$, R = lexicographic ordering on A.

(d) $A = \{0, 1\} \times Q \times Re$, R = lexicographic ordering on A.

(e) $A = Re \times \{0, 1\} \times Re$, R = lexicographic ordering on A.

(f) $A_1 = A_2 = Re$, $R_1 = >$, $R_2 = <$, R = lexicographic ordering on $A_1 \times A_2$.

(g) $A_1 = Re \times Re$, $A_2 = \{0, 1\}$, $(a, b)R_1(s, t) \Leftrightarrow a > s$, $R_2 = >$, R = lexicographic ordering on $A_1 \times A_2$ relative to R_1, R_2.

12. If (A, R) is homomorphic to $(Re, >)$, does it necessarily follow that (A, R) has a countable order-dense subset?

13. (Fishburn [1970, p. 27]) Let $A = [-1, 1]$ and define R on A by

$$aRb \Leftrightarrow \left[|a| > |b| \text{ or } (|a| = |b| \text{ and } a > b) \right].$$

Show that (A, R) has no countable order-dense subset.

14. For each (A, R) of Exer. 11, identify the sets G_1 and G_2, the lower and upper end points of gaps, respectively, and compute the cardinality of G_1 and G_2.

15. If f is an ordinal utility function on (A, R), which of the following assertions are meaningful?

(a) $f(a) > 2f(b)$.

(b) $f(a) \neq f(b)$.

(c) a has the largest utility of any element of A.

16. Suppose f is an ordinal utility function on (A, R) and $g: A \to Re$ is a derived scale defined by the condition

$$g(a) > g(b) \Leftrightarrow f(a) > f(b).$$

What is the scale type of g in the narrow sense?

17. Suppose R is a binary relation on a set A, S is a binary relation on Re, and f is a homomorphism from $\mathfrak{A} = (A, R)$ into $\mathfrak{B} = (Re, S)$ with $|f(A)| > 1$.

(a) If $(\mathfrak{A}, \mathfrak{B}, f)$ is an ordinal scale, show that S is $>$, $\geq$, $<$, or $\leq$ and $\mathfrak{A} \to \mathfrak{B}$ is regular.

(b) Show that $(\mathfrak{A}, \mathfrak{B}, f)$ could not be an interval scale.

(c) Show that the conclusion of part (a) is false without the hypothesis that $|f(A)| > 1$.

18. Prove that if (A, R) is an asymmetric, complete binary relation with a countable order-dense subset B and if G_1 is the set of lower end points of gaps, then we may map $G_1 - B$ into B in a one-to-one manner.

19. This exercise and the next two sketch an alternative proof of Theorem 3.4, which is due to David Radford (personal communication). Suppose (A, R) is an asymmetric binary relation. A subset O of A is called *left-order-dense* if, whenever aRb, then either $b \in O$ or there is $c \in O$ so that $aRcRb$. Show that if (A, R) is a strict simple order, then there is a countable left-order-dense subset of A if and only if there is a countable order-dense subset of A. (The proof uses the lemma on p. 114.)

20. Suppose (A, R) is a strict simple order and there is a countable left-order-dense subset O of A. Suppose $x_1, x_2, x_3, \ldots$ are the elements of O. Show that Eq. (3.7) defines a function f satisfying Eq. (3.1).

21. Suppose (A, R) is a strict simple order and there is a function $f: A \to Re$ satisfying (3.1). For each positive integer n and each integer k, $0 \leqq k < 2^{2(n+1)} - 1$, let

$$ J_{kn} = \left(-2^n + \frac{k}{2^{n+1}}, \ -2^n + \frac{k+1}{2^{n+1}} \right]. $$

If $f(x)$ is in J_{kn} for some x in A, let x_{kn} be that x for which $f(x)$ is maximum, if there is such an x, and be an arbitrary x in A for which $f(x)$ is in J_{kn}, if there is no x with $f(x)$ maximum. Let

$$ O = \{ x_{kn} : f(x) \in J_{kn} \quad \text{for some} \quad x \in A \}. $$

Show that O is a countable left-order-dense set in (A, R).

22. Suppose (A, R) is a preference relation. If there is some measure of closeness on A, then we might want to require that the utilities assigned to elements a and b in A be close whenever a and b are close. To make this idea precise, we introduce a topology on sets A with strict weak orders R on A. This topology, called the *R-order topology* or the *interval topology*, and denoted $\mathcal{T}(R)$, is defined to be the smallest system of subsets of A, closed under finite intersections and arbitrary unions, and containing all *open rays*, that is, subsets of the form

$$ \{ x \in A : xRa \} \quad \text{or} \quad \{ x \in A : aRx \} $$

for fixed a's in A. We search for ordinal utility functions continuous in the topology $\mathcal{T}(R)$.

(a) Show that if there is an ordinal utility function on (A, R) continuous in a topology $\mathcal{T}$ on A (and the usual topology on the reals), then $\mathcal{T}$ must contain all open rays. (Fishburn [1970, p. 36] proves that if $\mathcal{T}$ contains all open rays and there is an ordinal utility function, then there is an ordinal utility function continuous in $\mathcal{T}$. Hence, in particular, if there is an ordinal utility function, there is one that is continuous in the R-order topology. This result goes back to Debreu [1954].)

(b) If (A, R) is a strict simple order, show that (A, R) has a countable, order-dense subset if and only if the R-order topology $\mathcal{T}(R)$ has a countable base.

(c) Thus, show that a strict weak order (A, R) has a continuous ordinal utility function if and only if $\mathcal{T}(R^*)$ has a countable base, where R^* is the reduction of R. (This result is due to Debreu [1954].)

(d) By considering the open sets

$$ \{ (b, c) : (a, 1)R(b, c)R(a, 0) \}, $$

show that lexicographic preference on $Re \times Re$ could not have a countable base in the order topology. For further results on the order topology, see Pfanzagl [1968, Chapter 3].

3.2 Extensive Measurement

3.2.1 *Hölder's Theorem*

A second case of fundamental measurement which is of interest is that where the relational system is (A, R, o), where R is a binary relation on A and o is a binary operation. One seeks conditions on (A, R, o) (necessary and) sufficient for the existence of a real-valued function f on A satisfying (3.1) and

$$f(a \text{ o } b) = f(a) + f(b), \qquad (3.8)$$

or conditions on (A, R, o) (necessary and) sufficient for the existence of a real-valued function f on A satisfying (3.3) and (3.8). These representations arose in our study of mass and of preference where we want a utility function to "preserve" combinations of objects. To avoid redundancy, we shall state results carefully only for the former representation. That is, we seek conditions on (A, R, o) (necessary and) sufficient for the existence of a homomorphic map into $(Re, >, +)$.

Attributes that have additive properties, such as mass for example, have traditionally been called *extensive* in the literature of measurement, and so the problem of finding conditions on (A, R, o) (necessary and) sufficient for the existence of a homomorphic map into $(Re, >, +)$ is called the problem of *extensive measurement*.

The theory of extensive measurement will naturally overlap with abstract algebra, which often deals with relational systems of the form (A, o). We review a few of the relevant concepts dealing with such relational systems.* If o is an operation on the set A, the pair (A, o) is called a *group* if it satisfies the following axioms:†

AXIOM G1 (*Associativity*). *For all a, b, c in A, $(a \text{ o } b) \text{ o } c = a \text{ o } (b \text{ o } c)$.*

AXIOM G2 (*Identity*). *There is an (identity) element e in A such that for all a in A, $a \text{ o } e = e \text{ o } a = a$.*

AXIOM G3 (*Inverse*). *For all a in A, there is an (inverse) element b in A such that $a \text{ o } b = b \text{ o } a = e$.*

The reader is familiar with many examples of groups. $(Re, +)$ is an example, with the identity e of Axiom G2 being 0 and the inverse b of

---

*The reader familiar with group theory may wish to skip directly to the statement of Hölder's Theorem below.

†Often, an additional axiom called *Closure* is explicitly stated. This axiom asserts that for all a, b in A, $a \text{ o } b$ is in A. This is implicit in the definition of an operation.

Axiom G3 being $-a$. $(Re^+, \times)$ is a group, with the identity being 1 and the inverse of a being $1/a$. $(Re, \times)$ is not a group, since the only possible identity is 1, and the element 0 has no inverse b in Re such that $0 \times b = b \times 0 = 1$.

If $(A, \mathbf{o})$ is a group, we may define na for every positive integer n. The definition is inductive. We define $1a$ to be a. Having defined na, we define $(n + 1)a$ to be $a \mathbf{o} na$. The definition makes sense whenever $(A, \mathbf{o})$ is associative.

We have seen in the previous section that obtaining axiomatizations for homomorphisms into the reals can involve translating properties of the real number system into an abstract relational system. An example was the translation of the existence of a countable order-dense subset, the rationals, into an axiom for the representation of (A, R) into $(Re, >)$. Most axiomatizations in measurement theory try to capture an important property of the real numbers called the *Archimedean property*. This property can be defined as follows: If a and b are real numbers and $a > 0$, then there is a positive integer n such that $na > b$. That is, no matter how small a might be or how large b might be, if a is positive, then sufficiently many copies of a will turn out to be larger than b. This property of the real number system is what makes measurement possible; it makes it possible to roughly compare the relative magnitudes of any two quantities a and b, by seeing how many copies of a are required to obtain a larger number than b. We shall try to translate the Archimedean property of the reals into an Archimedean axiom in $(A, R, \mathbf{o})$ in order to axiomatize the representation $(A, R, \mathbf{o})$ into $(Re, >, +)$.

Sufficient conditions for extensive measurement, the representation $(A, R, \mathbf{o})$ into $(Re, >, +)$, were first given by Hölder [1901]. We state some conditions closely related to those originally given by Hölder, by giving the following definition. A relational system $(A, R, \mathbf{o})$ is an *Archimedean ordered group* if it satisfies the following axioms:

AXIOM A1. $(A, \mathbf{o})$ *is a group*.

AXIOM A2. (A, R) *is a strict simple order*.

AXIOM A3 (*Monotonicity*). *For all a, b, c in A,*

$$aRb \quad iff \quad (a \mathbf{o} c)R(b \mathbf{o} c) \quad iff \quad (c \mathbf{o} a)R(c \mathbf{o} b).$$

AXIOM A4 (*Archimedean*). *For all a, b in A, if aRe, where e is the identity for $(A, \mathbf{o})$, then there is a positive integer n such that $naRb$.*

The paradigm example of an Archimedean ordered group is of course $(Re, >, +)$. We know that $(Re, +)$ is a group and $(Re, >)$ is a strict simple order. Monotonicity follows, since

$$a > b \quad \text{iff} \quad a + c > b + c \quad \text{iff} \quad c + a > c + b.$$

The Archimedean axiom follows directly from the Archimedean property of the real numbers. $(Re^+, >, \times)$ is another example of an Archimedean ordered group. Hölder's Theorem is the following:

THEOREM 3.5 (Hölder).* *Every Archimedean ordered group is homomorphic to $(Re, >, +)$.*

Proof. Omitted. See Krantz *et al.* [1971] for a proof.

It should be noted that every homomorphism of an Archimedean ordered group into $(Re, >, +)$ is one-to-one, that is, is an isomorphism. This follows since (A, R) is strict simple.

The axioms for an Archimedean ordered group give a representation theorem for extensive measurement. As such, they should be tested for various examples. Let us begin with the case of mass. Here, A is a collection of objects, aRb is interpreted to mean that a is judged heavier than b, and $a \, o \, b$ is the combined object. Let us consider first the group axioms. To say (A, o) is a group requires that o be an operation, so in particular we have to make sense out of $a \, o \, a$, $a \, o \, (a \, o \, a)$, etc. This is the problem we discussed at the end of Section 1.7. Things work out in theory if we imagine an infinite number of ideal copies of each element in the group. However, it is still necessary to make sense out of complicated combinations like $a \, o \, [b \, o \, (c \, o \, d)]$. Of the explicit group axioms, certainly associativity seems to make sense: taking first the combination of a and b and then combining with c amounts to the same object (as far as mass is concerned) as first combining b and c and then combining with a. Also, at least ideally, there is an element with no mass, which could serve as the identity e. Speaking of inverses, however, does not make sense. Given an object a, the axiom requires that there be another object b, which, when a and b are combined, results in the identity, the object with no mass. There is no such b. Thus, to obtain a usable representation theorem for measurement of mass, it is necessary to modify this axiom.

Let us next consider the remaining axioms for an Archimedean ordered group. It might be reasonable to assume that (A, R) is a strict simple order, although we can run into problems with completeness: two different

---

*It should be remarked that from Hölder's Theorem it follows that (A, o) is *commutative*, that is, that $a \, o \, b = b \, o \, a$, all a, b in A. It is surprising that commutativity does not have to be assumed. (The second "iff" in Axiom A3 plays the role of a commutativity axiom, though this axiom may be weakened to read: if aRb, then $(a \, o \, c)R(b \, o \, c)$ and $(c \, o \, a)R(c \, o \, b)$.) The author knows of no simple direct proof of commutativity from the axioms for an Archimedean ordered group.

objects may be close enough in mass so that we cannot distinguish them. Monotonicity probably is reasonable; if you think a is heavier than b, then adding the same object c to each of a and b should not change your opinion, and taking c away similarly should not. Moreover, the order in which you add c to a and to b should not matter. Finally, it is probably reasonable to assume the Archimedean axiom: Given an object a that is heavier than the object with no mass, and given another object b, if we combine enough copies of a, it seems reasonable that, at least in principle, we can create an object that is heavier than b. The Archimedean axiom is not really empirically testable, in the sense that any finite experiment could not get enough data to refute it.* However, one could get very strong evidence against the Archimedean axiom, which would amount to a refutation. Still, if we accept this axiom, we often treat it as an idealization which seems reasonable.

Let us next consider the same axioms for the case of preference. Here, A is a set of objects or alternatives, o is again combination, and aRb means a is strictly preferred to b. Associativity is probably reasonable if the combination does not involve physical interaction between elements. However, if such interaction is allowed, then combining a with b first and then bringing in c might create a different object from that obtained when b and c are combined first. To give an example, if a is a flame, b is some cloth, and c is a fire retardant, then combining a and b first and then combining with c is quite different from combining b and c first and then combining with a. However, this result follows only if we allow interaction (for example, flame lights cloth) with our combination. Similar examples may be thought of from chemistry. It should be noted that in our discussion of associativity for the case of mass, we also tacitly assumed that combination did not involve physical or chemical interaction between alternatives being combined. The axioms of identity and inverse seem to be satisfied. For there is at least ideally an object with absolutely no worth at all. And given any object a, owing someone else such an object might be considered an inverse alternative. For having a and owing a amount to having nothing. (If future value is discounted over present value, we would have to look for an inverse to having a among the alternatives of owing objects of more worth than a.) The strict simple order axiom is one we have previously questioned for preference; we have even questioned whether preference is a strict weak order, and we probably can question completeness: it is possible to be indifferent between two alternatives. Let us turn next to monotonicity. Even this axiom might be questioned, if objects combined can be made more useful than individual objects. For

---

*The axiom could be refuted indirectly. In the presence of the other axioms for an Archimedean ordered group, the Archimedean axiom implies commutativity (see footnote on page 124). Hence, a violation of commutativity would provide evidence against the Archimedean axiom, given one believes the other axioms.

example, suppose a is black coffee, b is a candy bar, and c is sugar. You might prefer b to a (not liking black coffee), but prefer a o c to b o c. This is really an argument against the additive representation, not just the monotonicity axiom. Finally, perhaps one can even question the Archimedean axiom. For example, suppose a is a lamp and b is a long, healthy life. Will sufficiently many lamps ever be better than having a long healthy life?

The discussion above suggests that it is necessary to modify Hölder's axioms to obtain a satisfactory representation theorem for extensive measurement, even if it is only mass we wish to measure. Some early attempts to improve Hölder's Theorem can be found in Huntington [1902a,b, 1917], Suppes [1951], Behrend [1953, 1956], and Hoffman [1963]. All these improvements involve some axioms that are not necessary, as indeed does Hölder's Theorem. We shall present a set of axioms that are necessary as well as sufficient.

3.2.2 Necessary and Sufficient Conditions for Extensive Measurement

To find a set of axioms that are necessary as well as sufficient for extensive measurement, let us see which of Hölder's axioms are necessary. Let us suppose that f is a homomorphism from (A, R, o) into $(Re, >, +)$, and let us begin by looking at the group axioms. Certainly

$$[f(a) + f(b)] + f(c) = f(a) + [f(b) + f(c)].$$

However, the representation does not imply that

$$(a \text{ o } b) \text{ o } c = a \text{ o } (b \text{ o } c),$$

but only that

$$[(a \text{ o } b) \text{ o } c] E [a \text{ o } (b \text{ o } c)],$$

where E is defined by

$$xEy \Leftrightarrow \sim xRy \& \sim yRx. \tag{3.9}$$

Second, the representation does not imply that there is an identity element e, since there may not be any element e in A such that $f(e) = 0$. Third, the same is true of the inverse. Thus, of the group axioms, only a weak version of associativity is necessary.* Next, turning to the condition that (A, R) be

---

*Implicit in the group axioms is that o is an operation, that is, that a o b is defined for all a, b in A. This is not a necessary condition for the representation, and it might make sense to weaken it. In the case of mass, for example, we might limit combination only to objects whose combination fits into our lab! The approach to extensive measurement where combination is restricted can be found in Luce and Marley [1969] and Krantz et al. [1971, Section 3.4].

strict simple, we already know from the theorems of Section 3.1 that, since aRb iff $f(a) > f(b)$, (A, R) must be strict weak. It does not necessarily follow that (A, R) is strict simple. The next axiom, monotonicity, is necessary. For,

$$aRb \Leftrightarrow f(a) > f(b) \Leftrightarrow f(a) + f(c) > f(b) + f(c) \Leftrightarrow (a \mathbin{o} c)R(b \mathbin{o} c).$$

Similarly, $aRb \Leftrightarrow (c \mathbin{o} a)R(c \mathbin{o} b)$. Finally, the Archimedean axiom as stated cannot be necessary, for there is not necessarily an identity e in A. However, the Archimedean axiom can be restated as a necessary axiom. The assumption aRe amounted to saying that $f(a)$ was positive. This is the same as saying that $2f(a) > f(a)$, which, since f is additive and preserves R, is the same as saying that $2aRa$. Thus, the Archimedian axiom can be restated as follows: if a and b are in A and $2aRa$, then there is a positive integer n such that $naRb$. (The notation na makes sense so long as the weak form of associativity holds.) In this form, the Archimedean axiom is necessary. For $2aRa$ implies $2f(a) > f(a)$, or $f(a) > 0$. Thus, by the Archimedean property for the reals, there is a positive integer n such that $nf(a) > f(b)$. Since f is additive, $f(na) > f(b)$, and $naRb$. Summarizing, the following conditions are necessary for extensive measurement:

AXIOM E1 (*Weak Associativity*). *For all a, b, c in A,*

$$[a \mathbin{o} (b \mathbin{o} c)]E[(a \mathbin{o} b) \mathbin{o} c].$$

AXIOM E2. *(A, R) is a strict weak order.*

AXIOM E3 (*Monotonicity*). *For all a, b, c in A,*

$$aRb \Leftrightarrow (a \mathbin{o} c)R(b \mathbin{o} c) \Leftrightarrow (c \mathbin{o} a)R(c \mathbin{o} b).$$

AXIOM E4 (*Archimedean*). *For all a, b in A, if $2aRa$, then there is a positive integer n such that $naRb$.*

Unfortunately, Axioms E1 through E4 together are not sufficient for extensive measurement. Proof is left to the reader. (*Hint*: Let A be the negative reals supplemented with an element $-\infty$, let R be $>$, and define $\mathbin{o}$ to be $+$ except that $(-\infty) \mathbin{o} x$ and $x \mathbin{o} (-\infty)$ are always $-\infty$.) However, one obtains necessary and sufficient conditions by substituting for Axiom E4 a stronger Archimedean axiom:

AXIOM E4' (*Archimedean*). *For all a, b, c, d in A, if aRb, then there is a positive integer n such that $(na \mathbin{o} c)R(nb \mathbin{o} d)$.*

To see that Axiom E4' is indeed an Archimedean axiom (that is, it reflects the Archimedean properties of the reals) and that it is necessary, let us observe that aRb implies $f(a) > f(b)$, so $f(a) - f(b) > 0$. Thus, by the Archimedean property for the reals, there is a positive integer n such that $n[f(a) - f(b)] > f(d) - f(c)$, that is, $nf(a) + f(c) > nf(b) + f(d)$. From this, since f is additive, it follows that $f(na \circ c) > f(nb \circ d)$, or $(na \circ c)R(nb \circ d)$. A system $(A, R, \circ)$ satisfying Axioms E1, E2, E3, and E4' is called an *extensive structure*.

THEOREM 3.6 (Roberts and Luce [1968]). *Suppose A is a set, R is a binary relation on A and $\circ$ is a binary operation on A. Then there is a real-valued function f on A satisfying*

$$aRb \Leftrightarrow f(a) > f(b) \tag{3.1}$$

and

$$f(a \circ b) = f(a) + f(b) \tag{3.8}$$

if and only if $(A, R, \circ)$ is an extensive structure.

This theorem was stated as a corollary of a more general result in Roberts and Luce [1968]. We have already proved that $(A, R, \circ)$ is an extensive structure whenever a representation holds. A direct proof of the sufficiency of the conditions can be found in Krantz *et al.* [1971, Theorem 3.1]. The proof reduces this situation to Hölder's Theorem. We shall omit it. Other theorems giving necessary and sufficient conditions for extensive measurement can be found in Alimov [1950] and Holman [1969]. (See Exer. 12 below.)

Let us again consider the axioms for an extensive structure as axioms for measurement of mass and of preference. If R is the binary relation "heavier than," then the axioms for an extensive structure are probably satisfied, at least ideally. (The only axiom for an Archimedean ordered group with which we found serious fault for the case of mass was the existence of an inverse.) Suppose next that R is preference. Our discussion of the fire retardant casts doubts about weak associativity just as it did about associativity, if the combination allows interaction. The discussion in Section 3.1 casts doubts about (A, R) being strict weak. The example of the coffee, sugar, and candy bar casts doubts about monotonicity. To consider the Archimedean axiom, suppose a is one dollar, b is no dollars, c is life as a cripple, and d is a long and healthy life. It is conceivable that no amount of money will be enough to compensate for life as a cripple, and so there might be no n such that $(na \circ c)R(nb \circ d)$.

Extensive measurement is basic to the physical sciences. However, as Krantz *et al.* [1971, pp. 123, 124] point out, the attempt to apply extensive measurement to the social sciences usually meets with some sort of difficulty. Even if there is an operation available, the axioms for extensive measurement are usually not satisfied. The exceptions mentioned by Krantz *et al.* are risk and subjective probability. We discuss subjective probability in Chapter 8. Risk is discussed in a variety of places in the literature. One theory of risk, that of Pollatsek and Tversky [1970], can be formulated in terms of extensive measurement. The basic idea of the Pollatsek–Tversky theory is that one compares various probability distributions as to riskiness. Comparative riskiness defines the binary relation of the theory of extensive measurement. The operation o corresponds to convolution; that is,

$$(f \circ g)(t) = \int_{-\infty}^{+\infty} f(t - x)g(x) \, dx.$$

The Pollatsek–Tversky theory of risk has not been empirically verified.

The representation problem for extensive measurement has recently been treated from a probabilistic point of view. See Falmagne [1978].

3.2.3 Uniqueness

Before leaving the topic of extensive measurement, we ask for a uniqueness statement. Our earlier observations about measurement of mass suggest that the representation f should be unique up to a similarity transformation; that is, measurement should be on a ratio scale. We shall prove this. This result will have significance for preference as well. For it says that if a utility function f satisfies conditions (3.1) and (3.8), then it is meaningful to say that the utility of one alternative is twice the utility of a second, or half the utility of a second, and so on. If this is the case, we can begin to use utility functions to make "quantitative" decisions, rather than just "qualitative" ones.

THEOREM 3.7. *Suppose A is a non-empty set, R is a binary relation on A,* o *is a (binary) operation on A, and f is a real-valued function on A satisfying*

$$aRb \Leftrightarrow f(a) > f(b) \tag{3.1}$$

and

$$f(a \circ b) = f(a) + f(b). \tag{3.8}$$

Then $\mathfrak{A} = (A, R, o) \to \mathfrak{B} = (Re, >, +)$ *is a regular representation, and* $(\mathfrak{A}, \mathfrak{B}, f)$ *is a ratio scale.*

Proof. That $\mathfrak{A} \to \mathfrak{B}$ is a regular representation follows from Theorem 2.1.

To show that $(\mathfrak{A}, \mathfrak{B}, f)$ is a ratio scale, suppose first that for some $\alpha > 0$, $\phi(x) = \alpha x$ for all x in $f(A)$. Clearly $\phi \circ f$ satisfies (3.1) and (3.8), and so ϕ is an admissible transformation.

Conversely, suppose $\phi: f(A) \rightarrow Re$ is an admissible transformation. We shall show that for some $\alpha > 0$, $\phi(x) = \alpha x$, all $x \in f(A)$. Let $g = \phi \circ f$. We show first that $f(a) > 0$ iff $g(a) > 0$. For since f and g satisfy (3.1) and (3.8),

$$\begin{aligned}
f(a) > 0 \quad &\text{iff} \quad f(a) + f(a) > f(a), \\
&\text{iff} \quad f(a \circ a) > f(a), \\
&\text{iff} \quad (a \circ a) Ra, \\
&\text{iff} \quad g(a \circ a) > g(a), \\
&\text{iff} \quad g(a) + g(a) > g(a), \\
&\text{iff} \quad g(a) > 0.
\end{aligned}$$

A similar proof shows that $f(a) < 0$ iff $g(a) < 0$.

If $f(a) = 0$ for all a in A, any positive α will suffice to satisfy $\phi(x) = \alpha x$ for all x in $f(A)$. Thus, we may assume that for some e in A, $f(e) \neq 0$. We shall assume that $f(e) > 0$. The proof in the case that $f(e) < 0$ is similar. Since $f(e) > 0$, $g(e)$ must be > 0.

Pick α such that $g(e) = \alpha f(e)$. Since $f(e)$ and $g(e)$ are positive, so is α. We shall show that $g(a) = \alpha f(a)$, for all a in A. This proves that $\phi(x) = \alpha x$ for all x in $f(A)$. The proof that $g(a) = \alpha f(a)$ proceeds by contradiction, assuming first that $g(a) < \alpha f(a)$ and second that $g(a) > \alpha f(a)$. The proof is similar in the second case, and so is left to the reader.

If $g(a) < \alpha f(a)$, then

$$\frac{g(a)}{g(e)} = \frac{g(a)}{\alpha f(e)} < \frac{\alpha f(a)}{\alpha f(e)} = \frac{f(a)}{f(e)}.$$

Since between any pair of real numbers there is a rational, it follows that there are a pair of positive numbers m and n such that

$$\frac{g(a)}{g(e)} < \frac{m}{n} < \frac{f(a)}{f(e)}. \tag{3.10}$$

The second inequality of (3.10) implies that $mf(e) < nf(a)$, so $f(me) < f(na)$, so $naRme$. But then $g(na) > g(me)$, so $ng(a) > mg(e)$, so

$$\frac{g(a)}{g(e)} > \frac{m}{n}.$$

This contradicts the first inequality of (3.10) and shows that $g(a) < \alpha f(a)$ is impossible. ∎

3.2.4 *Additivity*

Before closing this section, we make two remarks about operations in general and additivity in particular. First, in spite of historical emphasis on additivity, there is nothing magic about the addition operation. Indeed, if f satisfies (3.1) and (3.8), then $g = e^f$ satisfies (3.1) and

$$g(a \mathbin{\mathbf{o}} b) = g(a)g(b). \tag{3.11}$$

Thus a multiplicative representation (3.1), (3.11) can also be obtained, with positive g. Conversely, if a multiplicative representation (3.1), (3.11) can be obtained with positive g, then $f = \ln g$ gives an additive representation (3.1), (3.8). Moreover, it is easy to see that the logarithm of the multiplicative representation gives rise to the same type of scale as the additive representation, so the same comparisons using $\ln g$ are meaningful as would be using f. The main point is that we make use of a representation to learn about empirical relations. We get just as much information out of the representation (3.1), (3.11) as we do out of the representation (3.1), (3.8).

As a second point, let us note that, as we suggested above, there are very few genuine empirical operations in the social sciences. Perhaps the most interesting one is the *bisection operation*, which arises when a subject is asked to produce a stimulus which he thinks is halfway between two given stimuli, for example, with respect to loudness or with respect to brightness. Suppose $a \mathbin{\mathbf{o}} b$ is defined to be the unique element that "bisects" a and b. Suppose R is a binary relation on the set of objects being compared, with aRb interpreted to mean that a is judged higher (louder, brighter, etc.) than b. Then we seek a real-valued function f on A such that for all a, b in A,

$$aRb \Leftrightarrow f(a) > f(b) \tag{3.1}$$

and

$$f(a \mathbin{\mathbf{o}} b) = \frac{f(a) + f(b)}{2}. \tag{3.12}$$

We shall study this representation in Exer. 15.

Exercises

1. (a) Suppose $A = \{0, 1\}$, $R = >$, and $\mathbf{o}$ is defined by

$$0 \mathbin{\mathbf{o}} 0 = 0,\ 0 \mathbin{\mathbf{o}} 1 = 1 \mathbin{\mathbf{o}} 0 = 1 \mathbin{\mathbf{o}} 1 = 1.$$

(The operation $\mathbf{o}$ is Boolean addition.) Show the following:
      (i) $(A, \mathbf{o})$ is associative.
      (ii) $(A, \mathbf{o})$ has an identity.

(iii) Not every element of A has an inverse.
(iv) (A, R) is a strict simple order.
(v) $(A, R, \mathbf{o})$ violates monotonicity.
(vi) $(A, R\ \mathbf{o})$ violates the Archimedean axiom, Axiom A4.
(vii) $(A, R, \mathbf{o})$ violates Axiom E4′.
(viii) $(A, R, \mathbf{o})$ satisfies Axiom E4.
(b) Make a similar analysis if $\mathbf{o}$ is modified as follows:
$$0 \mathbf{o} 0 = 0 \mathbf{o} 1 = 1 \mathbf{o} 0 = 0,\ 1 \mathbf{o} 1 = 1.$$

2. Show that if $(A, R, \mathbf{o})$ is homomorphic to $(Re, \geqq, +)$, then $(a \mathbf{o} b)R(b \mathbf{o} a)$ for all a, b in A.

3. (a) Neither of the following relational systems is an Archimedean ordered group. Which of the axioms for an Archimedean ordered group holds in each case?
(i) $(Re^+, <, +)$.
(ii) $(\{0, 1, 2\}, >, +(\mathrm{mod}\ 3))$.
(b) Which of the axioms for an Archimedean ordered group holds in each of the following relational systems? (Q^+ is the positive rationals, Re^- the negative reals.)
(i) $(Re^+, <, \times)$.
(ii) $(N, >, +)$.
(iii) $(Q, >, \times)$.
(iv) $(Q^+, >, \times)$.
(v) $(Re^-, <, +)$.

4. Suppose $A = Re \times Re$, R is the lexicographic ordering of the plane, and $(a, b) \mathbf{o} (c, d)$ is defined to be $(a + c, b + d)$.
(a) Show that it follows from Hölder's Theorem that $(A, R, \mathbf{o})$ is not an Archimedean ordered group.
(b) Thus, one of the axioms for an Archimedean ordered group must be violated. Which one?

5. (a) Show that $(Re^+, <, +)$ is an extensive structure.
(b) Which of the axioms for an extensive structure hold for each relational system in Exer. 3b?

6. (Krantz, et al. [1971, p. 77]) Which of the axioms for an extensive structure are satisfied by the following relational systems $(A, R, \mathbf{o})$?
(a) $A = N$, aRb iff $a + 1 > b$, $a \mathbf{o} b = a + b + 2$.
(b) $A = Re^+$, aRb iff $a > b$, $a \mathbf{o} b = \max\{a, b\} + \frac{1}{2}\min\{a, b\}$.
(c) $A = \{x\} \cup N$, aRb iff $[(a > b$ and $a, b \in N)$ or $(a = x$ and $b \in N)]$,

$$a \mathbf{o} b = \begin{cases} a + b & \text{if } a, b \in N, \\ 2 & \text{if } a = b = x, \\ 1 & \text{if } a = x, b \in N, \\ 1 & \text{if } a \in N, b = x. \end{cases}$$

(d) $(A, R, \mathbf{o})$ as in Exer. 4.

7. Show that Axioms E1 through E4 together are not sufficient for extensive measurement.

8. Suppose f satisfies extensive measurement. Show that the following statements are meaningful:
   (a) $f(a) > 7.6f(b)$.
   (b) $f(a) - f(b) > f(c) - f(d)$.
   (c) $f(a)/f(b) = 10$.

9. Suppose $f: A \to Re^+$ and $g: A \to Re^+$ satisfy (3.1) and (3.11). Show that

$$f(a) > 1 \quad \text{iff} \quad g(a) > 1,$$

10. Give examples to show that none of the group axioms are necessary for extensive measurement. Specifically, give examples of relational systems $(A, R, \mathbf{o})$ homomorphic to $(Re, >, +)$ but which violate
   (a) associativity (Axiom G1);
   (b) identity (Axiom G2);
   (c) inverse (Axiom G3).

11. Suppose (A, R) is a strict weak order and $(A, \mathbf{o})$ is associative. Suppose that every element of A satisfies $2aRa$ (such an element is called *positive*). Show that $4a \neq a$ and A must be infinite.

12. Suppose $(A, R, \mathbf{o})$ satisfies the first three axioms for an extensive structure. A *pair* of elements a and b in A is called *anomalous* if either aRb or bRa and either for all positive integers n,

$$naR(n + 1)b \quad \text{and} \quad nbR(n + 1)a$$

or for all positive integers n,

$$(n + 1)bRna \quad \text{and} \quad (n + 1)aRnb.$$

Show that if $(A, R, \mathbf{o})$ is homomorphic to $(Re, >, +)$, then there is no anomalous pair. (Alimov [1950] proved that Axioms E1, E2, and E3 of an extensive structure plus the assumption that there are no anomalous pairs provide necessary and sufficient conditions for extensive measurement.)

13. (Roberts and Luce [1968]) $(A, R, \mathbf{o})$ satisfies *weak solvability* if, for all $a, b \in A$,

$$\sim aRb \Rightarrow (\exists c \in A)[bR(a \mathbf{o} c)].$$

$(A, R, \mathbf{o})$ satisfies *positivity* if every a in A is positive, that is, it satisfies $2aRa$. Show that if $(A, R, \mathbf{o})$ satisfies Axioms E1, E2, and E3 of an extensive structure and positivity, then weak solvability and the standard Archimedean axiom (Axiom E4) imply the Archimedean Axiom E4' of an extensive structure.

14. Suppose $(A, R, \mathbf{o})$ is homomorphic to $(Re^+, >, +)$, via a homomorphism g. Fix e in A and define

$$S(a, e) = \{m/n: m, n \in N \quad \text{and} \quad \sim meRna\},$$

where N is the set of positive integers. Show the following:

(a) $S(a, e)$ is nonempty.

(b) $S(a, e)$ is bounded above.

(c) If $f(a)$ is the least upper bound of the set $S(a, e)$, then f is a homomorphism from (A, R, o) into $(Re^+, >, +)$. (To show this, consider the relation between f and g.)

Note: Under assumptions similar to Hölder's, Suppes [1951] and Suppes and Zinnes [1963] use f to give a constructive proof of the existence of an extensive measure. A key additional assumption needed for the proof is that (A, R, o) satisfies positivity (Exer. 13).

15. In this exercise we study the binary operation of bisection. We assume that the structure (A, R, o) is given, with R a binary relation on A and o a binary operation (bisection), and study the representation (3.1), (3.12). In point of fact, $a \mathrm{o} b$ might not be the same as $b \mathrm{o} a$, or even judged equal in strength (loudness, brightness, etc.), as has been observed by such writers as Stevens [1957]. Such order biases are called *hysteresis*. As a result of these biases, the representation (3.12) might not be appropriate, and we replace it with the representation

$$f(a \mathrm{o} b) = \delta f(a) + (1 - \delta)f(b), \qquad (3.13)$$

where $\delta \in (0, 1)$. If $\delta \neq \frac{1}{2}$, there is an order bias. The representation (3.1), (3.13) has been studied by Pfanzagl [1959a, b]. The specific representation (3.1), (3.12) is studied by Krantz, *et al.* [1971, Section 6.9.2], and a more general representation than (3.1), (3.13) is studied by Krantz *et al.* [1971, Section 6.9.1]. Pfanzagl's representation theorem involves four axioms. The following are three of Pfanzagl's axioms for the representation (3.1), (3.13). (The fourth axiom is a topological axiom, making precise the statement that $a \mathrm{o} b$ is continuous in both of its arguments.)

(a) Reflexivity: $a \mathrm{o} a = a$.

(b) Monotonicity: $aRb \Rightarrow (a \mathrm{o} c)R(b \mathrm{o} c)$.

(c) Bisymmetry: $(a \mathrm{o} b) \mathrm{o} (c \mathrm{o} d) = (a \mathrm{o} c) \mathrm{o} (b \mathrm{o} d)$.

For each axiom, determine if it is a necessary condition for the representation and, if not, modify the axiom so that it is necessary.

Note: For an alternative approach to bisection, see Exer. 12, Section 3.3.

16. (a) Show that if (A, R, o) is an extensive structure, then Eq. (1.24) of Chapter 1 is satisfied and hence (A, R, o) is shrinkable and the reduction (A^*, R^*, o^*) of Section 1.8 is well-defined.

(b) Construct an extensive structure (A, R, o) which is not isomorphic to its reduction.

3.3 Difference Measurement

3.3.1 *Algebraic Difference Structures*

Let A be a set and let D be a quaternary relation on A. In discussing temperature differences in Section 3.1, we interpreted $D(a, b, s, t)$ or, equivalently, $abDst$, to be the statement that the difference between the temperature of a and the temperature of b is judged to be greater than that between the temperature of s and the temperature of t. We then encountered the representation

$$abDst \Leftrightarrow f(a) - f(b) > f(s) - f(t). \tag{3.4}$$

This is called (*algebraic*) *difference measurement*. A similar representation makes sense in the case of preference, if *abDst* is interpreted to mean that I like a over b more than I like s over t. In this case, f is a different kind of utility function. In the present section, we shall state two representation theorems for the representation (3.4), and a uniqueness theorem. The uniqueness theorem is the focal point here, and the reader may wish to concentrate primarily on this theorem.

The relation D may be obtained by a variant of a pair comparison experiment, using pairs of pairs of alternatives. Alternatively, D may be obtained by asking a subject to make numerical estimates of absolute differences. For example, Table 3.9 shows such numerical estimates $\Delta(x, y)$. Suppose we define $\delta(x, y)$ as follows:

$$\delta(x, y) = \begin{cases} 0 & \text{if } x = y \text{ or } x \text{ and } y \text{ are judged equally warm.} \\ \Delta(x, y) & \text{if } x \text{ is judged warmer than } y, \\ -\Delta(x, y) & \text{if } y \text{ is judged warmer than } x. \end{cases}$$

Then D can be defined by

$$abDst \Leftrightarrow \delta(a, b) > \delta(s, t). \tag{3.14}$$

We now ask whether or not the numbers $\delta(x, y)$ or $\Delta(x, y)$ could have arisen as differences of temperatures, that is, whether or not there is a function f so that

$$\delta(a, b) > \delta(s, t) \Leftrightarrow f(a) - f(b) > f(s) - f(t). \tag{3.15}$$

This is the same as asking if there is a function f satisfying Eq. (3.4).

Table 3.9. Judgments of Absolute Temperature Difference

Objects Compared, x, y	Warmer Object	Estimated Absolute Temperature Difference $\Delta(x, y)$
a, b	a	4
a, c	a	2
a, d	a	12
b, c	b	4
b, d	b	8
c, d	c	4

To state a representation theorem for the representation (3.4), we introduce five defining axioms for an *algebraic difference structure* (A, D). We shall use the notation

$$abEst \Leftrightarrow [\sim abDst \ \& \ \sim stDab] \tag{3.16}$$

and

$$abWst \Leftrightarrow [abDst \ \text{or} \ abEst]. \tag{3.17}$$

The first three axioms are the following:

AXIOM D1. *Suppose R is defined on $A \times A$ by*

$$(a, b)R(s, t) \Leftrightarrow abDst.$$

Then $(A \times A, R)$ is strict weak.

AXIOM D2. *For all $a, b, s, t \in A$, if abDst, then tsDba.*

AXIOM D3. *For all $a, b, c, a', b', c' \in A$, if $abWa'b'$ and $bcWb'c'$, then $acWa'c'$.*

Axioms D1, D2, and D3 are clearly necessary conditions for the representation (3.4). To see that Axiom D1 is necessary, define $g: A \times A \to Re$ by

$$g(a, b) = f(a) - f(b).$$

Then

$$(a, b)R(s, t) \Leftrightarrow g(a, b) > g(s, t).$$

Proof of the necessity of Axioms D2 and D3 is left to the reader. Axiom D3 is sometimes called Weak Monotonicity. It is violated by the data of Table 3.9, in the sense that if D is defined from this data by Eq. (3.14), then (A, D) violates Axiom D3. This is the case because $abWbc$ and $bcWcc$ but not $acWbc$. [It is clear that for this data, there can be no function satisfying Eq. (3.15), because we have $\Delta(a, b) > 0$, $\Delta(b, c) > 0$, but $\Delta(a, c) < \Delta(b, c)$.]

The fourth axiom reads as follows:

AXIOM D4. *For all $a, b, s, t \in A$, if abWst holds and stWxx holds, then there are u, v in A such that auEst and vbEst.*

Axiom D4 is often called a *solvability condition*. The assumption $stWxx$ says that $f(s) - f(t) \geqq 0$. The assumption $abWst$ says that

$$f(a) - f(b) \geqq f(s) - f(t).$$

Then Axiom D4 says that we can find u and v which "solve" the equations

$$f(a) - f(u) = f(s) - f(t)$$

and

$$f(v) - f(b) = f(s) - f(t).$$

Axiom D4 is not a necessary condition. To give an example, let $A = \{0, 1, 3\}$, and let D on A be defined by

$$xyDuv \Leftrightarrow x - y > u - v. \tag{3.18}$$

Then $f(x) = x$ is a function satisfying Eq. (3.4). But (A, D) does not satisfy Axiom D4. For $3,1W1,0$ holds* and $1,0W0,0$ holds, but there are no u and v in A such that $3,uE1,0$ and $v,1E1,0$ both hold. For $3,uE1,0$ implies $u = 2$.

To state our fifth axiom, let $a_1, a_2, \ldots, a_i, \ldots$ be a sequence of elements from A. It is called a *standard sequence* if $a_{i+1}a_iEa_2a_1$ holds for all a_i, a_{i+1} in the sequence and $a_2a_1Ea_1a_1$ does not hold. The idea of a standard sequence is that the difference between two successive elements is the same nonzero amount. This follows from the representation (3.4), since

$$a_{i+1}a_iEa_2a_1 \Leftrightarrow f(a_{i+1}) - f(a_i) = f(a_2) - f(a_1),$$

so for all i and j,

$$f(a_{i+1}) - f(a_i) = f(a_{j+1}) - f(a_j),$$

and since $\sim a_2a_1Ea_1a_1$ implies $f(a_2) - f(a_1) \neq 0$.

A standard sequence is called *strictly bounded* if there are s, t in A such that $stDa_ia_1$ and a_ia_1Dts for all a_i in the sequence. That is,

$$f(t) - f(s) < f(a_i) - f(a_1) < f(s) - f(t)$$

for all i in the sequence. Our fifth axiom is:

AXIOM D5. *Every strictly bounded standard sequence is finite.*

---

*We have placed commas here purely to separate elements in the relation W.

Axiom D5 is an Archimedean condition, and it is a necessary condition for the representation (3.4). For if $a_1, a_2, \ldots$ is a strictly bounded standard sequence, then $f(a_{i+1}) - f(a_i) = f(a_2) - f(a_1)$, for all i. Thus,

$$f(a_i) - f(a_1) = (i - 1)[\, f(a_2) - f(a_1)\,].$$

Since $a_2 a_1 E a_1 a_1$ does not hold, $f(a_2) - f(a_1) \neq 0$. Thus, if the sequence is infinite, then, depending on whether $f(a_2) - f(a_1)$ is positive or negative, the Archimedean condition on the reals implies that for all s, t, there is an i such that $f(a_i) - f(a_1) > f(s) - f(t)$ or there is an i such that $f(t) - f(s) > f(a_i) - f(a_1)$. This violates either $stDa_i a_1$ or $a_i a_1 Dts$.

Remark: We shall encounter the idea of standard sequence on a number of occasions in this volume. It corresponds to a very practical idea in measurement. Before performing a measurement, one usually agrees on some degree of precision, say the difference between $f(a_2)$ and $f(a_1)$. Suppose this difference is positive. Then we are willing to make errors in measurement up to this degree of precision. We construct a standard sequence $a_1, a_2, \ldots$ by repeating this fixed difference over and over again. Given any "positive" difference $f(s) - f(t)$ larger than the degree of precision, that is, so that $stDa_2 a_1$, we find an i so that

$$stWa_i a_1 \quad \text{and} \quad a_{i+1} a_1 Wst.$$

Now we can assert that the difference $f(s) - f(t)$ is somewhere between the difference $f(a_i) - f(a_1)$ and the difference $f(a_{i+1}) - f(a_1)$, and this measurement is within the desired degree of precision. Standard sequences also arise in extensive measurement. Here, we fix any a so that $2aRa$. A standard sequence is a set

$$\{ na: n \in I \}, \tag{3.19}$$

where I is any consecutive set of integers. An Archimedean axiom for extensive measurement says that any strictly bounded standard sequence is finite, where the standard sequence (3.19) is *strictly bounded* if there is a b so that $bRna$ for all na in the sequence. An equivalent axiom says the following. Given $a \in A$ with $2aRa$ and given $b \in A$ so that bRa, there are n and $n + 1$ so that $bSna$ and $(n + 1)aSb$, where $xSy \Leftrightarrow\, \sim yRx$. Thus, we can assert that the measure $f(b)$ is between $nf(a)$ and $(n + 1)f(a)$. If a is chosen so that $f(a)$ is small enough, we can obtain $f(b)$ to within any desired degree of precision. In the case of measurement of mass, we have a set of standard weights in most laboratories. Combinations of these weights can be used to form standard sequences. Similarly, in the measurement of length, the ruler defines standard sequences (up to $\frac{1}{8}$ inch, $\frac{1}{16}$ inch, etc.)

A relational system (A, D) satisfying Axioms D1 through D5 is called an *algebraic difference structure*.

The reader should consider the "reasonableness" of the axioms for an algebraic difference structure, for the cases where D stands for comparison of temperature differences and where D stands for comparison of preferences. He also should consider the testability of the axioms, in particular Axioms D4 and D5.

We now state our first representation theorem for algebraic difference measurement.

THEOREM 3.8 (Krantz. *et al.* [1971]). *If (A, D) is an algebraic difference structure, then there is a real-valued function f on A so that for all $a, b, s, t \in A$,*

$$abDst \Leftrightarrow f(a) - f(b) > f(s) - f(t). \tag{3.4}$$

We omit the proof of this theorem. Earlier sets of sufficient conditions for algebraic difference measurement were given by Suppes and Winet [1955], Debreu [1958], Scott and Suppes [1958], Suppes and Zinnes [1963], and Kristof [1967]. The only known set of axioms necessary as well as sufficient for difference measurement is due to Scott [1964], and requires the assumption that A be finite. We state Scott's Theorem in the next subsection.*

3.3.2 Necessary and Sufficient Conditions

THEOREM 3.9 (Scott). *Suppose A is a finite set, D is a quaternary relation on A, and E and W are defined by Eqs. (3.16) and (3.17). Then the following conditions are necessary and sufficient for there to be a real-valued function f on A satisfying, for all $a, b, s, t \in A$,*

$$abDst \Leftrightarrow f(a) - f(b) > f(s) - f(t). \tag{3.4}$$

AXIOM SD1. $abWst$ *or* $stWab$, *all* $a, b, s, t \in A$.

AXIOM SD2. $abDst \Rightarrow tsDba$, *all* $a, b, s, t \in A$.

AXIOM SD3. *If* $n > 0$ *and* π *and* σ *are permutations of* $\{0, 1, \ldots, n - 1\}$, *and if* $a_i b_i W a_{\pi(i)} b_{\sigma(i)}$ *holds for all* $0 < i < n$, *then* $a_{\pi(0)} b_{\sigma(0)} W a_0 b_0$ *holds; this is true for all* $a_0, a_1, \ldots, a_{n-1}, b_0, b_1, \ldots, b_{n-1} \in A$.

Axioms SD1 and SD2 are clearly necessary. To illustrate Axiom SD3, let $A = \{a, b\}$ and let $n = 2$. Let $a_0 = b, a_1 = a, b_0 = a, b_1 = b$. Suppose π is

---

*A referee has pointed out that a variant of Scott's Theorem is valid for A of arbitrary cardinality, except that in the infinite case the representation is non-Archimedean in general.

the identity permutation, the permutation that takes 0 into 0 and 1 into 1, and σ is the permutation that takes 0 into 1 and 1 into 0. Axiom SD3 says that if $a_1 b_1 W a_{\pi(1)} b_{\sigma(1)}$ holds, then $a_{\pi(0)} b_{\sigma(0)} W a_0 b_0$ holds. Thus, $a_1 b_1 W a_1 b_0$ implies $a_0 b_1 W a_0 b_0$, or $abWaa$ implies $bbWba$. This result is clearly necessary, for

$$f(a) - f(b) \geq f(a) - f(a) = 0$$

implies

$$0 = f(b) - f(b) \geq f(b) - f(a).$$

Axiom SD3 is really an "infinite schema" of axioms. It is necessary to state one axiom for every n. (Even if A has only finitely many elements, there is need for infinitely many axioms, since the a_i and b_i do not need to be distinct.) To show that Axiom SD3 is necessary in general, let us observe that if f is any real-valued function on A, then

$$\sum_{i=0}^{n-1} f(a_i) - \sum_{i=0}^{n-1} f(b_i) = \sum_{i=0}^{n-1} f(a_{\pi(i)}) - \sum_{i=0}^{n-1} f(b_{\sigma(i)}).$$

Thus,

$$\sum_{i=0}^{n-1} \left[f(a_i) - f(b_i) \right] = \sum_{i=0}^{n-1} \left[f(a_{\pi(i)}) - f(b_{\sigma(i)}) \right]. \tag{3.20}$$

If f satisfies (3.4), then the hypothesis of Axiom SD3 says that

$$f(a_i) - f(b_i) \geq f(a_{\pi(i)}) - f(b_{\sigma(i)}),$$

for all $0 < i < n$. Now equality in (3.20) implies

$$f(a_{\pi(0)}) - f(b_{\sigma(0)}) \geq f(a_0) - f(b_0),$$

so $a_{\pi(0)} b_{\sigma(0)} W a_0 b_0$. Sufficiency of Axioms SD1 through SD3 involves a clever argument which uses the well-known separating hyperplane theorem. We refer the reader to Scott's paper for details.

Before leaving this section, let us observe how the data of Table 3.9 violates Axiom SD3. Let $a_0 = b$, $a_1 = a$, $a_2 = b$, $b_0 = c$, $b_1 = b$, and $b_2 = c$. Let $\pi(0) = 1$, $\pi(1) = 0$, $\pi(2) = 2$, $\sigma(0) = 0$, $\sigma(1) = 2$, and $\sigma(2) = 1$. Then, according to Axiom SD3,

$$abWbc \ \& \ bcWbb \Rightarrow acWbc.$$

This condition is violated by the data of Table 3.9.

3.3.3 *Uniqueness*

Next, we turn to a uniqueness theorem. We define a quaternary relation Δ on Re by

$$xy\Delta uv \Leftrightarrow x - y > u - v. \qquad (3.21)$$

THEOREM 3.10. *Suppose A is a nonempty set, D is a quaternary relation on A, and f is a real-valued function on A satisfying*

$$abDst \Leftrightarrow f(a) - f(b) > f(s) - f(t), \qquad (3.4)$$

for all a, b, s, t in A. Suppose (A, D) is an algebraic difference structure. Then $\mathfrak{A} = (A, D) \to \mathfrak{B} = (Re, \Delta)$ is a regular representation and $(\mathfrak{A}, \mathfrak{B}, f)$ is an interval scale.

In the case of preference and utility, this theorem says that if judgments of utility difference are sufficiently regular to give rise to a utility function satisfying Eq. (3.4), and if certain assumptions about these judgments— namely, that (A, D) be an algebraic difference structure—hold, then we can make more than just ordinal comparisons with utility. Some writers call f a cardinal utility function if f defines a scale as strong as an interval scale.

Before discussing the proof of this theorem, we note that it requires the special hypothesis that (A, D) be an algebraic difference structure. In particular, since Axioms D1, D2, D3, and D5 are necessary axioms for the representation (3.4), the required assumption is Axiom D4. To see that some assumption is needed, consider $A = \{0, 1, 3\}$ and D as defined on A by Eq. (3.18). Then two functions f and g satisfying Eq. (3.4) are given by $f(x) = x$ and $g(0) = 1, g(1) = 2, g(3) = 8$. If $\phi: f(A) \to Re$ satisfies $g = \phi \circ f$, then ϕ is not a positive linear transformation. For suppose there are α and β, with $\alpha > 0$, so that $\phi(x) = \alpha x + \beta$ for all x in $f(A)$. Then $\phi(0) = \beta$, so $\beta = 1$. Now $2 = \phi(1) = \alpha + \beta = \alpha + 1$, so $\alpha = 1$. Thus, $\phi(3) = \alpha \cdot 3 + \beta = 3 + 1 = 4 \neq g(3)$, so $g \neq \phi \circ f$.

The uniqueness problem for algebraic difference measurement is, therefore, not completely settled by Theorem 3.10. It would be interesting to find necessary and sufficient conditions on those (A, D) representable in the form (3.4) for the representation to be unique up to a positive linear transformation. It would also be helpful to have a systematic treatment of the possible admissible transformations that arise in difference measurement.

Turning to a proof of Theorem 3.10, we note first that by Theorem 2.1, the representation $\mathfrak{A} \to \mathfrak{B}$ is regular. For

$$f(a) = f(b) \quad \text{iff} \quad abEaa.$$

Next, suppose f satisfies (3.4) and $\phi(x) = \alpha x + \beta$, $\alpha > 0$, for all x in $f(A)$. Then it is easy to see that $g(a) = \phi \circ f(a)$ also satisfies (3.4). Finally, suppose $\phi : f(A) \to Re$ has the property that $g = \phi \circ f$ satisfies (3.4). It is necessary to show that $\phi(x)$ is of the form $\alpha x + \beta$, $\alpha > 0$.

We sketch a proof that works under the additional assumption that for all b, x, y in A, there is a c in A so that $bcExy$. This is a stronger solvability condition than Axiom D4. Krantz *et al.* [1971] have a proof that works without this additional assumption. The idea for the following proof goes back to Hölder [1901]. Let

$$B = \{(a, b) \in A \times A: \ abDaa\}$$

and let R on B be defined by

$$(a, b)R(c, d) \Leftrightarrow abDcd.$$

By Axiom D1 of an algebraic difference structure, (B, R) is a strict weak order. Let (B^*, R^*) be its reduction. Define an operation $\circ$ on B^* as follows. Given s and t in B^*, let (a, b) be in s, and let (x, y) be in t. Then by the additional solvability assumption, there is c in A so that $bcExy$, that is, (b, c) is in t. Define $s \circ t$ to be the equivalence class containing (a, c). It is not hard to show that $\circ$ is well-defined. (One part of the proof is to show that (a, c) is in B.) Having defined $\circ$, define $F: B^* \to Re$ by

$$F(s) = f(a) - f(b) \quad \text{if} \quad (a, b) \text{ is in } s.$$

F is well-defined and defines a homomorphism from $(B^*, R^*, \circ)$ into $(Re, >, +)$. Given the admissible transformation ϕ of f, let $\psi : F(B^*) \to Re$ be defined by $\psi(x - y) = \phi(x) - \phi(y)$. Then ψ is well-defined and an admissible transformation of F. It follows from the uniqueness theorem for extensive measurement (Theorem 3.7) that there is an $\alpha > 0$ such that $\psi(x) = \alpha x$, all x in $F(B^*)$. But then

$$\phi(x) - \phi(y) = \psi(x - y) = \alpha x - \alpha y.$$

Hence, for fixed x_0 in $f(A)$,

$$\phi(x) - \phi(x_0) = \alpha x - \alpha x_0$$

and so

$$\phi(x) = \alpha x + \beta,$$

with $\beta = \phi(x_0) - \alpha x_0$.

Exercises

1. Suppose $n \geq 2$, $A = \{0, 1, 2, \ldots, n\}$, and D is defined on A by Eq. (3.18). Show that (A, D) is an algebraic difference structure.

2. (a) If $k > 0$, let $A = \{nk: n \in N\}$. Let D be defined on A by Eq. (3.18). Show that (A, D) is an algebraic difference structure.

(b) Which of the following relational systems (A, D) are algebraic difference structures?

      (i) $A = Z$, D defined by Eq. (3.18).
      (ii) $A = Re^+$, $abDcd$ iff $a/b > c/d$.
      (iii) $A = Re^+$, $abDcd$ iff $a^2/b^2 > c^2/d^2$.
      (iv) $A = \{n^k: n \in N\}$, $abDcd$ iff $a/b > c/d$.

3. Suppose a subject in the laboratory is asked to estimate absolute temperature differences of objects he feels, and gives the data in Table 3.10. Show that the subject is inconsistent with the algebraic difference model in the sense that if δ is defined from Δ as in Section 3.3.1, there is no real-valued function f on the set A of alternative objects considered so that Eq. (3.15) is satisfied.

4. Show that the representation (3.15) is attainable for the subject of Table 3.11.

5. (a) Show that Axiom D2 for an algebraic difference structure is necessary for the representation (3.4).

Table 3.10. Judgments of Absolute Temperature Difference

Objects Compared, x, y	Warmer Object	Estimated Absolute Temperature Difference $\Delta(x, y)$
a, b	a	3
a, c	a	4
a, d	a	8
b, c	b	3
b, d	b	5
c, d	c	2

Table 3.11. Judgments of Absolute Temperature Difference

Objects Compared, x, y	Warmer Object	Estimated Absolute Temperature Difference $\Delta(x, y)$
a, b	a	2
a, c	a	3
a, d	a	6
b, c	b	2
b, d	b	3
c, d	c	2

(b) Show that Axiom D3 for an algebraic difference structure is necessary for the representation (3.4).

6. (a) Suppose $n \geq 2$, $A = \{0, 1, 2, \ldots, n, r\}$, where r is not in the set $\{-1, 0, 1, 2, \ldots, n, n+1\}$, and D is defined on A by Eq. (3.18). Show that (A, D) is not an algebraic difference structure.

(b) Show that $\mathfrak{A} = (A, D)$ is homomorphic to $\mathfrak{B} = (Re, \Delta)$, where Δ is defined by Eq. (3.21).

(c) Show that if $r > 2n$ and if f is any homomorphism from $\mathfrak{A}$ into $\mathfrak{B}$, then $(\mathfrak{A}, \mathfrak{B}, f)$ is not an interval scale.

(d) However, if $n + 1 < r \leq 2n$ and f is any homomorphism from $\mathfrak{A}$ into $\mathfrak{B}$, show that $(\mathfrak{A}, \mathfrak{B}, f)$ is an interval scale.

7. Suppose $A = \{0, 1, 2\}$ and D is a quaternary relation on A. Show that the following statements follow from Scott's axioms:

(a) If $0,0\,W\,0,1$ and $1,1\,W\,1,2$ then $2,0\,W\,2,2$.

(b) If $0,1\,W\,0,0$, then $0,0\,W\,1,0$.

(c) If $0,0\,W\,1,2$ and $1,2\,W\,1,0$, then $0,2\,W\,1,2$.

8. Let R be the lexicographic ordering of the plane, and define D on $A = Re$ by

$$xy\,Duv \Leftrightarrow (x, y)R(u, v).$$

(a) Show that (A, D) is not an algebraic difference structure.

(b) Determine which of the axioms D1 through D5 are violated.

9. Suppose (A, D) is an algebraic difference structure and f satisfies Eq. (3.4).

(a) Show that the statement $f(a) > f(b) + f(c)$ is not meaningful.

(b) Consider the meaningfulness of the following statements:

(i) $f(a) = 2f(b)$.

(ii) $f(a) > f(b) + 100$.

(iii) $f(a) - f(b)$ is a constant.

10. Suppose (A, D) is an algebraic difference structure. Show the following:

(a) There is a function $g: A \to Re^+$ so that for all a, b, c, d in A,

$$abDcd \Leftrightarrow g(a)/g(b) > g(c)/g(d).$$

(b) The function g defines a regular scale and the admissible transformations of g are all functions of the form $\phi(x) = \alpha x^\beta$, α, $\beta > 0$. (In the terminology of Section 2.3, g defines a log-interval scale.)

(c) The following statements are meaningful:

(i) $g(a) > g(b)$.

(ii) $g(a)/g(b) > g(c)/g(d)$.

11. (Krantz *et al.* [1971], Suppes and Zinnes [1963]) Suppose A is a nonempty set and D is a quaternary relation on A. Suppose (A, D) satisfies Axioms D1 through D3 of an algebraic difference structure. We say that a is an *immediate successor* of b, and write $a\,Jb$, if $abDaa$ and there

is no c so that $acDaa$ and $cbDaa$. A pair (A, D) is called an *equally spaced difference structure* if it satisfies Axioms D1 through D3 and the following condition:

$$(aJb \ \& \ uJv) \Rightarrow abEuv.$$

Define J^n inductively as follows:

$$aJ^1b \text{ is } aJb,$$
$$aJ^{n+1}b \text{ holds iff there is } c \text{ in } A \text{ so that } aJ^nc \text{ and } cJb.$$

We shall also use the notation aJ^0b, to mean $abEaa$.

(a) Show that if $A = \{0, 1, 3\}$, then (A, Δ) is not an equally spaced difference structure, where Δ is defined by Eq. (3.21).

(b) Which of the relational systems (i), (ii), (iii) or (iv) of Exer. 2b is an equally spaced difference structure?

(c) Show that in an equally spaced difference structure, aJ^nb and $n > 0$ implies $abDaa$.

(d) Show that in an equally spaced difference structure on a finite set A, $abDaa$ implies that aJ^nb, some $n > 0$.

(e) However, show that (d) is false if A is not finite.

(f) Show that in an equally spaced difference structure on a finite set A, if $abDaa$ and $cdDaa$, then $abDcd$ holds if and only if·there are m, n so that

$$n > m \geqq 1 \quad \text{and} \quad aJ^nb \quad \text{and} \quad cJ^md.$$

(g) In an equally spaced difference structure on a finite set A, let e be a minimal element in the sense that for all a in A, $\sim eaDaa$. Define $f: A \to Re$ by

$$f(a) = \begin{cases} 0 & \text{if } aeEaa, \\ n & \text{if } aeDaa \text{ and } aJ^ne. \end{cases}$$

Show that f is a homomorphism from (A, D) into (Re, Δ).

(h) Suppose (A, D) is an equally spaced difference structure on a finite set A, and f is a homomorphism from (A, D) into (Re, Δ). Show that f defines an interval scale. [*Hint*: Find e so that $f(e)$ is minimal, and consider the relation between $f(a) - f(e)$ and $f(a) - f(b)$.]

12. In Exer. 15 of Section 3.2, we discussed the bisection operation. Suppose we do not require that every pair of objects has a bisector. Rather, suppose we consider the ternary relation B on the set A defined by $B(a, b, c)$ iff b is a bisector of a and c. As Suppes and Zinnes [1963] point out, the ternary relation B is related to the quaternary difference relation D as follows:

$$B(a, b, c) \Leftrightarrow [\sim abDbc \ \& \ \sim bcDab]. \tag{3.22}$$

(a) Suppose Eq. (3.22) holds, suppose (A, D) is an equally spaced difference structure (Exer. 11), and suppose f is a homomorphism from (A, D) into (Re, Δ). Show that if $A = \{x, y, z, w, u\}$ and

$$f(x) > f(y) > f(z) > f(w) > f(u),$$

then $B(x, y, z)$.

(b) Identify other triples (a, b, c) for which $B(a, b, c)$ holds in part (a).

(c) Use Eq. (3.22) and the representation (3.4) to state some necessary axioms on the ternary relation B.

References

Adams, E. W., "Elements of a Theory of Inexact Measurement," *Phil. Sci.*, **32** (1965), 205–228.

Adams, E. W., and Carlstrom, I. F., "Representing Approximate Ordering and Equivalence Relations," *J. Math. Psychol.*, **19** (1979), 182–207.

Alimov, N. G., "On Ordered Semigroups," *Izv. Akad. Nauk. SSSR Ser. Mat.*, **14** (1950), 569–576.

Behrend, F. A., "A System of Independent Axioms for Magnitudes," *J. Proc. Roy. Soc. N. S. Wales*, **87** (1953), 27–30.

Behrend, F. A., "A Contribution to the Theory of Magnitudes and the Foundations of Analysis," *Math. Z.*, **63** (1956), 345–362.

Bernays, P., *Axiomatic Set Theory*, 2nd ed., North-Holland, Amsterdam, 1968.

Birkhoff, G., *Lattice Theory*, American Mathematical Society Colloquium Publication No. XXV, New York, 1948, 1967.

Cantor, G., "Beiträge zur Begründung der Transfiniten Mengenlehre," *Math. Ann.*, **46** (1895), 481–512.

Davidson, R. R., and Bradley, R. A., "Multivariate Paired Comparisons: The Extension of a Univariate Model and Associated Estimation and Test Procedures," *Biometrika*, **56** (1969), 81–95.

Debreu, G., "Representation of Preference Ordering by a Numerical Function," in R. M. Thrall, C. H. Coombs, and R. L. Davis (eds.), *Decision Processes*, Wiley, New York, 1954.

Debreu, G., "Stochastic Choice and Cardinal Utility," *Econometrica*, **26** (1958), 440–444.

Falmagne, J. C., "Random Conjoint Measurement and Loudness Summation," *Psychol. Rev.*, **83** (1976a), 65–79.

Falmagne, J. C., "Statistical Tests in Measurement Theory: Two Methods," mimeographed, New York University, Department of Psychology, paper presented at the Spring Meeting of the Psychometric Society, April 1976b.

Falmagne, J. C., "A Probabilistic Theory of Extensive Measurement," Tech. Rept. 78–7, Mathematical Studies in Perception and Cognition, Department of Psychology, New York University, New York, N.Y., 1978.

Falmagne, J. C., "On a Class of Probabilistic Conjoint Measurement Models: Some Diagnostic Properties," *J. Math. Psychol.*, **19** (1979), 73–88.

Falmagne, J. C., Iverson, G., and Marcovici, S., "Binaural Loudness Summation: Probabilistic Theory and Data," manuscript, Department of Psychology, New York University, New York, N.Y., 1978.

Fishburn, P. C., *Utility Theory for Decision Making*, Wiley, New York, 1970.

Hoffman, K. H., "Sur Mathematischen Theorie des Messens," *Rozprawy Mat. (Warsaw)*, **32** (1963), 1–31.

Hölder, O., "Die Axiome der Quantität und die Lehre vom Mass," *Ber. Verh. Kgl. Sächsis. Ges. Wiss. Leipzig, Math.-Phys. Klasse*, 53 (1901), 1–64.

Holman, E. W., "Strong and Weak Extensive Measurement," *J. Math. Psychol.*, 6 (1969), 286–293.

Huntington, E. V., "A Complete Set of Postulates for the Theory of Absolute Continuous Magnitude," *Trans. Amer. Math. Soc.*, 3 (1902a), 264–279.

Huntington, E. V., "Complete Sets of Postulates for the Theories of Positive Integral and of Positive Rational Numbers," *Trans. Amer. Math. Soc.*, 3 (1902b), 280–284.

Huntington, E. V., *The Continuum and Other Types of Serial Order*, Harvard University Press, Cambridge, Massachusetts, 1917.

Keeney, R. L., and Raiffa, H., *Decisions with Multiple Objectives: Preferences and Value Tradeoffs*, Wiley, New York, 1976.

Krantz, D. H., Luce, R. D., Suppes, P., and Tversky, A., *Foundations of Measurement*, Vol. I, Academic Press, New York, 1971.

Krantz, D. H., Luce, R. D., Suppes, P., and Tversky, A., *Foundations of Measurement*, Vol. II, Academic Press, New York, to appear.

Kristof, W., "A Foundation of Interval Scale Measurement," Research Bulletin RB 67-22, Educational Testing Service, Princeton, New Jersey, 1967.

Luce, R. D., and Marley, A. A. J., "Extensive Measurement When Concatenation Is Restricted and Maximal Elements May Exist," in S. Morgenbesser, P. Suppes, and M. G. White (eds.), *Philosophy, Science, and Method: Essays in Honor of Ernest Nagel*, St. Martin's Press, New York, 1969, pp. 235–249.

Milgram, A. N., "Partially Ordered Sets, Separating Systems and Inductiveness," in K. Menger (ed.), *Reports of a Mathematical Colloquium*, Second Series No. 1, University of Notre Dame, 1939.

Pfanzagl, J., *Die Axiomatischen Grundlagen einer Allgemeinen Theorie des Messens*, Schrift. Stat. Inst. Univ. Wien, Neue Folge Nr. 1, Physica-Verlag, Würzburg, 1959a.

Pfanzagl, J., "A General Theory of Measurement—Applications to Utility," *Naval Research Logistics Quarterly*, 6 (1959b), 283–294.

Pfanzagl, J., *Theory of Measurement*, Wiley, New York, 1968.

Pollatsek, A., and Tversky, A., "A Theory of Risk," *J. Math. Psychol.*, 7 (1970), 540–553.

Riker, W. H., and Ordeshook, P. C., *Positive Political Theory*, Prentice-Hall, Englewood Cliffs, New Jersey, 1973.

Roberts, F. S., *Discrete Mathematical Models, with Applications to Social, Biological, and Environmental Problems*, Prentice-Hall, Englewood Cliffs, New Jersey, 1976.

Roberts, F. S., and Luce, R. D., "Axiomatic Thermodynamics and Extensive Measurement," *Synthese*, 18 (1968), 311–326.

Scott, D., "Measurement Models and Linear Inequalities," *J. Math. Psychol.*, 1 (1964), 233–247.

Scott, D., and Suppes, P., "Foundational Aspects of Theories of Measurement," *J. Symbolic Logic*, 23 (1958), 113–128.

Stevens, S. S., "On the Psychophysical Law," *Psychol. Rev.*, 64 (1957), 153–181.

Suppes, P., "A Set of Independent Axioms for Extensive Quantities," *Portugal. Math.*, 10 (1951), 163–172.

Suppes, P., and Winet, M., "An Axiomatization of Utility Based on the Notion of Utility Differences," *Management Sci.*, 1 (1955), 259–270.

Suppes, P., and Zinnes, J., "Basic Measurement Theory," in R. D. Luce, R. R. Bush, and E. Galanter (eds.), *Handbook of Mathematical Psychology*, Vol. I, Wiley, New York, 1963, pp. 1–76.

CHAPTER 4 _____

Applications to Psychophysical Scaling

4.1 The Psychophysical Problem

4.1.1 *Loudness*

A sound has a variety of physical characteristics. For example, a pure tone can be described by its physical *intensity* (energy transported), its *frequency* (in cycles per second), its *duration*, and so on. The same sound has various psychological characteristics. For example, how *loud* does it seem? What *emotional meaning* does it portray? What *images* does it suggest? Since the middle of the nineteenth century, scientists have tried to study the relationships between the physical characteristics of stimuli like sounds and their psychological characteristics. Some psychological characteristics might have little relationship to physical ones. For example, emotional meaning probably has little relation to the physical intensity of a sound, but rather it may be related to past experiences, as, for example, with the sound of a siren. Other psychological characteristics seem to be related in fairly regular ways to physical characteristics. Such sensations as loudness are an example. *Psychophysics* is the discipline that studies various psychological sensations such as loudness, brightness, apparent length, and apparent duration, and their relations to physical stimuli. It attempts to scale or measure psychological sensations on the basis of corresponding physical stimuli. In this chapter, we shall describe some of the history of psychophysical scaling and its applications or potential applications to measurement of noise pollution,* of attitudes, of utility, etc., and we shall

*The loudness of a sound is different from its disturbing effect. It is this disturbing effect that is often called *noise*. Of course, noise has effects other than just disturbance. It is

ENCYCLOPEDIA OF MATHEMATICS and Its Applications, Gian-Carlo Rota (ed.). Vol. 7: Fred S. Roberts, Measurement Theory

discuss ways to put psychophysical scaling on a firm measurement-theoretic foundation. We shall concentrate on loudness.

The psychophysical approach to measuring sensations like loudness is very different from the fundamental measurement approach we spelled out in Chapters 2 and 3. That approach would start with an observed binary relation "sounds louder than," and seek a scale that preserves this relation. We shall return to that approach in Chapter 6.

4.1.2 *The Psychophysical Function*

Subjective judgments of loudness are certainly dependent on the intensity of a sound. Data suggests that such judgments are also dependent on the frequency of a sound. Figure 4.1 shows equal-loudness contours, which illustrate the fact that sounds at some intensities are judged equally as loud as sounds at other intensities at different frequencies. Presumably, the duration of a sound also affects judgments of loudness. So does the *rise time*, the time for a sound to rise to maximum intensity. To simplify matters, one tries to eliminate all physical factors but one. For example, we shall study the relationship between intensity and loudness. To do so, we deal with *pure tones*, sounds of constant intensity at one fixed frequency (often taken to be 1000 cycles per second, cps), and we consider the case where these tones are presented for a fixed length of time. Then, only the intensity is varied. (Alternatively, we could deal with *white noise*, sounds with the same intensity at all frequencies.)

In principle, the scaling of the loudness of any sound, no matter how complex, can be reduced to the scaling of loudness of 1000-cps pure tones. For we simply find such a pure tone that gives rise to an equal sensation as the original sound (Stevens [1955, p. 825]). In practice, this procedure is difficult to carry out. The general case of loudness scaling involves fairly complicated procedures which are not a straightforward generalization of the procedure we have described. See Kryter [1970] or Stevens [1969] for a detailed discussion.

Suppose $I(a)$ denotes the intensity of a pure tone a. $I(a)$ is proportional to the square of the root-mean-square (rms) pressure $p(a)$. The common unit of measurement of intensity is the decibel (dB). This is $10 \log_{10}(I/I_0)$,

---

becoming increasingly evident that noise has numerous physiological effects. It obviously can affect hearing. More subtly, it has been linked to ulcers, changes in the cardiovascular system, possible decrease in fertility, etc. Noise also has psychological effects. Many of these are closely related to perceived loudness. However, measurement of noise pollution is different from measurement of loudness. For a survey of effects of noise on people, see Environmental Protection Agency [1971] or Kryter [1970]. For a summary of alternative ways of measuring noise or noisiness, see Kryter [1970].

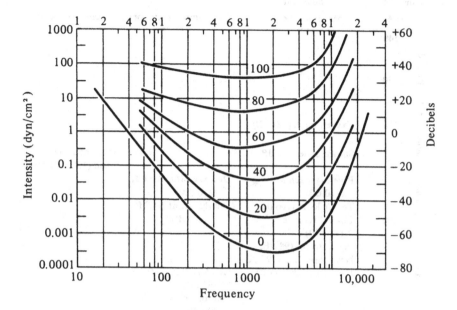

Figure 4.1. Equal-loudness contours. All the points on a given curve represent tones whose loudness is judged equal to that of a particular 1000-cps tone. The number on a given curve represents the "sensation level" of the given 1000-cps tone, where a sensation level of 0 is threshold, a sensation level of 20 is 20 dB above threshold, etc. This figure is adapted from Wever [1949, p. 307] and Krantz *et al.* [1971, p. 255] with permission of Academic Press and the authors. The data was due to Fletcher and Munson [1933].

where I_0 is the intensity of a reference sound. Thus,

$$dB(a) = 10 \log_{10}[I(a)/I_0] = 10 \log_{10}[p(a)/p_0]^2,$$

where p_0 is the rms pressure of the reference sound. (Cf. Kryter [1970] or Sears and Zemansky [1955, Section 23-3].) A sound of 1 dB is essentially the lowest audible sound.* For reference, some typical environmental noises measured in the decibel scale are given in Table 4.1.

---

*The reader of the acoustical literature will see notations like dBA, dBC, etc. If sounds occur over several frequencies, and we would like to get one number representing their sound pressure, we take the average pressure by summing (integrating) over different frequencies. Sometimes a weighted average is obtained, with some frequencies weighted more than others. For example, since the human ear is more sensitive to certain frequencies, it is considered reasonable to use a weighted average with a frequency weighted relative to the human ear's sensitivity to it. This weighting procedure leads to a decibel measure on the so-called A scale, denoted dBA. The decibel measure dBC corresponds to yet another weighting procedure, etc. (Cf. Kryter [1970, p. 13].) We do not have the problem of distinguishing the different decibel scales, since our discussion is limited to sounds of constant frequency.

Table 4.1. Sound Level in dBA of Some Typical Environmental Noises.*

	Sound Level, dBA	Industrial or Machine Operator†	Community–Outdoors	Home–Indoors
Painful	140			
	130			
Uncomfortably loud	120	Oxygen torch (121 dB)		
		Snowmobile (113 dB)		Rock-and-roll band (108–114 dB)
	110	Riveting machine (110 dB)	Jet take-off at 1000 ft (110 dB)	
		Textile loom (106 dB)	Jet flyover at 1000 ft (103 dB)	
	100	Electric furnace (100 dB)		
		Farm tractor (98 dB)		
		Newspaper press (97 dB)		
Very loud		Power mower (96 dB)	Rock drill at 50 ft (95 dB)	
			Motorcycle at 50 ft (90 dB)	
	90		Compressor at 50 ft (90 dB)	
			Snowmobile at 50 ft (90 dB)	Food blender (88 dB)
		Milling machine (85 dB)	Power mower at 50 ft (85 dB)	
			Diesel truck at 50 ft (85 dB)	
			Diesel train at 50 ft (85 dB)	
Moderately loud	80	Lathe (81 dB)		Garbage disposal (80 dB)
				Clothes washer (78 dB)
			Passenger car at 50 ft (75 dB)	Dishwasher (75 dB)
	70		Air-conditioning unit at 50 ft (60 dB)	Conversation (60 dB)
Quiet	60		Large transformer at 50 ft (60 dB)	
	50			
	40			
Very quiet	30			
	20			
	10			
	0			

*For meaning of A scale, see footnote, p.151. Data from Department of Public Health, State of California [1971].

†*Note:* Unless otherwise specified, listed sound levels prevail at *typical operator–listener distance from source.*

152

Let us call the scale of loudness the *psychological scale*. (Its unit is known as the *sone*, a term due to S. S. Stevens.*) This is the scale we are trying to derive from the *physical scale*, the scale I of intensity in our case. We shall denote the loudness of a sound a by $L(a)$. The measurement of loudness now reduces to the following question: What is the relation between the psychological scale L and the physical scale I? This relation is called the *psychophysical law*. It is usually stated as a function ψ which satisfies the equation

$$L(a) = \psi[I(a)].$$

The function ψ used to calculate psychological values from physical ones is called the *psychophysical function*. In the general situation, if $f(a)$ is a physical scale and $g(a)$ is a corresponding psychological scale, and if ψ is the psychophysical function, then for all a,

$$g(a) = \psi[f(a)].$$

We are in a situation of derived measurement, trying to derive one scale from another. The only difference between our present situation and the situations studied in Sections 2.5 and 2.6 is that we do not know the function ψ relating one scale to another.

A basic goal of psychophysics is to find the general form of the psychophysical function ψ which applies in many different cases of physical and psychological scales. This general form is often guessed at from large amounts of data, or even derived on the basis of some general assumptions. Then specific parameters needed to determine the exact form of the function ψ in a particular application are estimated in context.

The first attempt to specify the psychophysical function for a large class of psychological variables was made by Fechner [1860]. He argued that, under reasonable assumptions, the psychophysical law was logarithmic, that is, of the form

$$\psi(x) = \alpha \log x + \beta, \tag{4.1}$$

for α and β constant.* Equation (4.1) is called *Fechner's Law*. In the case of loudness, Fechner's Law says that $L(a) = \alpha \log I(a) + \beta$. The decibel scale of loudness arises from a special case of Fechner's Law, where the base of log is 10, $\alpha = 10$, and $\beta = -10 \log_{10} I_0$. If $L(a) = dB(a)$, then a

---

*A sone corresponds to the loudness of a pure tone of 40 dB at 1000 cps (Stevens [1955]).

*The base of log may be any number, since the change of base can be incorporated in the constant α. A critical assumption Fechner used in arguing for the logarithmic law was that you could scale sensations on the basis of variability or confusion or error: if two pairs of stimuli are equally often confused, then psychologically they are equally far apart. This assumption, which is embodied in the *Fechnerian utility model* of Section 6.2, has been questioned by writers such as S. S. Stevens (see below).

doubling of the dB level of a sound should lead to a doubling of the perceived loudness. It follows, for example, that a sound of 100 dB should sound twice as loud as a sound of 50 dB. Not long after the introduction of the dB scale, acoustical engineers noted that this seemed to be false (Stevens [1955, pp. 815, 816; 1957, p. 163]). Other data violated even the genera form of Fechner's Law (see below).

One of the earliest attempts at measuring loudness was that of Fletcher and Munson [1933].[†] They assumed that loudness was proportional to the number of auditory nerve impulses reaching the brain. Thus, a sound delive ed to two ears should appear to be twice as loud as it is when presented to only one ear.[‡] Fletcher and Munson discovered that a tone presented to only one ear had to be about 10 dB higher in energy (intensity) level than the level of a tone presented to both ears and judged equally loud. Thus, they concluded that *subjective loudness doubles for each 10-dB increase in sound pressure.*[*] Hence, Fletcher and Munson suggested that an increase from 50 dB to 60 dB should double perceived loudness, whereas if decibels measure loudness, the increase would have to be from 50 dB to 100 dB! The Fletcher–Munson observation has been confirmed (at least approximately) by many experiments. See Stevens [1955] for a summary of data and experiments. The Fletcher–Munson observation implies that there are no α and β so that for all sounds (pure tones) a,

$$L(a) = \alpha \log_{10} I(a) + \beta.$$

That is, the Fletcher-Munson observation implies that the general form of Fechner's Law could not hold. For suppose this law were to hold. Then certainly $\alpha \neq 0$; otherwise $L(a)$ is constant for all a. Moreover, we have

---

[†] The approach of Fletcher and Munson is discussed in some detail in Kryter [1970, pp. 247, 255].

[‡] This is a special case of the general hypothesis sometimes called the *loudness summation hypothesis*, which says the following. Suppose the loudness of a plus the loudness of x equals the loudness of b plus the loudness of y. Suppose a is presented to the left ear and simultaneously x to the right ear. This should produce a sound equally as loud as when b is presented to the left ear and simultaneously y to the right ear. We discuss this summation or additive hypothesis in our discussion of conjoint measurement in Section 5.4. For reviews of this hypothesis, see Hirsch [1948], Reynolds and Stevens [1960], Békésy [1960], Treisman and Irwin [1967], Scharf [1969], Tobias [1972], Levelt, Riemersma, and Bunt [1972], and Falmagne, Iverson, and Marcovici [1978].

[*] It has been observed that noise levels in urban areas in the United States have been increasing at the rate of approximately 1 dB a year. Thus, it is fair to conclude that, on the basis of results like those of Fletcher and Munson, noise levels in our urban areas are doubling every ten years (Rienow and Rienow [1967, p. 179].)

for all a,

$$
\begin{aligned}
L(a) &= \alpha \log_{10} I(a) + \beta \\
&= \alpha \log_{10}[I(a)/I_0] + (\beta + \alpha \log_{10} I_0) \\
&= \alpha \log_{10}[I(a)/I_0] + \beta' \\
&= \alpha \, \mathrm{dB}(a) + \beta'.
\end{aligned}
$$

Suppose $\mathrm{dB}(b) = \mathrm{dB}(a) + 10$. By the Fletcher–Munson observation, $L(b) = 2L(a)$. Thus,

$$
\alpha \, \mathrm{dB}(b) + \beta' = 2[\alpha \, \mathrm{dB}(a) + \beta'],
$$

or

$$
\alpha[\mathrm{dB}(a) + 10] + \beta' = 2[\alpha \, \mathrm{dB}(a) + \beta'],
$$

or, using $\alpha \neq 0$,

$$
\mathrm{dB}(a) = \frac{10\alpha - \beta'}{\alpha}.
$$

That is, for all a, $\mathrm{dB}(a)$ is a constant. This is a contradiction.

Stevens [1957, 1960, 1961a,b,c, and elsewhere] has argued that instead of a logarithmic law, the fundamental psychophysical law for loudness and many other psychological variables is a power law,

$$
\psi(x) = \alpha x^{\beta}, \tag{4.2}
$$

for α, β constant, $\alpha > 0$. This idea goes back to Plateau [1872]. We shall see in Section 4.3 that the power law is consistent with data like that of Fletcher and Munson, at least if the exponent β is chosen properly. The data of Stevens and his colleagues [1960 and elsewhere] seems to suggest that, at least to a first approximation, and sometimes only for limited intervals of values of the physical stimuli, the power law holds for more than two dozen psychological variables. These psychological variables are shown in Table 4.2. In general, these variables seem to be concerned with quantitative judgments, like "how much"? Stevens calls such variables *prothetic continua*. Other psychological variables are more concerned with qualitative judgments, like "what kind" or "where"? Examples of such variables are pitch, apparent azimuth, and apparent inclination. These variables are called *metathetic continua*. The power law may or may not hold on metathetic continua (Stevens [1968]). For example, it fails for pitch as a function of frequency (Stevens and Volkmann [1940]). However, for

Table 4.2. Some Prothetic Psychological Continua and Their Exponents under the Power Law*

Psychological Continuum	Name of Psychological Unit	Exponent	Conditions
Loudness	Sone	0.3	Binaural, 1000-cps tone
Loudness		0.27	Monaural, 1000-cps tone
Brightness	Bril	0.33	5° target—dark-adapted eye
Brightness		0.5	Point source—dark-adapted eye
Lightness		1.2	Reflectance of gray papers
Smell		0.55	Coffee odor
Smell		0.6	Heptane
Taste	Gust	0.8	Saccharine
Taste		1.3	Sucrose
Taste		1.3	Salt
Temperature		1.0	Cold—on arm
Temperature		1.6	Warm—on arm
Vibration		0.95	60 cps—on finger
Vibration		0.6	250 cps—on finger
Duration	Chron	1.1	White noise stimulus
Repetition rate		1.0	Light, sound, touch, and shocks
Finger span		1.3	Thickness of wood blocks
Pressure on palm		1.1	Static force on skin
Heaviness	Vog	1.45	Lifted weights
Force of handgrip		1.7	Precision hand dynamometer
Vocal effort		1.1	Sound pressure of vocalization
Electric shock		3.5	60 cps through fingers

*Table adapted from Stevens [1957, p. 166; 1960, p. 236; 1961b].

prothetic continua, the power law is accepted quite widely. According to Ekman and Sjöberg [1965]: "As an experimental fact, the power law is established beyond any reasonable doubt, possibly more firmly established than anything else in psychology." Still, there is some conflicting evidence.*

In summary, the literature of psychophysics has not been and still is not in agreement on the general form of the psychophysical law. In the next

*Pradhan and Hoffman [1963] found violations in the power law by individuals (though not by the whole group of subjects, if the data was averaged). Others have found tremendous variability in data, both between different individual subjects and for an individual subject being retested. (See Luce and Mo [1965], Schneider and Lane [1963], and Stevens and Guirao [1964].) Stevens [1957, 1959c, 1961b,c, 1971] argues that this variation is primarily due to randomness in the data, and it averages out over subjects. However, Pradhan and Hoffman suggest that the power law is simply an "artifact" of group averaging. And Green and Luce [1974] try to make the case that the variations might be "intimately related" to the decisionmaking process underlying sensory judgments. Luce [1972] argues that the large variation in data makes it hard to speak of psychophysical measurement as analogous to physical measurement or, indeed, any sort of fundamental measurement. For other criticisms of the power law, see Savage [1970].

section, we discuss a theory that allows us to determine the possible forms of the psychophysical law. In Section 4.3, we introduce assumptions that allow us to derive the power law, and we discuss some implications of the power law, and some of its applications. In Section 4.4, we introduce a measurement axiomatization for the key assumption used to justify the power law in Section 4.3.

Exercises

1. A 40-dB pure tone at 1000 cps receives a loudness rating of 1 sone. According to the Fletcher–Munson observation, what sone rating does a 60-dB pure tone at 1000 cps receive?

2. (Stevens [1960]) It is possible to introduce a decibel scale for light as well as sound. We would define

$$N_{dB} = 10 \log_{10}(E/E_0),$$

where E is the light energy and E_0 is a reference light energy. The brightness of a light a, $B(a)$, is a function $\psi(E(a))$. The unit of brightness is the bril. If $B(a) = N_{dB}(a)$, a halving in the number of brils would correspond to a halving in the decibel level. Yet, experiments suggest that, as with sound, a halving in the number of brils corresponds to a 10-dB decrease. Show that this observation even violates the general form of Fechner's Law:

$$B(a) = \alpha \log_{10} E(a) + \beta.$$

3. The decibel scale could also be used for vibration. The decibel level v_{dB} would be taken as $10 \log_{10}(E/E_0)$, with E physical energy transmitted and E_0 a reference energy level. Then a $5v_{dB}$ increase corresponds to a doubling of sensation. (See Exer. 6 of Section 4.3.) Show that Fechner's Law again fails to hold.

4. If a plot of physical versus psychological scales is made in log–log coordinates, show that the power law predicts the plot will be a straight line.

5. (Stevens [1960, p. 235]) To most observers, the apparent length of a straight line of 100 cm is about twice that of a straight line of 50 cm; that of a straight line of 80 cm is about twice that of a straight line of 40 cm; and so on. Thus, if the physical scale is length and the psychological scale is apparent length, and if the psychophysical function ψ is a power function, show that $\psi(x) = \alpha x$. (We make a stronger observation in Exer. 1, Section 4.2.)

6. Since sound pressure p defines a ratio scale, so does sound intensity. However, show that the decibel scale is an absolute scale in the wide sense. (It is not a difference scale.)

Table 4.3. Median Threshold of Audibility in Decibels*

	Frequency (cps)						
	500	1000	1500	2000	3000	4000	6000
Farmworkers	4	5	4	4	18	33	30
Office workers	3	5	7	8	15	22	26
Factory workers	8	9	11	16	35	43	42

*Data from Glorig *et al.* [1957].

7. The *threshold of audibility* or *hearing level* (in decibels) is measured for various individuals, and it depends on frequency. (A lower threshold means more acute hearing.) In a study reported in Kryter [1970, p. 116], Glorig *et al.* [1957] measured the median threshold of audibility for farmworkers, office workers, and factory workers, and reported the data shown in Table 4.3.

(a) Is it meaningful to assert that the median threshold of hearing of farmworkers at 1000 cps is better than that of factory workers?

(b) Is it meaningful to assert that the median threshold of hearing of office workers at 2000 cps is twice as high as that of farmworkers, that is, that the threshold has been increased (worsened) by 100%?

(c) If arithmetic means had been calculated instead of medians, would the statements of (a) and (b) have been meaningful?

8. The American Academy of Ophthalmology and Otolaryngology of the American Medical Association proposed in Lierle [1959] that the percentage of hearing impairment suffered by an individual should be measured as follows. (This description disregards how to average in the effect of differential hearing loss in two ears.) Measure the threshold of audibility (Exer. 7) at the three frequencies 500, 1000, and 2000 cps. Subtract a fixed number of dB (15 dB) from each. Average the three numbers, and then multiply by 1.5%. If $t_i(f)$ is the threshold of audibility of individual i at frequency f, then the impairment of i is given by

$$\text{Imp}_i = \frac{t_i(500) - 15 + t_i(1000) - 15 + t_i(2000) - 15}{3} \times 1.5.$$

If Imp is to be considered a percentage of hearing impairment, it should be meaningful to say that an individual with an Imp of 60 has only 60% of the impairment of an individual with an Imp of 100. Is this a meaningful assertion? (For a further discussion of measurement of impairment, see Kryter [1970, pp. 125ff].)

4.2 The Possible Psychophysical Laws

In trying to derive a psychological scale g from a physical scale f, we need to determine the psychophysical function ψ relating g to f. The domain and range of ψ are usually taken to be all of Re or Re^+, though real intervals can be used. Thus, it is assumed that all real numbers or positive real numbers are (potential) physical and psychological scale values. Although the function ψ may not be known, some of its properties may be. For example, we are usually willing to assume that the psychophysical function is continuous. We might also discover that ψ is additive, that is, it satisfies

$$\psi(x + y) = \psi(x) + \psi(y), \tag{4.3}$$

for all x, y in the domain of ψ. Equation (4.3) gives a so-called *functional equation* involving the function ψ. In the next subsection, we indicate the continuous solutions to four simple functional equations called the *Cauchy equations* (Cauchy [1821]). Our approach follows that of Aczél [1966]. We shall apply the solutions to several of these Cauchy equations to derive possible forms of the psychophysical function.

4.2.1 *Excursis: Solution of the Cauchy Equations*

Equation (4.3) is the first Cauchy equation. The remaining Cauchy equations are

$$\psi(x + y) = \psi(x)\psi(y), \tag{4.4}$$

$$\psi(xy) = \psi(x) + \psi(y), \tag{4.5}$$

and

$$\psi(xy) = \psi(x)\psi(y). \tag{4.6}$$

THEOREM 4.1. *Let* $Re' = Re$ *or* Re^+. *Suppose* $\psi: Re' \to Re$ *satisfies the first Cauchy equation*

$$\psi(x + y) = \psi(x) + \psi(y) \tag{4.3}$$

for all x *and* y *in* Re', *and suppose that* ψ *is continuous. Then there is a real number* c *such that*

$$\psi(x) = cx, \tag{4.7}$$

all x.

Proof. By induction from Eq. (4.3), $\psi(nx) = n\psi(x)$, all positive integers n. Next, let $x = (m/n)t$, m, n positive integers. Then $nx = mt$ so $\psi(nx) = \psi(mt)$, whence $n\psi(x) = m\psi(t)$. We conclude that for all positive integers m

and n and for all t real,

$$\psi\left(\frac{m}{n}t\right) = \frac{m}{n}\psi(t). \tag{4.8}$$

Suppose $\psi(1) = c$. Since t can be any positive real number, we let $t = 1$ in (4.8) and conclude that

$$\psi(r) = cr \tag{4.9}$$

for all positive rationals r. Since (4.9) holds for all positive rational numbers r, (4.7) follows for all positive real numbers x by taking limits on both sides of (4.9). (More precisely, by density of the rationals, we find a sequence of rationals $r_1, r_2, \ldots, r_n, \ldots$ such that $r_n \to x$. Then continuity of ψ implies that $\psi(r_n) \to \psi(x)$. But we know that $\psi(r_n) = cr_n$, so $\psi(r_n) \to cx$.) This proves the theorem if $Re' = Re^+$.

If $Re' = Re$, then note that $\psi(0) = 0 = c0$ follows immediately from (4.3). If x is negative, then $\psi(x) = \psi(0) - \psi(-x) = -\psi(-x) = -c(-x) = cx$. ∎

Remark: This theorem (and the next three) hold if $\psi(x)$ is assumed continuous at only one point x_0. This observation is due to Darboux [1875]. To see why Darboux's observation holds in the present case when $Re' = Re$, we suppose that $\psi(x)$ satisfies (4.3) and is continuous at x_0. Then $\lim_{t \to x_0} \psi(t) = \psi(x_0)$. For any other x, we have

$$\begin{aligned}
\lim_{u \to x} \psi(u) &= \lim_{u - x + x_0 \to x_0} \psi[(u - x + x_0) + (x - x_0)] \\
&= \lim_{t \to x_0} \psi[t + (x - x_0)] \\
&= \lim_{t \to x_0} [\psi(t) + \psi(x - x_0)] \\
&= \psi(x_0) + \psi(x - x_0) \\
&= \psi(x_0 + x - x_0) \\
&= \psi(x).
\end{aligned}$$

Thus, ψ is continuous at all x. If $Re' = Re^+$, the argument is slightly more complicated.

LEMMA. *Let $Re' = Re$ or Re^+. If $\psi: Re' \to Re$ satisfies the second Cauchy equation*

$$\psi(x + y) = \psi(x)\psi(y) \tag{4.4}$$

for all x and y in Re', then either $\psi(x) = 0$, all x, or $\psi(x) > 0$, all x.

Proof. Suppose $\psi(x_0) = 0$. If $Re' = Re$, then for all $x \in Re'$,

$$\psi(x) = \psi[(x - x_0) + x_0] = \psi(x - x_0)\psi(x_0) = 0. \qquad (4.10)$$

If $Re' = Re^+$, then for all $x > x_0$, (4.10) holds, and so $\psi(x) = 0$. If $y \in Re^+$ and $y < x_0$, then $ny > x_0$, some positive integer n. Hence, $\psi(ny) = 0$. But

$$\psi(ny) = \psi(y)^n,$$

so $\psi(y) = 0$.

Suppose next that for all x in Re', $\psi(x) \neq 0$. Then for all x in Re',

$$\psi(x) = \psi\left(\frac{x}{2} + \frac{x}{2}\right) = \psi\left(\frac{x}{2}\right)\psi\left(\frac{x}{2}\right) > 0.$$

∎

THEOREM 4.2. *Let $Re' = Re$ or Re^+. Suppose $\psi: Re' \to Re$ satisfies the second Cauchy equation*

$$\psi(x + y) = \psi(x)\psi(y) \qquad (4.4)$$

for all x and y in Re', and suppose that ψ is continuous. Then either

$$\psi \equiv 0$$

or

$$\psi(x) = e^{cx},$$

some real constant c.

Proof. Suppose $\psi \neq 0$. Let $f(x) = \ln \psi(x)$. By the lemma, f is well-defined, since $\psi(x) > 0$, all x. Then $f(x + y) = f(x) + f(y)$ and so f satisfies the first Cauchy equation, Eq. (4.3). By Theorem 4.1, $f(x) = cx$, some real c. We conclude

$$\psi(x) = e^{f(x)} = e^{cx},$$

as desired.

∎

THEOREM 4.3. *Suppose $\psi: Re^+ \to Re$ satisfies the third Cauchy equation*

$$\psi(xy) = \psi(x) + \psi(y) \qquad (4.5)$$

for all positive reals x and y, and suppose that ψ is continuous. Then

$$\psi(x) = c \ln x,$$

some real constant c.

Proof. Let $f(x) = \psi(e^x)$. Then

$$f(x + y) = \psi(e^{x+y}) = \psi(e^x e^y) = \psi(e^x) + \psi(e^y) = f(x) + f(y).$$

We conclude that f satisfies the first Cauchy equation, Eq. (4.3), for all real x and y. Thus, $f(x) = cx$, some real constant c. It follows that for all positive x, $\psi(x) = f(\ln x) = c \ln x$. ∎

COROLLARY. *Suppose $\psi: Re \to Re$ satisfies the third Cauchy equation*

$$\psi(xy) = \psi(x) + \psi(y) \tag{4.5}$$

for all reals x and y, and suppose that ψ is continuous. Then $\psi \equiv 0$.

Proof. $\psi(x) = c \ln x$ for all $x \in Re^+$. But

$$\psi(0) = \psi(0 \cdot 0) = \psi(0) + \psi(0),$$

so $\psi(0) = 0$. By continuity of ψ, $c \ln x \to 0$ as $x \to 0$, so $c = 0$. Therefore, $\psi(x) = 0$, all $x \geq 0$.

Finally, if x is negative,

$$\psi(x) = \psi[(-1)(-x)] = \psi(-1) + \psi(-x) = \psi(-1).$$

Thus, $\psi(x) = \psi(-1)$, all $x < 0$. By continuity, $\psi(-1) = \psi(0) = 0$. ∎

THEOREM 4.4. *Suppose $\psi: Re^+ \to Re$ satisfies the fourth Cauchy equation*

$$\psi(xy) = \psi(x)\psi(y) \tag{4.6}$$

for all positive reals x and y, and suppose that ψ is continuous. Then either

$$\psi \equiv 0$$

or

$$\psi(x) = x^c,$$

some real c.

Proof. As in the proof of Theorem 4.3, let $f(x) = \psi(e^x)$. Then f satisfies the second Cauchy equation (4.4). Proceed from there.

COROLLARY. *Suppose* $\psi: Re \to Re$ *satisfies the fourth Cauchy equation*

$$\psi(xy) = \psi(x)\psi(y) \tag{4.6}$$

for all real x and y, and suppose that ψ *is continuous. Then either*

(a) $$\psi \equiv 0,$$

or for some real c,

(b) $$\psi(x) = \begin{cases} x^c & \text{if } x > 0, \\ (-x)^c & \text{if } x < 0, \\ 0 & \text{if } x = 0, \end{cases}$$

or for some real c,

(c) $$\psi(x) = \begin{cases} x^c & \text{if } x > 0, \\ -(-x)^c & \text{if } x < 0, \\ 0 & \text{if } x = 0. \end{cases}$$

Proof. By Theorem 4.4, $\psi(x) \equiv 0$ or $\psi(x) = x^c$ holds for all positive reals. If the former, then x negative implies that

$$\psi(x) = \psi(-1)\psi(-x) = 0,$$

so $\psi \equiv 0$ for all negative x. By continuity, $\psi \equiv 0$.

If $\psi(x) = x^c$, all x positive, and if x is negative, then

$$\psi(x) = \psi(-1)(-x)^c.$$

Moreover,

$$\psi(-1)\psi(-1) = \psi(1) = \psi(1)\psi(1) = 1^c 1^c = 1.$$

Thus, $\psi(-1) = 1$ or $\psi(-1) = -1$. In the former case, using continuity, one concludes that ψ has the form (b). In the latter case, one concludes that ψ has the form (c). ∎

Remark: This Corollary does not hold if f is only continuous at a point. For example, the following function ψ is continuous everywhere but at 0, and satisfies the fourth Cauchy equation:

$$\psi(x) = \begin{cases} 1 & \text{if} \quad x \neq 0, \\ 0 & \text{if} \quad x = 0. \end{cases}$$

4.2.2 *Derivation of the Possible Psychophysical Laws*

Luce [1959] observed that we can sometimes derive the possible forms of psychophysical functions ψ if we assume that they must be continuous and we know (on some grounds) what types of scales f and g form. We present Luce's results here.[*] We derive the possible psychophysical functions in four cases, where f is taken to be either a ratio scale or an interval scale and g is taken to be a ratio or an interval scale *in the wide sense*.[†] (Additional cases are handled by Luce.) The results are summarized in Table 4.4.

Table 4.4. Possible Psychophysical Laws

Physical Scale	Psychological Scale	Functional Equation	Psychophysical Function
$f : A \to Re^+$ ratio scale	$g : A \to Re^+$ ratio scale in the wide sense	$\psi(kx) = K(k)\psi(x)$ $k > 0,\ K(k) > 0$	$\psi(x) = \alpha x^\beta$
$f : A \to Re^+$ ratio scale	$g : A \to Re$ interval scale in the wide sense	$\psi(kx) =$ $K(k)\psi(x) + C(k),$ $k > 0,\ K(k) > 0$	$\psi(x) = \alpha \ln x + \beta$ or $\psi(x) = \alpha x^\beta + \delta$
$f : A \to Re$ interval scale	$g : A \to Re^+$ ratio scale in the wide sense	$\psi(kx + c) =$ $K(k, c)\psi(x),$ $k > 0,\ K(k, c) > 0$	ψ constant
$f : A \to Re$ interval scale	$g : A \to Re$ interval scale in the wide sense	$\psi(kx + c) =$ $K(k, c)\psi(x) + C(k, c),$ $k > 0,\ K(k, c) > 0$	$\psi(x) = \alpha x + \beta$

It should be mentioned that the results that follow have much wider applicability than just to the determination of the possible psychophysical laws. Indeed, given any two scales of known scale type that are related by some unknown law, we can derive the possible forms of this law by Luce's methods. In Exer. 6, we explore the application of this idea to laws such as Newton's Law and Ohm's Law relating physical variables to each other

---

[*]For a criticism of this approach and a reply, see Rozeboom [1962a, b] and Luce [1962].
[†]We use the term "in the wide sense" because we are thinking of derived measurement.

and laws relating geometrical variables such as volume and radius of a sphere.

We shall always assume that an interval scale has a range of all real numbers, but that a ratio scale can be limited to a range of positive reals. If the physical scale f has range Re', where Re' is Re or Re^+, then we shall assume that f attains all possible values in its range; that is, every real (positive real) number is a (potential) scale value for some stimulus. Thus, ψ will have domain all of Re' (and range the range of the psychological scale g). Hence, if A is the set of objects being measured both physically and psychologically, our assumptions can be summarized as follows:

If f is a ratio scale, $f(A) = Re^+$.
If f is an interval scale, $f(A) = Re$.
If g is a ratio scale, $g(A) \subseteq Re^+$.
If g is an interval scale, $g(A) \subseteq Re$.

The general procedure of Luce is to use the observation, based on our discussion of Section 2.5, that an admissible transformation of f leads to an admissible transformation of g (in the wide sense).* Using Luce's observation, we shall obtain a functional equation for the psychophysical function. For example, suppose both f and g are ratio scales, the latter in the wide sense. Then multiplication by a positive constant k is an admissible transformation of f and so must result in an admissible transformation of g, that is, multiplication by a positive constant $K(k)$. Thus, for all a in A, the set of objects being measured both physically and psychologically, $\psi[kf(a)] = K(k)\psi[f(a)]$. Thus, for all x in the range of f, we have the equation

$$\psi(kx) = K(k)\psi(x). \tag{4.11}$$

By our conventions, all positive reals x are attained as scale values $f(a)$, so (4.11) holds for all positive reals x. To solve Eq. (4.11), we reduce it to one of the Cauchy equations solved in the previous section. The procedure under other assumptions about the scale types of f and g is similar.

THEOREM 4.5 (Luce). *Suppose the psychophysical function ψ is continuous and suppose $f:A \to Re^+$ and $g:A \to Re^+$ are both ratio scales, the latter in the wide sense. Then*

$$\psi(x) = \alpha x^\beta,$$

where $\alpha > 0$.

---

*It is this observation that Rozeboom [1962a] criticizes. Luce [1962] admits that his procedure is subject to difficulties if the psychophysical law involves "dimensional parameters" that can be transformed only by transformations that depend on the transformation of the physical parameter f. (Compare the Remark at the end of Section 2.5.)

Proof. By our convention, $\psi: Re^+ \to Re^+$. We have already shown that ψ satisfies the functional equation (4.11), for $k > 0$ and $K(k) > 0$. Setting $x = 1$ in (4.11), we obtain

$$\psi(k) = K(k)\psi(1), \qquad (4.12)$$

all $k > 0$. Since the range of ψ is contained in Re^+, $\psi(1) > 0$. Thus, $K(k) = \psi(k)/\psi(1)$. Now Eq. (4.11) becomes

$$\psi(kx) = \psi(k)\psi(x)/\psi(1).$$

For all $x \in Re^+$, let $\gamma(x) = \ln[\psi(x)/\psi(1)]$. This function is well-defined, since the range of ψ is Re^+ and so $\psi(x)/\psi(1) > 0$. We have

$$\gamma(kx) = \ln[\psi(kx)/\psi(1)]$$

$$= \ln\left[\frac{\psi(k)\psi(x)}{\psi(1)\psi(1)}\right]$$

$$= \ln\left[\frac{\psi(k)}{\psi(1)}\right] + \ln\left[\frac{\psi(x)}{\psi(1)}\right]$$

$$= \gamma(k) + \gamma(x).$$

Since ψ is continuous, so is γ, and thus by Theorem 4.3,

$$\gamma(x) = \beta \ln x = \ln x^\beta.$$

It follows that $\psi(x) = \alpha e^{\gamma(x)} = \alpha x^\beta$, where $\alpha = \psi(1)$. Finally, note that $\alpha > 0$, since $\psi(1) > 0$. ∎

The next result is obtained immediately from the proof of the preceding theorem.

COROLLARY 1. *Suppose $\psi: Re^+ \to Re^+$ is continuous and satisfies the functional equation*

$$\psi(kx) = K(k)\psi(x) \qquad (4.11)$$

for $k > 0$ and $\dot{K}(k) > 0$. Then $\psi(x) = \alpha x^\beta$, where $\alpha > 0$.

COROLLARY 2. *Suppose $\psi: Re \to Re^+$ is continuous and satisfies the functional equation*

$$\psi(kx) = K(k)\psi(x), \qquad (4.11)$$

for $k > 0$ and $K(k) > 0$. Then $\psi(x) \equiv c$, for some constant c.

Proof. In the proof of Theorem 4.5, $\gamma(x)$ is well-defined for all $x \in Re$. Use the Corollary to Theorem 4.3 to conclude that $\gamma(x) \equiv 0$. Thus,

$$0 \equiv \ln[\psi(x)/\psi(1)],$$

so $\psi(x)/\psi(1) \equiv 1$, so $\psi(x) = \psi(1)$, all x. ∎

THEOREM 4.6 (Luce). *Suppose the psychophysical function ψ is continuous, suppose $f:A \to Re^+$ is a ratio scale and $g:A \to Re$ is an interval scale in the wide sense. Then either*

$$\psi(x) = \alpha \ln x + \beta,$$

or

$$\psi(x) = \alpha x^\beta + \delta.$$

Proof. * By our convention, $\psi:Re^+ \to Re$. Since admissible transformations of f (similarity transformations) lead to admissible transformations of g (positive linear transformations), we derive the functional equation

$$\psi(kx) = K(k)\psi(x) + C(k), \tag{4.13}$$

where $k > 0$ and $K(k) > 0$.

Case 1. $K(k) \equiv 1$. Here, we define $\gamma(x) = e^{\psi(x)}$. Thus, $\gamma:Re^+ \to Re^+$ and γ is continuous. Since $K(k) = 1$, Eq. (4.13) becomes $\gamma(kx) = D(k)\gamma(x)$, where $D(k) = e^{C(k)} > 0$.

By Corollary 1 to Theorem 4.5, we conclude that $\gamma(x) = \delta x^\alpha$, where $\delta > 0$. Taking logarithms, we obtain $\psi(x) = \alpha \ln x + \beta$, where $\beta = \ln \delta$.

Case 2. $K(k) \not\equiv 1$. We shall assume that $\psi(x)$ satisfies (4.13), and $\psi(x)$ is not constant. For if ψ is constant, then $\psi(x) = 0 \cdot x^\beta + \delta$, some δ, and we are done. Using Eq. (4.13), we find that

$$\psi(k_1 k_2 x) = K(k_1 k_2)\psi(x) + C(k_1 k_2). \tag{4.14}$$

Also using Eq. (4.13),

$$\psi(k_1 k_2 x) = K(k_1)\psi(k_2 x) + C(k_1)$$
$$= K(k_1)[K(k_2)\psi(x) + C(k_2)] + C(k_1),$$

*The author thanks J. Rosenstein for some of the ideas of this proof.

so

$$\psi(k_1 k_2 x) = K(k_1)K(k_2)\psi(x) + K(k_1)C(k_2) + C(k_1). \qquad (4.15)$$

Similarly,

$$\psi(k_1 k_2 x) = K(k_1)K(k_2)\psi(x) + K(k_2)C(k_1) + C(k_2). \qquad (4.16)$$

Note that if

$$a\psi(x) + b = c\psi(x) + d$$

for all $x \in Re^+$, then

$$(a - c)\psi(x) = d - b$$

for all $x \in Re^+$. Thus, if $a - c \neq 0$,

$$\psi(x) = \frac{d - b}{a - c}$$

is a constant, contrary to assumption. Thus, we conclude $a = c$ and therefore $b = d$. By this line of reasoning, Eqs. (4.14) and (4.15) imply

$$K(k_1 k_2) = K(k_1)K(k_2) \qquad (4.17)$$

for all $k_1, k_2 \in Re^+$, and Eqs. (4.15) and (4.16) imply

$$K(k_1)C(k_2) + C(k_1) = K(k_2)C(k_1) + C(k_2) \qquad (4.18)$$

for all $k_1, k_2 \in Re^+$.

Since $K(k) \not\equiv 1$ by the assumption of Case 2, there is k_1 so that $K(k_1) \neq 1$. Then for all k, Eq. (4.18) implies that

$$C(k_1)[1 - K(k)] = C(k)[1 - K(k_1)].$$

Thus, since $K(k_1) \neq 1$,

$$C(k) = C(k_1)\left[\frac{1 - K(k)}{1 - K(k_1)} \right]. \qquad (4.19)$$

Using this value of $C(k)$ in Eq. (4.13), we have

$$\psi(kx) = K(k)\psi(x) + C(k_1)\left[\frac{1 - K(k)}{1 - K(k_1)} \right]$$

$$= K(k)\left[\psi(x) - \frac{C(k_1)}{1 - K(k_1)} \right] + \frac{C(k_1)}{1 - K(k_1)}.$$

Since ψ was assumed not constant, there is x_0 so that

$$\psi(x_0) \neq \frac{C(k_1)}{1 - K(k_1)}.$$

Thus,

$$K(k) = \frac{\psi(kx_0) - \dfrac{C(k_1)}{1 - K(k_1)}}{\psi(x_0) - \dfrac{C(k_1)}{1 - K(k_1)}}.$$

It follows that K is a continuous function of k. Thus, by Theorem 4.4, Eq. (4.17) implies that either

$$K(k) \equiv 0$$

or

$$K(k) = k^\beta,$$

some β. If $K(k) \equiv 0$, then Eq. (4.13) implies that

$$\psi(x) = \psi(1 \cdot x) = C(1),$$

so ψ is a constant, contrary to assumption. Thus, $K(k) = k^\beta$, some β.
We next claim that if

$$\delta = \frac{C(k_1)}{1 - k_1{}^\beta},$$

and if

$$\psi^*(x) = \gamma x^\beta + \delta,$$

then ψ^* satisfies Eq. (4.13) for every γ. (The number δ is well-defined, since $1 \neq K(k_1) = k_1{}^\beta$.) The claim follows, since

$$\begin{aligned}
\psi^*(kx) &= \gamma k^\beta x^\beta + \delta \\
&= k^\beta(\gamma x^\beta + \delta) + \delta(1 - k^\beta) \\
&= k^\beta(\gamma x^\beta + \delta) + \frac{C(k_1)}{1 - k_1{}^\beta}(1 - k^\beta) \\
&= k^\beta\psi^*(x) + C(k),
\end{aligned}$$

using Eq. (4.19).

Finally, we show that any continuous solution ψ of Eq. (4.13) is of the form $\alpha x^\beta + \delta$. For suppose $\psi^*(x) = \gamma x^\beta + \delta$ and $\psi(x_0) \neq \psi^*(x_0)$, some $x_0 > 0$. Suppose $\psi(x_0) > \psi^*(x_0)$. Now given $x > 0$, $x = kx_0$ some $k > 0$. Thus,

$$
\begin{aligned}
\psi(x) &= \psi(kx_0) \\
&= K(k)\psi(x_0) + C(k) \\
&> K(k)\psi^*(x_0) + C(k) \\
&= \psi^*(kx_0) \\
&= \psi^*(x).
\end{aligned}
$$

Let $\lambda(x) = \psi(x) - \psi^*(x)$. Then $\lambda(x) > 0$, all $x \in Re^+$, so λ is a function from Re^+ to Re^+. Moreover, by Eq. (4.13), λ must satisfy the functional equation

$$\lambda(kx) = K(k)\lambda(x), \tag{4.20}$$

$k > 0$, $K(k) > 0$. Since both ψ and ψ^* are continuous, so is λ. Thus, Corollary 1 to Theorem 4.5 applies to λ. We conclude that $\lambda(x) = ax^b$, $a > 0$. Substituting this into Eq. (4.20), we find that $b = \beta$. Thus, $\lambda(x) = ax^\beta$, and

$$
\begin{aligned}
\psi(x) &= \psi^*(x) + \lambda(x) \\
&= \gamma x^\beta + \delta + ax^\beta \\
&= (\gamma + a)x^\beta + \delta.
\end{aligned}
$$

Thus, ψ has the form $\alpha x^\beta + \delta$, for $\alpha = \gamma + a$.

A similar proof applies if $\psi(x_0) < \psi^*(x_0)$. ■

THEOREM 4.7 (Luce). *Suppose the psychophysical function ψ is continuous, $f:A \to Re$ is an interval scale, and $g:A \to Re^+$ is a ratio scale in the wide sense. Then ψ is constant.*

Proof. By our convention, $\psi: Re \to Re^+$. Once again, by using admissible transformations, we derive a functional equation:

$$\psi(kx + c) = K(k, c)\psi(x), \tag{4.21}$$

$k > 0$, $K(k, c) > 0$. Let $c = 0$ in Eq. (4.21), note that ψ is nonzero, since its range is contained in Re^+, and apply Corollary 2 to Theorem 4.5.

THEOREM 4.8 (Luce). *Suppose the psychophysical function ψ is continuous and $f:A \to Re$ and $g:A \to Re$ are both interval scales, the latter in the wide* *sense. Then*

$$\psi(x) = \alpha x + \beta.$$

Proof. By our convention, $\psi: Re \to Re$. Once again, we derive a functional equation

$$\psi(kx + c) = K(k, c)\psi(x) + C(k, c), \tag{4.22}$$

where $k > 0$, $K(k, c) > 0$. If we set $c = 0$, then Eq. (4.22) reduces to Eq. (4.13) and so by the proof of Theorem 4.6, at least for $x > 0$, $\psi(x) = \alpha \ln x + \beta$ or $\psi(x) = \alpha x^\beta + \delta$. In the former case, set $k = c = 1$ in Eq. (4.22). It follows that for $x > 0$,

$$\alpha \ln(x + 1) + \beta = K(1, 1)\left[\alpha \ln x\right] + K(1, 1)\beta + C(1, 1).$$

Differentiating with respect to x, we obtain

$$\frac{\alpha}{x + 1} = \frac{K(1, 1)\alpha}{x}.$$

If $\alpha \neq 0$, we obtain

$$K(1, 1) = \frac{x}{x + 1}, \tag{4.23}$$

all positive x. Setting $x = 1$ and $x = 2$ in Eq. (4.23), we obtain $K(1, 1) = \frac{1}{2}$ and $K(1, 1) = \frac{2}{3}$, respectively, a contradiction. The conclusion is that α must be 0. Thus, $\psi(x) = \alpha \ln x + \beta = \beta$, for all x positive.

Now given $k > 0$ and c, find $x > 0$ so that $kx + c > 0$. Then $\psi(kx + c) = \beta$ and $\psi(x) = \beta$, so Eq. (4.22) implies that

$$\beta = K(k, c)\beta + C(k, c), \tag{4.24}$$

all $k > 0$, all c. Now given $y \leq 0$, choose c so that $y = x + c$, some $x > 0$. Then, using (4.22) and (4.24), we have

$$\begin{aligned}
\psi(y) &= K(1, c)\psi(x) + C(1, c) \\
&= K(1, c)\beta + C(1, c) \\
&= \beta.
\end{aligned}$$

Thus, $\psi \equiv \beta$.

In the case that $\psi(x) = \alpha x^\beta + \delta$, $x > 0$, assume $\alpha \neq 0$. For otherwise, $\psi(x) \equiv \delta$, all x, follows as in the previous case. Using $\psi(x) = \alpha x^\beta + \delta$,

$x > 0$, set $k = c = 1$ in Eq. (4.22). It follows that for $x > 0$,

$$\alpha(x + 1)^\beta + \delta = K(1, 1)\alpha x^\beta + K(1, 1)\delta + C(1, 1).$$

Differentiating with respect to x, we obtain

$$\alpha\beta(x + 1)^{\beta - 1} = \alpha\beta K(1, 1)x^{\beta - 1}, \tag{4.25}$$

all positive x. Since $\alpha \neq 0$, setting $x = 1$ and $x = 2$ in (4.25) gives $\beta 2^{\beta - 1} = \beta K(1, 1) = \beta(3/2)^{\beta - 1}$. We conclude that $\beta = 0$ or $\beta = 1$, so $\psi(x) = \alpha + \delta$ or $\psi(x) = \alpha x + \delta$, all x positive.

Thus in all cases, we have, for all x positive, $\psi(x) = \alpha x + \beta$, some α, β. We wish to show this for all real x. Given any $x > 0$ and $c > 0$, we know that

$$\psi(x + c) = \alpha(x + c) + \beta = \alpha x + (\alpha c + \beta). \tag{4.26}$$

Also, using Eq. (4.22), we have

$$\psi(x + c) = K(1, c)\psi(x) + C(1, c),$$

so

$$\psi(x + c) = K(1, c)\alpha x + K(1, c)\beta + C(1, c). \tag{4.27}$$

Thus, equating the right-hand sides of Eqs. (4.26) and (4.27), we conclude that for all $x > 0$ and $c > 0$,

$$\alpha x + (\alpha c + \beta) = K(1, c)\alpha x + [K(1, c)\beta + C(1, c)].$$

Now as in the proof of Theorem 4.6,

$$ax + b = cx + d$$

for all $x > 0$ implies that $a = c$ and $b = d$. Thus, $\alpha = K(1, c)\alpha$, or, since $\alpha \neq 0$, $K(1, c) = 1$, all $c > 0$. Moreover,

$$\alpha c + \beta = K(1, c)\beta + C(1, c) = \beta + C(1, c),$$

or $C(1, c) = \alpha c$, all $c > 0$. Now, given any x, choose $c > 0$ such that $x + c > 0$. Then

$$\begin{aligned}
\alpha x + \alpha c + \beta &= \alpha(x + c) + \beta \\
&= \psi(x + c) \\
&= K(1, c)\psi(x) + C(1, c) \\
&= \psi(x) + \alpha c.
\end{aligned}$$

It follows that $\psi(x) = \alpha x + \beta$.   ∎

Exercises

1. In the measurement of apparent length of straight lines, a doubling of physical length leads to (essentially) a doubling of apparent length. More generally, suppose multiplication of physical length by a (positive) rational amount r leads to (essentially) multiplication of apparent length by an amount r. Show that this is enough to conclude that, if the psychophysical function ψ is continuous, then $\psi(x) = \alpha x$, some α.

2. (Aczél [1966, p. 42]) Suppose $\psi: Re^+ \to Re$ satisfies

$$\psi(x + y) = \psi(x) + \psi(y)$$

and

$$\psi(xy) = \psi(x)\psi(y),$$

for all positive x and y, and suppose ψ is continuous. Show that $\psi(x) = x$ or $\psi(x) \equiv 0$, all positive x.

3. (Aczél [1966, p. 43]) The functional equation

$$\psi\left(\frac{x + y}{2}\right) = \frac{\psi(x) + \psi(y)}{2}.$$

is called Jensen's equation, after J. L. W. V. Jensen [1905, 1906]. If $\psi: Re \to Re$ satisfies Jensen's equation for all real x and y, and ψ is continuous, show that $\psi(x) = cx + a$. [*Hint:* Set $\gamma(x) = \psi(x) - \psi(0)$.]

4. (Aczél [1966, pp. 46, 47]) (a) Suppose $m \geq 2$ and $\psi: Re \to Re$ is continuous and satisfies

$$\psi(x_1 + x_2 + \cdots + x_n) = \psi(x_1) + \psi(x_2) + \cdots + \psi(x_n),$$

all $x_1, x_2, \ldots, x_n \in Re$. Show from the result of Theorem 4.1 that $\psi(x) = cx$.

(b) Suppose $n \geq 2$ and $\psi: Re \to Re$ is continuous and satisfies

$$\psi\left(\frac{x_1 + x_2 + \cdots + x_n}{n}\right) = \frac{\psi(x_1) + \psi(x_2) + \cdots + \psi(x_n)}{n},$$

all $x_1, x_2, \ldots, x_n \in Re$. Show that $\psi(x) = cx + a$. Use the result of Exer. 3.

5. (Aczél [1966, pp. 49, 50]) Suppose $\psi: Re \to Re$ is continuous and satisfies

$$\psi(x + y) = \psi(x) + \psi(y) + \psi(x)\psi(y),$$

all $x, y \in Re$. In Chapter 5, we shall call such ψ *quasi-additive*. If ψ is quasi-additive, show that either $\psi(x) \equiv -1$ or $\psi(x) = e^{cx} - 1$.

(a) Give a quick proof by setting $\gamma(x) = \psi(x) + 1$ and reducing to the second Cauchy equation.

(b) An alternative proof goes as follows:

(i) Show that $\psi(nx) = [1 + \psi(x)]^n - 1$.

(ii) Using part (i) with $n = 2$, prove that $\psi(x) \geqq -1$, all x.

(iii) Using quasi-additivity, prove that $\psi(1) = -1$ implies $\psi(x) = -1$, all x.

(iv) If $\psi(1) \neq -1$, let $c = \ln[1 + \psi(1)]$. Using part (i), prove that $\psi(m/n) = e^{cm/n} - 1$, all $m/n \geqq 0$.

(v) Using the result of part (iv) and continuity, conclude that if $\psi(1) \neq -1$, then $\psi(x) = e^{cx} - 1$, all $x \geqq 0$.

(vi) Show that part (v) implies that if $\psi(1) \neq -1$, then $\psi(0) = 0$. Then use quasi-additivity with $y = -x$ to prove from this that $\psi(x) = e^{cx} - 1$, all $x \in Re$.

6. (Luce [1959]) Theorems 4.5 through 4.8 hold for derived measurement in general, not just for psychophysical scaling. Thus, for example, if the independent variable f and the derived variable g are both ratio scales, and $g = \psi(f)$ for ψ continuous, then Theorem 4.5 implies that ψ is a power function. Show that the following physical and geometrical laws are of the forms called for by Theorems 4.5 through 4.8:[*]

(a) For a sphere, $V = \frac{4}{3}\pi r^3$, where $V =$ volume, $r =$ radius.

(b) *Ohm's Law*: Under fixed resistance, voltage is proportional to current. (Voltage and current are ratio scales.)

(c) *Newton's Law of Gravitation*: $F = G(mm'/r^2)$, where F is the force of attraction, G is the gravitational constant, m and m' are the masses of two bodies being attracted, and r is the distance between the bodies.

(d) If a body of constant mass is moving at velocity v, then its energy is $\alpha v^2 + \delta$, α, δ constant. (Energy is an interval scale.)

(e) If the temperature of a perfect gas is constant, then as a function of pressure p, the entropy of the gas is $\alpha \log p + \beta$, α, β constant. (Entropy defines an interval scale.)

(f) For a square, $A = l^2$, where A is area and l is length.

7. (a) Show that if the variables on both the independent (physical) and dependent (psychological) scales are dimensionless (that is, define absolute scales), then there is no restriction on the psychophysical function.

---

[*]Luce [1959] and Rozeboom [1962a] give examples of physical laws that seem to violate the conclusions of Theorems 4.5 through 4.8, for example the exponential law of radioactive decay. Luce [1959, 1962] argues that such violations occur if the independent variable enters the equation in a "dimensionless fashion," and argues that the conclusions of Theorems 4.5 through 4.8 hold only if there are no "dimensional parameters" present. The admissible transformations of such dimensional parameters are not determined by a measurement theory, but are included in the statement of a law. See the Remark at the end of Section 2.5 for a more detailed discussion.

(b) Dimensionless scales can be constructed rather easily. For example, suppose a variable x defines a ratio scale, but some value x_0 is taken as a reference value and the scale used is x/x_0. Show that x/x_0 defines an absolute scale.

8. (a) In Theorem 4.5, show that β is independent of the units of f and of g; that is, β doesn't change if admissible transformations are applied to either f or g. For example, show that if ψ relates kf to g, then $\psi(x) = \alpha'x^\beta$, for the same β as in the relation of f to g.

(b) In Theorem 4.6, if $\psi(x) = \alpha \ln x + \beta$, show that α is independent of the unit of f. What about β?

(c) In Theorem 4.6, if $\psi(x) = \alpha x^\beta + \delta$, show that β is independent of the units of both f and g. What about δ? What about α?

(d) In Theorem 4.8, what is the dependence of α and of β on the units of f and g?

9. (Luce [1959]) Derive a functional equation for the psychophysical function ψ in each of the following cases:

(a) f is a ratio scale, g is a log-interval scale (cf. Section 2.3).

(b) f is an interval scale, g is a log-interval scale.

(c) f is a log-interval scale, g is a ratio scale.

(d) f is a log-interval scale, g is an interval scale.

(e) f is a log-interval scale, g is a log-interval scale.

10. (Luce [1959]) (a) In case (a) of Exer. 9, if ψ is continuous, show that either $\psi(x) = \delta e^{\alpha x^\beta}$ or $\psi(x) = \alpha x^\beta$. [*Hint*: Take $\ln$ of the functional equation, let $\gamma = \ln \psi$, and reduce to one of the functional equations in Table 4.4.]

(b) In case (b) of Exer. 9, if ψ is continuous, show that $\psi(x) = \alpha e^{\beta x}$. [*Hint*: The method of proof for (a) applies.]

(c) In case (c) of Exer. 9, if ψ is continuous, show that $\psi(x)$ is constant. [*Hint*: Take $\gamma(\ln x) = \psi(x)$ and reduce to one of the functional equations in Table 4.4.]

(d) In case (d) of Exer. 9, if ψ is continuous, show that $\psi(x) = \alpha \ln x + \beta$. [*Hint*: The method of proof of part (c) applies.]

(e) In case (e) of Exer. 9, if ψ is continuous, show that $\psi(x) = \alpha x^\beta$. [*Hint*: Take $\ln$ of the functional equation, let $\gamma = \ln \psi$, and reduce to one of the functional equations of Table 4.4.]

11. (a) Suppose $f: Re \times Re \to Re$ satisfies

$$f(x_1 + y_1, x_2 + y_2) = f(x_1, x_2) + f(y_1, y_2)$$

for all x_1, x_2, y_1, y_2 in Re, and suppose f is continuous. Find all such functions f.

(b) Generalize to continuous $f: Re^n \to Re$ satisfying

$$f(x_1 + y_1, x_2 + y_2, \ldots, x_n + y_n) = f(x_1, x_2, \ldots, x_n) + f(y_1, y_2, \ldots, y_n).$$

12. Show that the solutions given to Cauchy's equations in Theorems 4.1 through 4.4 hold for functions ψ which are assumed monotone increasing

rather than continuous, for such functions are continuous at at least one point.

13. (Aczél [1966, pp. 105–106], Eichhorn [1978, pp. 3–4, 10–11]) Functional equations have a wide variety of applications in economics, as illustrated in the recent book by Eichhorn [1978]. This exercise and Exer. 15 present some of these applications. Suppose $I(K, T)$ represents the compound interest earned by a capital K during a time interval of length t. Thus, $I: Re^+ \times Re^+ \to Re^+$. If the interest accrued is not changed if K is divided into two separate capital investments K_1 and K_2, we have

(i) $I(K_1 + K_2, t) = I(K_1, t) + I(K_2, t),$ $K_1, K_2, t > 0.$

Also, if the interest rate doesn't change over the length of an account, the amount of interest on $I(K, t_1)$ during a time interval of length t_2 is equal to the amount of interest on K during a time interval of length $t_1 + t_2$. Hence,

(ii) $I[I(K, t_1), t_2] = I(K, t_1 + t_2),$ $K, t_1, t_2 > 0.$

Finally, note that I is monotone increasing in each variable.

(a) Since I is monotone increasing, Exer. 12 implies that the results of Theorems 4.1 through 4.4 apply. Show from the above assumptions that there is $h(t) > 0$ so that

$$I(K, t) = h(t)K.$$

(b) Show from (ii) that

$$h(t_1 + t_2) = h(t_1)h(t_2),   t_1, t_2 > 0.$$

(c) Show that $I(K, t) = Kq^t$, some $q > 1$. Hence, the standard way of computing compound interest follows from the simple assumptions we have made.

14. (Eichhorn [1978, pp. 51–52]) (a) Suppose $\psi: Re^n \to Re$ is monotone increasing in each variable and satisfies

$$\psi(x + y) = \psi(x) + \psi(y),$$

all $x, y \in Re^n$. Show that there is a vector $c = (c_1, c_2, \ldots, c_n)$ in Re^n such that

$$\psi(x) = c \cdot x,$$

where $c \cdot x$ is the dot product $c_1 x_1 + c_2 x_2 + \cdots + c_n x_n$.

(b) Show that the conclusion of (a) still holds if the domain of ψ is changed to all nonnegative real vectors of length n. (Hint: Note that the proof of Theorem 4.1 can be modified to hold for the domain of ψ equal to the nonnegative reals.)

(c) Show that the conclusion of (a) still holds if the domain of ψ is changed to all nonnegative real vectors of length n except the vector $\mathbf{0}$.

15. (Eichhorn [1978, pp. 53–58]) In Section 2.6.2, we studied index numbers. We can look at an index number (consumer price index, consumer confidence index, etc.) as a function $I{:}A \times A \to Re$, where A is the set of all nonnegative real vectors of length n except the vector $\mathbf{0}$. One often takes

$$I(\mathbf{x}, \mathbf{u}) = \alpha \cdot \mathbf{u}/\beta \cdot \mathbf{x} = \Sigma\alpha_i u_i/\Sigma\beta_i x_i,$$

where $\mathbf{x} = (x_1, x_2, \ldots, x_n)$ and $\mathbf{u} = (u_1, u_2, \ldots, u_n)$ are in A, $\alpha = (\alpha_1, \alpha_2, \ldots, \alpha_n)$ and $\beta = (\beta_1, \beta_2, \ldots, \beta_n)$ are vectors of positive reals, and $\cdot$ means dot product. For price indices, $x_i = p_i(t)$, the price of good i in year t, and $u_i = p_i(0)$. The Paasche price index (Exer. 5, Section 2.6.2) uses $\alpha_i = \beta_i = q_i(t)$, the quantity of good i in an average market basket in year t, and the Laspeyres price index (Exer. 5, Section 2.6.2) uses $\alpha_i = \beta_i = q_i(0)$. From $I(\mathbf{x}, \mathbf{u}) = \alpha \cdot \mathbf{u}/\beta \cdot \mathbf{x}$ and $\alpha_i > 0$, $\beta_i > 0$, all i, it follows that

(i) $\qquad\qquad I(\mathbf{x}, \mathbf{u} + \mathbf{v}) = I(\mathbf{x}, \mathbf{u}) + I(\mathbf{x}, \mathbf{v});$

(ii) $\qquad\qquad \dfrac{1}{I(\mathbf{x} + \mathbf{y}, \mathbf{u})} = \dfrac{1}{I(\mathbf{x}, \mathbf{u})} + \dfrac{1}{I(\mathbf{y}, \mathbf{u})};$

(iii) $\qquad\qquad I(\mathbf{x}, \mathbf{u}) > 0, \qquad$ all $\mathbf{x}, \mathbf{u} \in A.$

This exercise presents a sketch of the proof that if I is monotone decreasing in its first variable and monotone increasing in its second variable, then these three conditions essentially determine the general formula

$$I(\mathbf{x}, \mathbf{u}) = \alpha \cdot \mathbf{u}/\beta \cdot \mathbf{x}.$$

(a) Use Exer. 14 to show from monotonicity, (ii), and (iii) that there is a vector $\mathbf{b}(\mathbf{u}) = [b_1(\mathbf{u}), b_2(\mathbf{u}), \ldots, b_n(\mathbf{u})]$, with each $b_i(\mathbf{u}) > 0$, such that

$$\frac{1}{I(\mathbf{x}, \mathbf{u})} = \mathbf{b}(\mathbf{u}) \cdot \mathbf{x}.$$

(b) Using the results of (a) and (i), show that

$$\frac{1}{\mathbf{b}(\mathbf{u} + \mathbf{v}) \cdot \mathbf{x}} = \frac{1}{\mathbf{b}(\mathbf{u}) \cdot \mathbf{x}} + \frac{1}{\mathbf{b}(\mathbf{v}) \cdot \mathbf{x}}.$$

(c) Show that for all i,

$$\frac{1}{b_i(\mathbf{u} + \mathbf{v})} = \frac{1}{b_i(\mathbf{u})} + \frac{1}{b_i(\mathbf{v})}.$$

(d) Show that for all i, there is a vector $\mathbf{a}^i$ with all entries positive such that

$$\frac{1}{b_i(\mathbf{u})} = \mathbf{a}^i \cdot \mathbf{u} = \sum_j a_j^i u_j.$$

(See the hint to Exer. 14b.)

(e) Use parts (a) and (d) to show that for all x, u,

$$\frac{1}{I(x, u)} = \frac{x_1}{a^1 \cdot u} + \frac{x_2}{a^2 \cdot u} + \cdots + \frac{x_n}{a^n \cdot u}.$$

(f) Let $e^i = (0, \ldots, 0, 1, 0, \ldots, 0)$, with a 1 in the ith component. Show that the result of (e) implies that for all u in A,

$$I(e^1 + e^2 + \cdots + e^n, u) = \left(\frac{1}{a^1 \cdot u} + \frac{1}{a^2 \cdot u} + \cdots + \frac{1}{a^n \cdot u} \right)^{-1}.$$

(g) Show that for some vector α with each component $\alpha_i > 0$,

$$I(e^1 + e^2 + \cdots + e^n, u) = \alpha \cdot u,$$

all u in A.

(h) Show from the results of (f) and (g) that

$$(a^1 \cdot u)\left(1 + \frac{a^1 \cdot u}{a^2 \cdot u} + \cdots + \frac{a^1 \cdot u}{a^n \cdot u} \right)^{-1} = \alpha \cdot u.$$

(i) Show that

$$\left[a^1 - \psi(u)\alpha \right] \cdot u = 0,$$

where

$$\psi(u) = 1 + \frac{a^1 \cdot u}{a^2 \cdot u} + \cdots + \frac{a^1 \cdot u}{a^n \cdot u},$$

and $\psi(u) > 0$ for all u in A.

(j) It follows from the result of part (i) that $a^1 = \lambda_1 \alpha$, $\lambda_1 > 0$. This is because if $a^1 > 0$ and $\alpha > 0$ are linearly independent, the result of (i) cannot hold for all u, since it cannot hold for any u not perpendicular to the plane spanned by a^1 and α. Similarly, we conclude that for all i, $a^i = \lambda_i \alpha$, $\lambda_i > 0$. Using this conclusion, show from the result of part (e) that for all x, u in A,

$$\frac{1}{I(x, u)} = \frac{\beta_1 x_1 + \beta_2 x_2 + \cdots + \beta_n x_n}{\alpha \cdot u} = \frac{\beta \cdot x}{\alpha \cdot u},$$

for some $\beta_i > 0$.

(k) Show that if in addition $I(x, x) = 1$ for all x in A, a reasonable assumption for price indices, then

$$I(x, u) = \frac{\alpha \cdot u}{\alpha \cdot x}.$$

Note: For further applications of functional equations to the study of price indices, see Eichhorn [1978, Chapter 8].

4.3 The Power Law

4.3.1 *Magnitude Estimation*

In this section, we shall present a justification of the power law and then investigate some consequences of this law.

Judgments of subjective loudness are made in the laboratory in various ways. Stevens [1957] classifies four different methods. One of the most common is the method of *magnitude estimation*, which we encountered in our discussion of choice of most important variables in Section 2.6. The subject hears a reference sound (or light or other stimulus) and is told to assign it a fixed number, say 100. Then he is presented other sounds (or lights, etc.) and asked to assign them numbers proportionate to the apparent loudness (or brightness, etc.). For example, if a sound seems twice as loud as the reference sound, it should be assigned a value of 200; if it sounds one half as loud, it should be assigned a value of 50. Magnitude estimation may also be performed without a reference stimulus, letting the subject pick out his own reference stimulus. The data of Fig. 4.2 shows the results of one magnitude estimation for loudness, plotted in

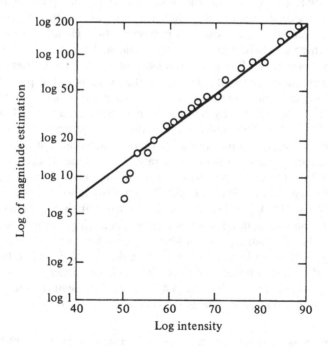

Figure 4.2. Magnitude estimation judgments of loudness in log–log coordinates, fitted with a straight line of slope 0.3. Adapted from Galanter and Messick [1961, p. 366]. Copyright 1961 by American Psychological Association. Reprinted by permission.

log–log coordinates. In log–log coordinates, the psychophysical function, if it were a power function, would appear as a straight line whose slope is the exponent in the power law. Figure 4.2 shows a straight line which has been fitted to the data. This line has a slope of .3, and the number .3 has been used by Stevens as an estimate of the exponent of the power function for loudness.

A variant of magnitude estimation is *magnitude production*. An experimenter names magnitudes and asks the subject to adjust stimuli to match those magnitudes. In *ratio production*, a subject is asked to select a stimulus that is one-half as loud as a given stimulus, or one that is three times as loud, etc. Finally, in *ratio estimation*, the experimenter presents two stimuli and asks the subject to estimate the ratio between them.

Subjects seem to feel quite comfortable with all these procedures. Stevens argues that the results of such scaling procedures are ratio scales of loudness and other psychological variables. He doesn't "prove" that magnitude estimation, for example, leads to a ratio scale, because he develops no representation and uniqueness theorems. Rather, he uses the principle we have discussed in Chapter 2, that an admissible transformation is one that keeps "intact the empirical information depicted by the scale" (Stevens [1968, p. 850]). With this idea, it seems plausible that the only admissible transformations of judgments obtained using methods like magnitude estimation are those that preserve ratios. Hence, it is suggested that magnitude estimation gives rise to a ratio scale.

Now measurement of physical intensity of a sound (or intensity of a light, etc.) is on a ratio scale. Hence, if the corresponding psychological scale of loudness (or brightness, etc.) is also a ratio scale (in the wide sense), and if the psychophysical function is continuous, it follows from Theorem 4.5 that the psychophysical function ψ is a power function.*

It should be remarked that not all psychophysicists accept the argument that magnitude estimation leads to a ratio scale. Indeed, there are some who claim that the entire procedure of magnitude estimation is nonsense. (See, for example, Final Report of the British Association for the Advancement of Science [1940] or Moon [1936] for early criticisms.) Perhaps more important, there are those who feel that specifying uniqueness of the derived scale of loudness (or brightness or subjective duration) before specifying the psychophysical law is begging the question, and so the whole approach described above is useless. We shall not attempt to settle this issue here. However, in Section 4.4, we shall present axioms that, if satisfied, imply that magnitude estimation leads to a ratio scale.

---

*If the psychological scale is only an interval scale (in the wide sense), then it follows from Theorem 4.6 that the possible psychophysical laws are essentially either the logarithmic law or the power law.

4.3.2 Consequences of the Power Law

The usual way of testing a proposed scientific law is to derive predictions from the law and subject these to test. These predictions can be new, or they can be previously observed results, testable by previously gathered data, or requiring new experiments or observations to be tested. In this subsection and the next, we shall derive some predictions or consequences of using the power law as the psychophysical law for the case of loudness, and discuss tests of these consequences.

In particular, assuming the power law, it is possible to estimate the parameter β experimentally. In the case of loudness, Stevens [1955, and elsewhere] has estimated β for 1000-cps pure tones by having subjects use the method of magnitude estimation. As we pointed out above, he estimates that $\beta \approx 0.3 \approx \log_{10} 2$. A sample of such a magnitude estimation was given in Fig. 4.2. Exponents for other psychophysical variables are shown in Table 4.2. These were all determined by use of the method of magnitude estimation.

Using a value of $\beta = \log_{10} 2$, we find that the formula

$$L(x) = \psi(I(x))$$

for loudness becomes

$$L(x) = \alpha I(x)^{\log_{10} 2}. \tag{4.28}$$

Recall that

$$dB(x) = 10 \log_{10}\left[I(x)/I_0 \right].$$

Hence,

$$I(x) = I_0 10^{\frac{1}{10}dB(x)}.$$

It follows from Eq. (4.28) that

$$L(x) = \alpha I_0^{\log_{10} 2}\left[10^{\frac{1}{10}dB(x)} \right]^{\log_{10} 2}. \tag{4.29}$$

It is interesting to demonstrate that from Eq. (4.29) we may derive as a conclusion the observation made by Fletcher and Munson (Section 4.1.2) that an increase of 10 dB in intensity is equivalent to a doubling of loudness. For, if $dB(b) = dB(a) + 10$, then by (4.29),

$$L(b) = \alpha I_0^{\log_{10} 2} \left[10^{\frac{1}{10}(dB(a) + 10)} \right]^{\log_{10} 2}$$

$$= \alpha I_0^{\log_{10} 2} \left[10^{\frac{1}{10} dB(a)} 10 \right]^{\log_{10} 2}$$

$$= \alpha I_0^{\log_{10} 2} \left[10^{\frac{1}{10} dB(a)} \right]^{\log_{10} 2} 10^{\log_{10} 2}$$

$$= L(a) \cdot 2.$$

Similar results hold for pure tones other than the 1000-cps tones, and for other kinds of noises. Indeed, Stevens [1955, p. 825] says that he is increasingly convinced by data that the Fletcher–Munson observation holds for all continuous noises of engineering interest.

4.3.3 Cross-Modality Matching

The power law has passed a basic test required of any such law: it accounts for known empirical data. An additional test may be provided by asking for new predictions. Stevens has used the idea of "cross-modality matching" to test the power law. Observers are apparently able to match the strengths of the sensations produced in two different modalities, for example, loudness and brightness. For instance, a person can adjust the brightness of a light to match the perceived loudness of a 50-dB pure tone at 1000 cps. In general, suppose two different psychological quantities such as loudness and brightness are each related to physical quantities by a power law. Suppose A_1 and A_2 are sets, $f_1: A_1 \rightarrow Re$ and $f_2: A_2 \rightarrow Re$ represent physical scales, and $g_1: A_1 \rightarrow Re$ and $g_2: A_2 \rightarrow Re$ represent the corresponding psychological scales. Then we have

$$g_1 = \alpha_1 f_1^{\beta_1} \tag{4.30}$$

and

$$g_2 = \alpha_2 f_2^{\beta_2}. \tag{4.31}$$

Subjects are now asked to match a given $a \in A_1$ to an $a' \in A_2$ of "equal sensation." If the two power laws (4.30) and (4.31) are in effect, then the equal sensation function $g_1(a) = g_2(a')$ is given by

$$\alpha_1 \left[f_1(a) \right]^{\beta_1} = \alpha_2 \left[f_2(a') \right]^{\beta_2}$$

or

$$\ln f_1(a) = \gamma + (\beta_2/\beta_1) \ln f_2(a'),$$

where

$$\gamma = \frac{\ln \alpha_2 - \ln \alpha_1}{\beta_1}.$$

Thus if for all (a, a') with a judged equal in sensation to a', $f_1(a)$ is plotted against $f_2(a')$ in log–log coordinates, the graph should be a straight line whose slope is given by the ratio of the exponents. This prediction has been tested for various pairs of modalities, and the results come out quite well (Stevens [1959a, 1960, 1962]). See Figs. 4.3 and 4.4 for typical "equal sensation graphs."

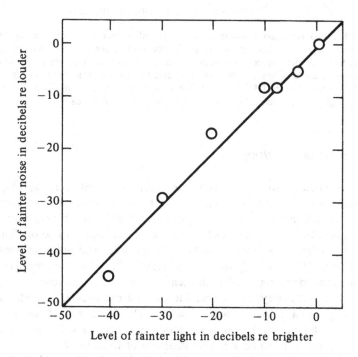

Figure 4.3. Equal sensation function for loudness versus brightness. Results of adjusting a loudness ratio to match an apparent brightness ratio defined by a pair of luminous circles. One of the circles was made dimmer than the other by the amount shown on the abscissa. The observer produced white noises by pressing on one or the other of two keys, and he adjusted the level of one noise (ordinate) to make the loudness ratio seem equal to the brightness ratio. From Stevens [1960, p. 241]. Reprinted with permission of *American Scientist*, journal of Sigma Xi, the Scientific Research Society of North America.

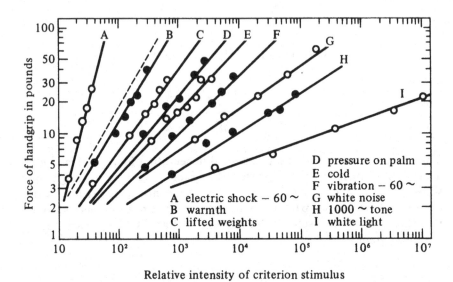

Figure 4.4. Equal sensation functions. Data obtained by matching force of handgrip to various criterion stimuli. Each point stands for the median force exerted by ten or more observers to match the apparent intensity of a criterion stimulus. The relative position of a function along the abscissa is arbitrary. The dashed line shows a slope of 1.0 in these coordinates. From Stevens [1960, p. 246]. Reprinted with permission of *American Scientist*, journal of Sigma Xi, the Scientific Research Society of North America.

4.3.4 *Attitude Scaling*

The procedures such as magnitude estimation that were developed to help scale sensory variables like loudness and brightness have also been applied in scaling attitudes or opinions, judgments of pleasantness of musical selections, judgments of seriousness of crimes, and so on. The idea of using psychophysical methods to study attitudes and opinions goes back to Thurstone [1927, 1959]. See Stevens [1966, 1968] for a survey. Here, the stimuli (corresponding to the physical scales in the discussion of earlier sections) cannot always be assumed to be on a ratio scale. Indeed, they are often only on a nominal scale (Stevens [1966, 1968]). However, when stimuli can be measured on what looks like a ratio scale, many of the attitude scales seem to be a power function of the stimuli.

To mention some references, Indow [1961] scaled desirability for wrist watches by asking subjects to match desirability of a watch to length of a line segment. This scale of preference, it can be argued, is a ratio scale. Then Indow had subjects state what they would regard as a fair price for each watch. The judged fair price (averaged over the group) turned out to be close to a power function of (average) desirability.

Other experimenters have had subjects indicate their opinions of the prestigiousness of various occupations by adjusting the intensity of a light or the level of a sound so that levels observed are proportional to judged prestigiousness (Cross [1976]).

Sellin and Wolfgang [1964] considered criminal offenses consisting of thefts of varying amounts of money, $5, $20, $50, $1000, $5000. They then asked subjects, using the magnitude estimation procedure, to scale the seriousness of each crime. The judged seriousness (geometric mean of group estimates) turned out to be a power function of the amount of money stolen, with exponent .17. The exponent suggests that, although it is worse to steal $2 than to steal $1, it is not twice as bad, which seems reasonable.

It is possible that the judged seriousness of the crimes in an experiment like Sellin and Wolfgang's could be a measure of the subjective value or utility of money. Most economists believe that the utility of money is not linearly related to the dollar amount of money, but rather that fixed increments become less and less important.* Bernoulli [1738] hypothesized that the utility of money is a logarithmic function of the dollar amount. Arguments that utility is a power function of dollar amount go back to the mathematician Gabriel Cramer in 1728 (Bernoulli [1738]). Some more modern arguments to this effect can be found in Stevens [1959b] and Galanter [1962]. We return to the utility of money in Chapters 5 and 7.

Exercises

1. Show that, according to the data of Table 4.2, if a magnitude estimation were performed for the subjective number of repetitions of a sound, and the results were plotted in log–log coordinates, the data would fit a 45° straight line.

2. Suppose we know (from data) that the psychophysical function is a power function, and we assume that the physical scale is a ratio scale. If the psychological scale is either a ratio scale, an interval scale, or a log-interval scale in the wide sense, show (by Theorems 4.5 and 4.6 and the results of Exer. 10a of Section 4.2) that the psychological scale is either a ratio scale or a log-interval scale.

3. Show that if the two physical scales f_1 and f_2 are both interval scales and the two corresponding psychological scales g_1 and g_2 are both interval scales in the wide sense, then the equal sensation graph is a straight line (in regular coordinates).

4. (Pfanzagl [1968, p. 128]) In cross-modality matching experiments, Stevens [1959a] reports that if $g_1 = \psi_1(f_1)$ and $g_2 = \psi_2(f_2)$, then $\psi_2^{-1}(\psi_1)$ turns out to be a power function.

(a) Show that this result follows if both ψ_1 and ψ_2 are power functions.

*This is the principle known as *decreasing marginal utility*.

(b) Show, however, that it also follows if

$$\psi_i(x) = \alpha_i \ln x + \beta_i, \quad i = 1, 2.$$

5. (Luce and Galanter [1963, p. 291]) An early experiment with painted disks performed by Plateau [1872] suggested that equal stimulus ratios must induce equal sensation ratios. If true, this observation implies the power law. For, suppose $\psi: Re^+ \to Re^+$ is continuous, and for s, s', t, t',

$$\frac{s}{t} = \frac{s'}{t'} \Rightarrow \frac{\psi(s)}{\psi(t)} = \frac{\psi(s')}{\psi(t')}.$$

Letting $u(x) = \psi(x)/\psi(1)$, derive Cauchy's fourth equation and hence show that $\psi(x) = \alpha x^\beta$.

6. According to Stevens [1960], vibration (of 250 cps) on the finger gives rise to sensations that satisfy a power law with exponent .6. The law is

$$V(a) = \alpha \left[\frac{E(a)}{E_0} \right]^{.6},$$

where $E(a)$ is the physical energy level of the vibration, E_0 a reference energy level, and $V(a)$ the psychological sensation. If the decibel scale $v_{dB}(a)$ is defined by

$$v_{dB}(a) = 10 \log_{10} \left[\frac{E(a)}{E_0} \right],$$

show that a 5-dB increase leads to a doubling of sensation.

4.4 A Measurement Axiomatization for Magnitude Estimation and Cross-Modality Matching

4.4.1 Consistency Conditions

In this section, we discuss axioms under which magnitude estimation leads to a ratio scale. Such axioms would put Stevens' argument that magnitude estimation leads to such a scale on a firm measurement-theoretic foundation—provided the axioms are satisfied. A similar approach might help settle a variety of disputes about the scale type resulting from empirical procedures. We follow Krantz [1972] and Krantz et al. [1971, Section 4.6]. Suppose $A_1, A_2, \ldots, A_n$ are different physical continua, one for sounds, one for lights, etc. Suppose we are performing a magnitude estimation on the ith continuum. We fix x_i in A_i and assign to x_i the psychological magnitude p. (We shall confuse x_i with its physical magnitude and assume that all magnitudes are positive numbers.) Next, we ask the subject to assign to each y_i in A_i a magnitude q, with q depending on x_i and p. That is,

$$N_i(y_i|x_i, p) = q,$$

where N_i is the magnitude estimate for y_i given that x_i is assigned a magnitude p. In particular,

$$N_i(x_i|x_i, p) = p.$$

The psychophysical law gives q/p as a function of y_i/x_i. The power law states that

$$N_i(y_i|x_i, p) = \alpha_i p(y_i/x_i)^{\beta_i},$$

where α_i and β_i are constants, α_i positive.

In the variant of magnitude estimation called ratio estimation or pair estimation, a pair of stimuli x_i and y_i from A_i are presented, and the subject is asked to provide a numerical estimate of the "sensation ratio" of x_i to y_i. We denote this estimate

$$P_i(x_i, y_i).$$

In scaling, pair estimates and magnitude estimates are often assumed to satisfy the following *magnitude–pair consistency condition*: for all z_i, p,

$$P_i(x_i, y_i) = \frac{N_i(x_i|z_i, p)}{N_i(y_i|z_i, p)}. \tag{4.32}$$

Moreover, it is often assumed that pair estimates act like ratios; that is, they satisfy the following *pair consistency condition*:

$$P_i(x_i, y_i) \cdot P_i(y_i, z_i) = P_i(x_i, z_i). \tag{4.33}$$

In cross-modality matching, we usually fix $x_i \in A_i$ and $x_j \in A_j$ and say they match. We then ask the subject to find a stimulus $y_i \in A_i$ that matches a given stimulus $y_j \in A_j$. We write

$$C_{ji}(y_j|x_j, x_i) = y_i.$$

It is often assumed that magnitude estimation and cross-modality matching are also related, by the following *magnitude–cross-modality consistency condition*:

$$\frac{N_i\left[C_{ji}(y_j|x_j, x_i)|z_i, p\right]}{N_i(x_i|z_i, p)} = \frac{N_j(y_j|z_j, q)}{N_j(x_j|z_j, q)}. \tag{4.34}$$

That is, if y_j is matched with y_i in the cross-modality matching, where x_j is given as matched with x_i, then the ratio of the magnitude estimate of y_i to the magnitude estimate of x_i on the ith modality equals the ratio of the magnitude estimate of y_j to the magnitude estimate of x_j on the jth modality for any reference estimates p for z_i and q for z_j. If $C_{ji}(y_j|x_j, x_i) = y_i$ and if $x_i = z_i$, then Eq. (4.34) and Eq. (4.32) with j replacing i and y and x reversed yield

$$\frac{N_i(y_i|x_i, p)}{p} = P_j(y_j, x_j), \tag{4.35}$$

using $N_i(x_i|x_i, p) = p$. Equation (4.35) is a second magnitude–pair consistency condition.

4.4.2 *Cross-Modality Ordering*

Let us introduce a binary relation $\succsim$, the *cross-modality ordering relation*, as follows. Suppose x_i and y_i are in A_i and u_j and v_j are in A_j. Then

$$(x_i, y_i) \succsim (u_j, v_j)$$

if and only if the sensation ratio of x_i to y_i is judged greater than or equal to the sensation ratio of u_j to v_j. Technically, $\succsim$ is a relation on

$$(A_1 \times A_1 \cup A_2 \times A_2 \cup \cdots \cup A_n \times A_n).$$

We would hope that $\succsim$ is related to the pair estimates as follows:

$$(x_i, y_i) \succsim (u_j, v_j) \Leftrightarrow P_i(x_i, y_i) \geqq P_j(u_j, v_j). \tag{4.36}$$

The cross-modality ordering relation $\succsim$ is closely related to the matching relation of cross-modality matching. Indeed, suppose $\sim$ is defined from $\succsim$ by

$$(x_i, y_i) \sim (u_j, v_j) \Leftrightarrow (x_i, y_i) \succsim (u_j, v_j) \,\&\, (u_j, v_j) \succsim (x_i, y_i).$$

Then it is reasonable to assume that

$$C_{ji}(y_j|x_j, x_i) = y_i \Rightarrow (y_j, x_j) \sim (y_i, x_i). \tag{4.37}$$

This assumption says that if x_j is matched by x_i and y_j by y_i, then the corresponding sensation ratios are judged equal. The cross-modality ordering relation $\succeq$ is closely related to the quaternary relation W of algebraic difference measurement (Section 3.3). Indeed, by modifying the axioms for algebraic difference measurement, Krantz *et al.* [1971, p. 165] present axioms sufficient to prove the following *representation theorem* (see Exers. 4 through 9 below for the axioms):

REPRESENTATION THEOREM FOR CROSS-MODALITY ORDERING. *There exist functions* $\phi_i : A_i \to Re^+$ *so that for all i, j, and for all* $x_i, y_i \in A_i$ *and* $u_j, v_j \in A_j$,

$$(x_i, y_i) \succeq (u_j, v_j) \Leftrightarrow \phi_i(x_i)/\phi_i(y_i) \geq \phi_j(u_j)/\phi_j(v_j). \qquad (4.38)$$

If ϕ_i' *are any other such functions, there are positive numbers* $\alpha_1, \ldots, \alpha_m$ *and* β *so that*

$$\phi_i' = \alpha_i \phi_i^\beta,$$

all i.

The functions ϕ_i define psychological scales. Equation (4.38) says that one sensation ratio is judged greater than or equal to a second if and only if the ratio of the corresponding psychological magnitudes for the first pair is greater than or equal to the ratio for this second pair. The representation (4.38) is very close to that for algebraic difference measurement. The critical axiom needed for the representation theorem is the monotonicity axiom:

$$\left[(x_i, y_i) \succeq (u_j, v_j) \ \& \ (y_i, z_i) \succeq (v_j, w_j) \right] \Rightarrow (x_i, z_i) \succeq (u_j, w_j). \qquad (4.39)$$

This is closely related to the monotonicity axiom (Axiom D3) for algebraic difference structures (Section 3.3).

4.4.3 *Magnitude Estimation as a Ratio Scale*

Suppose that in addition to the Krantz *et al.* axioms sufficient for the representation (4.38), we assume that the functions N_i, P_i, and C_{ji} satisfy the following conditions:

(a) Pair consistency on the first continuum [Eq. (4.33) for $i = 1$]:

$$P_1(x_1, y_1) \cdot P_1(y_1, z_1) = P_1(x_1, z_1).$$

(b) $\qquad (x_i, y_i) \succeq (u_j, v_j) \Leftrightarrow P_i(x_i, y_i) \geq P_j(u_j, v_j). \qquad (4.36)$

(c) $\qquad C_{ji}(y_j | x_j, x_i) = y_i \Rightarrow (y_j, x_j) \sim (y_i, x_i). \qquad (4.37)$

(d) Equation (4.35) for $j = 1$ whenever $(y_i, x_i) \sim (y_1, x_1)$:

$$(y_i, x_i) \sim (y_1, x_1) \Rightarrow \frac{N_i(y_i|x_i, p)}{p} = P_1(y_1, x_1).$$

Then Krantz [1972] proves the following representation theorem:

REPRESENTATION THEOREM FOR MAGNITUDE ESTIMATION. *There is a power function* $\phi: Re^+ \rightarrow Re^+$, *so that if* ϕ_i *are the functions satisfying Eq.* (4.38), *then*

$$N_i(y_i|x_i, p) = q \Leftrightarrow \phi_i(y_i)/\phi_i(x_i) = \phi(q)/\phi(p), \qquad (4.40)$$

$$P_i(x_i, y_i) = r \Leftrightarrow \phi_i(x_i)/\phi_i(y_i) = \phi(r), \qquad (4.41)$$

and

$$C_{ji}(y_j|x_j, x_i) = y_i \Rightarrow \phi_i(y_i)/\phi_j(y_j) = \phi_i(x_i)/\phi_j(x_j). \qquad (4.42)$$

Moreover, if ϕ_i' *and* ϕ' *also satisfy Eqs.* (4.38) *and* (4.40) *through* (4.42), *then there are positive numbers* $\alpha_1, \ldots, \alpha_m$ *and* β *so that*

and
$$\left.\begin{array}{c} \phi_i' = \alpha_i\phi_i^\beta \\ \phi' = \phi^\beta \end{array}\right\} \qquad (4.43)$$

In this theorem we can think of the ϕ_i and ϕ as psychological scales. Then Eq. (4.40) says that our magnitude estimate of y_i is q if and only if the ratio of the psychological magnitudes of y_i and x_i is the same as the ratio of the psychological magnitudes of the numbers q and p.

We shall show that a function N_i satisfying Eq. (4.40) defines a ratio scale in the wide sense. Thus, the magnitude estimates lead to a ratio scale. To proceed, we first rewrite Eq. (4.40).

Let us fix x_i and p, and denote by $N_i(y_i)$ the number $N_i(y_i|x_i, p)$. It follows from Eq. (4.40) that

$$\frac{\phi_i(y_i)}{\phi_i(x_i)} = \frac{\phi[N_i(y_i)]}{\phi[N_i(x_i)]}. \qquad (4.44)$$

We shall show that a function N_i satisfying Eq. (4.44) defines a ratio scale in the wide sense. Since ϕ is a power function, every similarity transformation $N_i'(y_i) = \lambda N_i(y_i)$ still satisfies (4.44). Conversely, suppose N_i satisfies (4.44), and we allow admissible transformations of ϕ and ϕ_i. Let us see what the corresponding transformation N_i' of N_i must be. We have

$$\phi_i'(y_i)/\phi_i'(x_i) = \frac{\phi'[N_i'(y_i)]}{\phi'[N_i'(x_i)]}.$$

Using the uniqueness results of Eq. (4.43), we have

$$\frac{\alpha_i[\phi_i(y_i)]^{\beta}}{\alpha_i[\phi_i(x_i)]^{\beta}} = \frac{\{\phi[N_i'(y_i)]\}^{\beta}}{\{\phi[N_i'(x_i)]\}^{\beta}}.$$

Since α_i and β are positive, we have

$$\frac{\phi_i(y_i)}{\phi_i(x_i)} = \frac{\phi[N_i'(y_i)]}{\phi[N_i'(x_i)]},$$

and so by (4.44),

$$\frac{\phi[N_i(y_i)]}{\phi[N_i(x_i)]} = \frac{\phi[N_i'(y_i)]}{\phi[N_i'(x_i)]}. \tag{4.45}$$

Let λ be $N_i'(x_i)/N_i(x_i)$. Since magnitude estimates are assumed to be positive numbers, λ is positive. We shall see that

$$N_i'(y_i) = \lambda N_i(y_i), \quad \text{all} \quad y_i. \tag{4.46}$$

Thus, in the wide sense, all admissible transformations of N_i are similarity transformations. It follows that N_i is a ratio scale in the wide sense. To demonstrate (4.46), note that since ϕ is a power function, we have from (4.45)

$$\frac{\mu[N_i(y_i)]^{\nu}}{\mu[N_i(x_i)]^{\nu}} = \frac{\mu[N_i'(y_i)]^{\nu}}{\mu[N_i'(x_i)]^{\nu}},$$

$\mu > 0$. Now we may assume that $\nu \neq 0$, for otherwise $\phi(x)$ is a constant, and $N_i(y_i|x_i, p) = q$ implies $N_i(y_i|x_i, p) = q'$ for any other q', which is nonsense. It follows from $\mu > 0$, $\nu \neq 0$, that

$$\frac{N_i(y_i)}{N_i(x_i)} = \frac{N_i'(y_i)}{N_i'(x_i)}.$$

Thus,

$$N_i'(y_i) = \frac{N_i'(x_i)}{N_i(x_i)} \cdot N_i(y_i) = \lambda N_i(y_i),$$

as desired.

Remark: Quite different formal theories of magnitude estimation can be found in Green and Luce [1974], Luce and Green [1974 a,b], and Marley [1972].

Exercises

1. Suppose loudness is modality $i = 1$ and the magnitude estimates are as in Fig. 4.2.

(a) Let $\log y_1 = 60$ and $\log z_1 = 80$. Suppose $N_1(y_1|x_1, p) \approx 60$ and $N_1(z_1|x_1, p) \approx 80$. If magnitude–pair consistency holds, show that $P_1(y_1, z_1) \approx \frac{1}{4}$.

(b) Repeat the computation of $P_1(y_1, z_1)$ for the following cases:
   (i) $N_1(y_1|x_1, p) \approx 40$,
        $N_1(z_1|x_1, p) \approx 50$.
   (ii) $N_1(y_1|x_1, p) \approx 70$,
        $N_1(z_1|x_1, p) \approx 90$.

2. Suppose modality $i = 1$ is cold and modality $i = 2$ is force of handgrip, and suppose cross-modality matchings are given as in Fig. 4.4 for fixed x_1 and x_2.

(a) Show that if $\log y_1 \approx 10^3$, $C_{12}(y_1|x_1, x_2)$ is that handgrip force y_2 whose logarithm is ≈ 15.

(b) Compute $C_{32}(y_3|x_3, x_2)$ if modality $i = 3$ is warmth and if $\log y_3 \approx 10^2$.

3. Suppose that for all i, $N_i(y_i|z_i, p) = \alpha_i p(y_i/z_i)^{\beta_i}$, where β_i is the exponent given in Table 4.2. Assume the conditions (4.32) and (4.36) hold. Given the following i, j, x_i, y_i, and u_j, determine v_j so that

$$(x_i, y_i) \sim (u_j, v_j):$$

(a) $i =$ brightness under 5° target.
    $j =$ temperature (cold on arm).
    $x_i = 100$.
    $y_i = 200$.
    $u_j = 100$.
(b) $i =$ taste (salt).
    $j =$ vocal effort.
    $x_i = 100$.
    $y_i = 200$.
    $u_j = 100$.

4. Show that if Eq. (4.38) holds, then the cross-modality ordering $\gtrsim$ defines a weak order.

5. Show that if Eq. (4.38) holds, then $\gtrsim$ satisfies

$$(x_i, y_i) \gtrsim (u_j, v_j) \Rightarrow (v_j, u_j) \gtrsim (y_i, x_i).$$

6. Show that the monotonicity condition of Eq. (4.39) follows from the representation of Eq. (4.38).

7. Show that the following condition is not a necessary condition for the representation (4.38): Given $x_i, y_i \in A_i$, there are $x_1, y_1 \in A_1$ so that

$$(x_i, y_i) \sim (x_1, y_1).$$

8. Show that even if the conditions of Exers. 4 through 7 hold, the following solvability condition on the cross-modality ordering is not a necessary condition for the representation (4.38): If

$$(x_1, y_1) \gtrsim (u_1, v_1) \gtrsim (x_1, x_1),$$

then there are z_1' and z_1'' so that

$$(x_1, z_1') \sim (u_1, v_1) \sim (z_1'', y_1).$$

9. Suppose the conditions of Exers. 4 through 8 hold. The elements $x_1^{(1)}, x_1^{(2)}, \ldots, x_1^{(i)}, \ldots$ from A_1 form a *strictly bounded standard sequence* if

$$(x_1^{(i+1)}, x_1^{(i)}) \sim (x_1^{(2)}, x_1^{(1)}), \quad \text{for all } i \text{ in the sequence,}$$
$$\text{not} \quad (x_1^{(2)}, x_1^{(1)}) \sim (x_1^{(1)}, x_1^{(1)}),$$

and there exist y_1' and y_1'' in A_1 so that

$$(y_1', y_1'') > (x_1^{(i)}, x_1^{(1)}) > (y_1'', y_1'), \quad \text{for all } i \text{ in the sequence,}$$

where

$$(z_1, z_1') > (w_1, w_1') \Leftrightarrow \sim [(w_1, w_1') \gtrsim (z_1, z_1')].$$

Show from the representation (4.38) that every strictly bounded standard sequence is finite. [*Note*: Krantz *et al.* [1971, p. 165] show that the conditions studied in Exers. 4 through 9 are sufficient to prove the representation (4.38).]

References

Aczél, J., *Lectures on Functional Equations and Their Applications*, Academic Press, New York, 1966.

Békésy, G. von, *Experiments in Hearing*, edited and translated by E. G. Wever, McGraw-Hill, New York, 1960.

Bernoulli, D., "Specimen Novae de Mensura Sortis," *Comentarii Academiae Scientiarum Imperiales Petropolitanae*, **5** (1738), 175–192; translated by L. Sommer in *Econometrica*, **22** (1954), 23–36.

Cauchy, A. L., "Cours d'analyse de l'École Polytechnique," *Analyse Algébrique*, Vol. I, V. Paris, 1821. ("Oeuvres," Ser. 2, Vol. 3, pp. 98–113, 220, Paris, 1897.)

Cross, D. V., "Multivariate Psychophysical Scaling," mimeographed, State University of New York, Stony Brook, paper presented at the Spring Meeting of the Psychometric Society, April 1976.

Darboux, G., "Sur la Composition des Forces en Statistique," *Bull. Sci. Math.*, **9** (1875), 281–288.

Department of Public Health, State of California, *A Report to the 1971 Legislature on the Subject of Noise Pursuant to Assembly Concurrent Resolution 165, 1970*, Sacramento, California, 1971.

Eichhorn, W., *Functional Equations in Economics*, Addison-Wesley, Reading, Massachusetts, 1978.

Ekman, G., and Sjöberg, L., "Scaling," *Ann. Rev. Psychol.*, **16** (1965), 451–474.

Environmental Protection Agency, "Effects of Noise on People," Report NT1D300.7, Washington, D.C., December 1971.

Falmagne, J. C., Iverson, G., and Marcovici, S., "Binaural Loudness Summation: Probabilistic Theory and Data," manuscript, Department of Psychology, New York University, New York, New York, 1978.

Fechner, G. T., *Elemente der Psychophysik*, Breitkopf und Hartel, Leipzig, 1860.

Final Report of the British Association for the Advancement of Science, *Advan. Sci.*, **2** (1940), 331–349.

Fletcher, H., and Munson, W. A., "Loudness, Its Definition, Measurement and Calculation," *J. Acoust. Soc. Amer.*, **5** (1933), 82–108.

Galanter, E., "The Direct Measurement of Utility and Subjective Probability," *Amer. J. Psychol.*, **75** (1962), 208–220.

Galanter, E., and Messick, S., "The Relation between Category and Magnitude Scales of Loudness," *Psychol. Rev.*, **68** (1961), 363–372.

Glorig, A., Wheeler, D., Quiggle, R., Grings, W., and Summerfield, A., "1954 Wisconsin State Fair Hearing Survey: Statistical Treatment of Clinical and Audiometric Data," American Academy Ophthalmology and Otolaryngology and Research Center Subcommittee on Noise in Industry, Los Angeles, California, 1957.

Green, D. M., and Luce, R. D., "Variability of Magnitude Estimates: A Timing Theory Analysis," *Perception and Psychophysics*, **15** (1974), 291–300.

Hirsch, I. J., "Binaural Summation: A Century of Investigation," *Psychol. Bull.*, **45** (1948), 193–206.

Indow, T., ["An Example of Motivation Research Applied to Product Design,"], published in Japanese in *Chosa to gijutsu*, **102** (1961), 45–60.

Jensen, J. L. W. V., "Om Konvekse Functioner og Uligheder Imellem Middelvaerdier," *Mat. Tidssler.*, **B** (1905), 49–68.

Jensen, J. L. W. V., "Sur les Fonctions Convexes et les Inégalités entre les Valeurs Moyennes," *Acta Math.*, **30** (1906), 175–193.

Krantz, D. H., "A Theory of Magnitude Estimation and Cross-Modality Matching," *J. Math. Psychol.*, **9** (1972), 168–199.

Krantz, D. H., Luce, R. D., Suppes, P., and Tversky, A., *Foundations of Measurement*, Vol. I, Academic Press, New York, 1971.

Kryter, K. D., *The Effects of Noise on Man*, Academic Press, New York, 1970.

Levelt, W. J. M., Riemersma, J. B., and Bunt, A. A., "Binaural Additivity of Loudness," *Brit. J. Math. Stat. Psychol.*, **25** (1972), 51–68.

Lierle, D. M., "Guide for the Evaluation of Hearing Impairment: A Report of the Committee

Luce, R. D., "On the Possible Psychophysical Laws," *Psychol. Rev.*, **66** (1959), 81–95.
on Conservation of Hearing," *Trans. Amer. Acad. Ophthalmol. Otolaryngol.*, 1959, 236–238.

Luce, R. D., "Comments on Rozeboom's Criticisms of 'On the Possible Psychophysical Laws'," *Psychol. Rev.*, **69** (1962), 548–551.

Luce, R. D., "What Sort of Measurement Is Psychophysical Measurement?" *Amer. Psychologist*, **27** (1972), 96–106.

Luce, R. D., and Galanter, E., "Psychophysical Scaling," in R. D. Luce, R. R. Bush, and E. Galanter (eds.), *Handbook of Mathematical Psychology*, Vol. I, Wiley, New York, 1963, pp. 245–307.

Luce, R. D., and Green, D. M., "Ratios of Magnitude Estimates," in H. R. Moskowitz *et al.* (eds.), *Sensation and Measurement*, Reidel, Dordrecht, Holland, 1974a, pp. 99–111.

Luce, R. D., and Green, D. M., "The Response Ratio Hypothesis for Magnitude Estimation," *J. Math. Psychol.*, **11** (1974b), 1–14.

Luce, R. D., and Mo, S. S., "Magnitude Estimation of Heaviness and Loudness by Individual Subjects: A Test of a Probabilistic Response Theory," *Brit. J. Math. Stat. Psychol.*, **18** (1965), 159–174.

Marley, A. A. J., "Internal State Models for Magnitude Estimation and Related Experiments," *J. Math. Psychol.*, **9** (1972), 306–319.

Moon, P., *The Scientific Bases of Illuminating Engineering*, McGraw-Hill, New York, 1936.

Pfanzagl, J., *Theory of Measurement*, Wiley, New York, 1968.

Plateau, J. A. F., "Sur la Mesure des Sensations Physiques, et sur la Loi Qui Lie l'Intensité de la Cause Excitante," *Bull. Acad. Roy. Belg.*, **33** (1872), 376–388.

Pradhan, P. L., and Hoffman, P. J., "Effect of Spacing and Range of Stimuli on Magnitude Estimation Judgments," *J. Exp. Psychol.*, **66** (1963), 533–541.

Reynolds, G. S., and Stevens, S. S., "Binaural Summation of Loudness," *J. Acoust. Soc. Amer.*, **32** (1960), 1337–1344.

Rienow, R., and Rienow, L. T., *Moment in the Sun*, Ballantine, New York, 1967.

Rozeboom, W. W., "The Untenability of Luce's Principle," *Psychol. Rev.*, **69** (1962a), 542–547.

Rozeboom, W. W., "Comment," *Psychol. Rev.*, **69** (1962b), 552.

Savage, C. W., *Measurement of Sensation*, University of California Press, Berkeley, California, 1970.

Scharf, B., "Dichotic Summation of Loudness," *J. Acoust. Soc. Amer.*, **45** (1969), 1193–1205.

Schneider, B., and Lane, H., "Ratio Scales, Category Scales, and Variability in the Production of Loudness and Softness," *J. Acoust. Soc. Amer.*, **35** (1963), 1953–1961.

Sears, F. W., and Zemansky, M. W., *University Physics*, Addison-Wesley, Reading, Massachusetts, 1955.

Sellin, T., and Wolfgang, M. E., *The Measurement of Delinquency*, Wiley, New York, 1964.

Stevens, J. C., and Guirao, M., "Individual Loudness Functions," *J. Acoust. Soc. Amer.*, **36** (1964), 2210–2213.

Stevens, S. S., "The Measurement of Loudness," *J. Acoust. Soc. Amer.*, **27** (1955), 815–829.

Stevens, S. S., "On the Psychophysical Law," *Psychol. Rev.*, **64** (1957), 153–181.

Stevens, S. S., "Cross-Modality Validation of Subjective Scales," *J. Exp. Psychol.*, **57** (1959a), 201–209.

Stevens, S. S., "Measurement, Psychophysics, and Utility," in C. W. Churchman and P. Ratoosh (eds.), *Measurement: Definitions and Theories*, Wiley, New York, 1959b.

Stevens, S. S., "On the Validity of the Loudness Scale," *J. Acoust. Soc. Amer.*, **31** (1959c), 995–1003.

Stevens, S. S., "The Psychophysics of Sensory Function," *Amer. Scientist*, **48** (1960), 226–253.

Stevens, S. S., "The Psychophysics of Sensory Function," in W. A. Rosenblith (ed.), *Sensory Communication*, Wiley, New York, 1961a, pp. 1–33.

Stevens, S. S., "To Honor Fechner and Repeal His Law," *Science*, 133 (1961b), 80–86.

Stevens, S. S., "Toward a Resolution of the Fechner–Thurstone Legacy, *Psychometrika*, 26 (1961c), 35–47.

Stevens, S. S., "The Surprising Simplicity of Sensory Metrics," *Amer. Psychologist*, 17 (1962), 29–39.

Stevens, S. S., "A Metric for the Social Consensus," *Science*, 151 (1966), 530–541.

Stevens, S. S., "Ratio Scales of Opinion," in D. K. Whitla (ed.), *Handbook of Measurement and Assessment in Behavioral Sciences*, Addison-Wesley, Reading, Massachusetts, 1968.

Stevens, S. S., "Assessment of Noise: Calculation Procedure Mark VII," Paper 355-128, Laboratory of Psychophysics, Harvard University, Cambridge, Massachusetts, December 1969.

Stevens, S. S., "Issues in Psychophysical Measurement," *Psychol. Rev.*, 78 (1971), 426–450.

Stevens, S. S., and Volkmann, J., "The Relation of Pitch to Frequency: A Revised Scale," *Amer. J. Psychol.*, 53 (1940), 329–353.

Thurstone, L. L., "Psychophysical Analysis," *Amer. J. Psychol.*, 38 (1927), 368–389.

Thurstone, L. L., *The Measurement of Values*, University of Chicago Press, Chicago, Illinois, 1959.

Tobias, J. V., "Curious Binaural Phenomena," in J. V. Tobias (ed.), *Foundations of Modern Auditory Theory*, Vol. 2, Academic Press, New York, 1972.

Treisman, M., and Irwin, R. J., "Auditory Intensity and Discrimination Scale I. Evidence Derived from Binaural Intensity Summation," *J. Acoust. Soc. Amer.*, 42 (1967), 586–592.

Wever, E. G., *Theory of Hearing*, Wiley, New York, 1949.

CHAPTER 5

Product Structures

5.1 Obtaining a Product Structure

In this chapter we study a variant on the measurement problems we have considered so far. Specifically, we consider measurement when the underlying set can be expressed as a Cartesian product. In making choices, we often consider alternatives with a variety of attributes or dimensions or from several points of view. We talk about *multidimensional* or *multiattributed* alternatives. For example, in choosing a job, we might consider salary, job security, possibility for advancement, geographical location, and so on. In buying a house, we might consider price, location, school system, availability of transportation, and the like. In designing a rapid transit system, we might consider power source, vehicle design, right of way design, and so on. In such a situation, each alternative **a** in the set of alternatives can be thought of as an *n*-tuple $(a_1, a_2, \ldots, a_n)$, where a_i is some rating of alternative **a** on the *i*th attribute or dimension. For example, in the case of a job, a_1 might be salary, a_2 might be fringe benefits, a_3 might be some measure of job security (for example, amount of notice required), and so on.

Multidimensional alternatives arise in a different way in economics. If there are *n* commodities in consideration, a_i might be a quantity of the *i*th commodity, and $(a_1, a_2, \ldots, a_n)$ then is a *commodity bundle* or *market basket*. Preferences are expressed among alternative market baskets. (We have previously encountered market baskets in our study of the consumer price index in Section 2.6.)

Let A_i be the set of all possible a_i. We think of the set of alternatives A as a Cartesian product $A_1 \times A_2 \times \ldots \times A_n$, and say that A has a *product structure.* If each A_i is a set of real numbers, we say A has a *numerical product structure.*

A product structure that is nonnumerical can arise in a variety of ways. For example, A_i can be a set of possibilities for the ith attribute, and a_i can be chosen as an element of A_i. Thus, in the rapid transit situation, A_1 might be a set of alternative power sources, A_2 a set of vehicle designs, A_3 a set of right of way designs, and so on. To choose an alternative, we pick one member from each A_i, that is, one power source, one vehicle design, one right of way design, and so on.

Product structures arise in numerous applications where we are trying to explain a response or a dependent variable on the basis of a number of factors or independent variables. In studying response strength, psychologists often consider two factors, drive and incentive. A given situation is defined by some measure d of drive and some measure k of incentive, and the set of situations corresponds to the set $D \times K$, where D is a set of different (levels of) drives and K a set of different (levels of) incentives. Sometimes habit strength is also considered. In that case, if H is a set of strengths of habits, one considers situations corresponding to ordered triples in the set $D \times K \times H$.

In studying binaural loudness in auditory perception, psychologists present sounds of different intensity to each ear. The set A of alternative stimuli is $L \times R$, where L is the set of sounds (sound intensities) presented to the left ear and R the set presented to the right ear.

In a mental testing situation, it is important to study the interaction between a subject and an item on a test. If S is the set of subjects and T is the set of test items, then $S \times T$ is the set that is often studied.

In studying discomfort under different weather conditions, the factors temperature t and humidity h play a principal rule. Relative discomfort is considered under various combinations of temperature and humidity. If T is the set of temperatures of interest and H the set of humidities, then $T \times H$ is the set of alternative weather conditions that can be compared as to discomfort.

Often a product structure will be presented ahead of time. However, especially if we are out to calculate a utility function, the first step is often to structure the set of alternatives A. In this section, we make some remarks on how to find a product structure. In the next section, we discuss, in the context of preferences, how to reduce one product structure to another with fewer dimensions, an important reduction in practice. In Section 5.3, we consider sets of alternatives with numerical product structures, in particular sets of alternative commodity bundles, and discuss the calculation of ordinal utility functions over sets of alternatives with such product structures. In Section 5.4, we seek functions (utility functions and other order-preserving functions) that can be calculated by reducing the computation to each dimension separately, and then adding. In Section

5.5, we continue the discussion of reduction of computation to dimension by dimension computation, but consider ways other than addition of combining results from different dimensions. In Section 5.6 we consider quite different measurement problems, where there are two dimensions and the set of individuals making judgments is one of the dimensions. Throughout the chapter, we shall keep in mind the applicability of the results to the drive and incentive problem, the binaural loudness problem, the mental testing problem, and the temperature–humidity problem. A variety of other applications of product structures, and in particular of utility functions over product structures, is described in Keeney and Raiffa [1976]. See also Farquhar [1977] and Cochrane and Zeleny [1973] for surveys of multidimensional utility theory. Other applications of utility functions over product structures include decisionmaking about educational priorities, choice of air pollution abatement strategies, development of water quality indices, choice of medical treatments, and siting for major new facilities such as airports. See Section 7.3.1 for references.

The question of how to introduce a product structure on A is a very important one, and one for which there is no precise theory. There are, at best, rules of thumb. Often, we shall simply be given relevant dimensions, and there is nothing to do. However, the problem we consider briefly is this: Given an unstructured set of alternatives A, how do we give it a product structure? The first step is to define the set of aspects or dimensions. Perhaps the most natural procedure for doing this in many decisionmaking contexts is that described by Manheim and Hall [1968], Miller [1969], and Raiffa [1969]: build up the structure hierarchically. Namely, start by listing an inclusive set of first-level attributes or objectives or facets. Then subdivide each of these into an inclusive list of second-level objectives, more precise than these. And so on. For an extensive discussion of the problem of introducing a product structure, see Keeney and Raiffa [1976, Chapter 2].

To illustrate this hierarchical procedure, we present the Manheim–Hall hierarchical structuring of attributes useful in comparing alternative transportation systems in the Northeast Corridor of the United States. The "super-goal" Manheim and Hall begin with is "The Good Life." They subdivide this goal into four dimensions: convenience, safety, aesthetics, and economic considerations. Each of these dimensions is further subdivided. And so on. The subdivisions are shown in a "tree diagram" in Fig. 5.1. The boxes at the bottom end of each branch represent the final collection of dimensions; there are twenty in all.

Once a final collection of dimensions or attributes has been obtained, these define a product structure. The next step is often to obtain some numerical assignment $a_i = f_i(a)$ for each alternative $a \in A$ on each dimension i. That is, we want to scale each element on each of the final list of dimensions, such as travel time, probability of delay, and noise. These

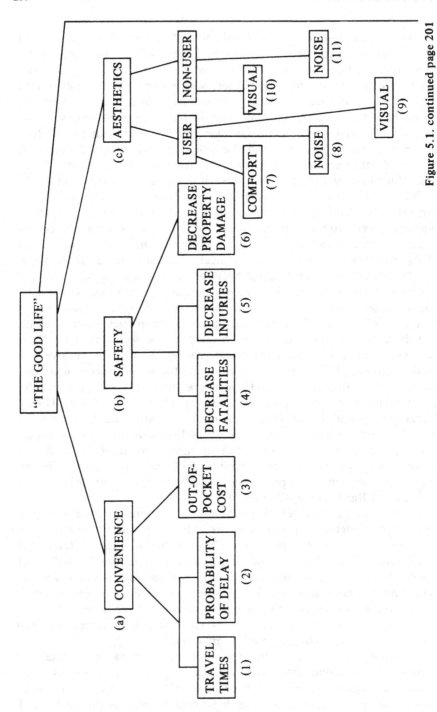

Figure 5.1. continued page 201

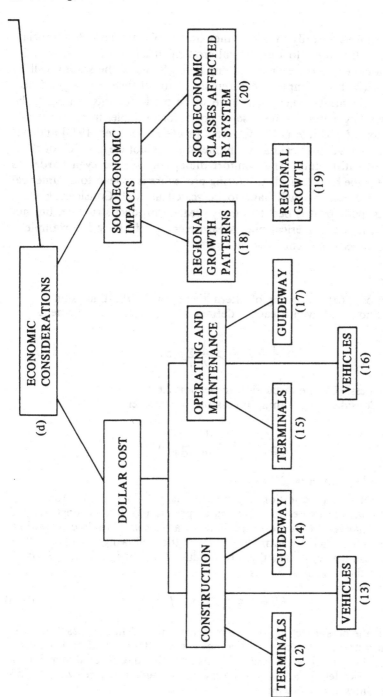

Figure 5.1. Manheim–Hall hierarchical structuring of attributes for comparison of alternative Northeast Corridor transportation systems. Adapted from Raiffa [1969, pp. 19, 20] with permission of the RAND Corporation.

scales are not necessarily utilities—they are simply numbers that translate the product structure into a numerical product structure and make it easier for us to define our preferences among them. Some of the scaling will be relatively easy, for example, measurement of travel time. But a good deal of progress in measurement will have to be made for this scaling to be carried out in general. In the case of noise, for example, there is a whole proliferation of possible noise measures in use (cf. Kryter [1970] and our discussion of loudness scaling in Chapter 4). It is not clear which of these is the most satisfactory. The comfort dimension will be even harder to scale, and if the hierarchical structuring procedure is to lead to a numerical product structure, this will have to be scaled as well. Considering these difficulties, perhaps it is best to simply create a product structure, but not translate it into a numerical one. Measurement—of utility for example—can still proceed, as Section 5.4 illustrates.

Exercises

1. Suppose $f_i(a)$ is a rating of alternative a on the ith dimension.
   (a) Show that even if each f_i defines a ratio scale, the statement

$$\frac{1}{n} \sum_i f_i(a) > \frac{1}{n} \sum_i f_i(b)$$

is meaningless. (Cf. the discussion in Section 2.6.)
   (b) Consider the meaningfulness of the statement

$$\frac{1}{n} \sum_{a \in X} f_i(a) > \frac{1}{m} \sum_{b \in Y} f_j(b),$$

where $n = |X|$ and $m = |Y|$.

2. An interesting question in measurement theory is the following: Suppose someone expresses or exercises preferences (on an unstructured set of alternatives). Can we judge if he is acting *as if* he had a product structure? One way to formalize this is the following: Suppose R is preference on the set A. Are there real-valued functions $f_1, f_2, \ldots, f_n$ on A, some n, so that for all a, b in A,

$$aRb \Leftrightarrow (\forall i)[\, f_i(a) > f_i(b) \,]. \tag{5.1}$$

That is, is the person acting as if he measures "preference" on each of a set of dimensions, and expresses preference for an alternative if and only if it receives a higher score on each dimension? The theory of dimension of strict partial orders (Section 1.6, Exers. 12 through 18) is relevant to this problem. Show the following:
   (a) If there are functions $f_1, f_2, \ldots, f_n$ satisfying Eq. (5.1), then (A, R) is the intersection of n strict weak orders (A, R_i) and (A, R) is a strict partial order.

(b) If A is finite and (A, R) is the intersection of a set of n strict weak orders on A, then (A, R) is the intersection of a set of n strict simple orders on A, provided $n > 1$. Hence, if A is finite and there are functions $f_1, f_2, \ldots, f_n$ satisfying (5.1), with $n > 1$, then the dimension of the strict partial order (A, R) is at most n.

(c) Conversely, if A is finite and (A, R) is a strict partial order of dimension at most n, $n > 1$, there are functions $f_1, f_2, \ldots, f_n$ satisfying (5.1).

(d) Hence, for every strict partial order on a finite set, there are functions $f_1, f_2, \ldots, f_n$ satisfying (5.1), for some n. (That is, every person's preferences can be given a product structure, provided they satisfy the strict partial order assumptions.) The smallest n for which there are such functions is the dimension of the strict partial order (A, R), except that some two-dimensional strict partial orders can be represented in the form (5.1) with $n = 1$.

For a further measurement-theoretic discussion of dimension of strict partial orders, see Baker, Fishburn, and Roberts [1971] and Roberts [1972].

Note: Variants of the representation (5.1) are of interest in measurement theory. We might ask for functions f_i such that

$$aRb \Leftrightarrow [f_1(a) > f_1(b)] \text{ or } [f_1(a) = f_1(b) \ \& \ f_2(a) > f_2(b)] \text{ or } \ldots$$
$$\text{or } [f_1(a) = f_1(b) \ \& \ \ldots \ \& \ f_{n-1}(a) = f_{n-1}(b) \ \& \ f_n(a) > f_n(b)].$$

This is a lexicographic representation. In the footnote on page 269, we shall mention a representation that has as a special case

$$aRb \Leftrightarrow (\forall i)\big[|f_i(a) - f_i(b)| \leqq \delta\big].$$

These are all representations into a product structure rather than into the reals, with R corresponding to some relation on the product structure. Not much progress has been made in studying any of these representations. However, they are potentially very useful.

5.2 Calculating Ordinal Utility Functions by Reducing the Dimensionality

In this chapter, we shall study a binary relation R on a set of alternatives A which has a product structure. Often R will be interpreted as preference. However, sometimes it will have the interpretation "responds more strongly than," "sounds louder than," "is scored higher than," etc. We shall use the notation

$$aSb \Leftrightarrow \sim bRa \tag{5.2}$$

and

$$aEb \Leftrightarrow \sim aRb \ \& \ \sim bRa. \tag{5.3}$$

If R is strict preference, then S is weak preference and E is indifference.*
In this section, we mention some tools for calculating a function f on A
which preserves the relation R, that is, a function f on A so that for all
$(a_1, a_2, \ldots, a_n)$ and $(b_1, b_2, \ldots, b_n)$ in A,

$$(a_1, a_2, \ldots, a_n)R(b_1, b_2, \ldots, b_n) \Leftrightarrow f(a_1, a_2, \ldots, a_n) > f(b_1, b_2, \ldots, b_n). \tag{5.4}$$

In case R is preference, f is an ordinal utility function. In Section 5.4, we
shall ask more of our utility function, namely that it be additive, that is,
that there be real-valued functions f_1 on A_1, f_2 on A_2, $\ldots$, up to f_n on A_n
so that

$$f(a_1, a_2, \ldots, a_n) = f_1(a_1) + f_2(a_2) + \cdots + f_n(a_n) \tag{5.5}$$

or

$$(a_1, a_2, \ldots, a_n)R(b_1, b_2, \ldots, b_n)$$
$$\Leftrightarrow \tag{5.6}$$
$$f_1(a_1) + f_2(a_2) + \cdots + f_n(a_n) > f_1(b_1) + f_2(b_2) + \cdots + f_n(b_n).$$

If such functions f_i exist, then we may calculate utility by calculating it
separately on each component, and adding. In Section 5.5, we shall again
ask for an f that can be calculated separately on each component, but we
shall not assume that it is obtained from the individual component values
by addition, but rather by other "composition rules."

There are times when the number of dimensions used in a decision
problem might number in the thousands.† In order to calculate a utility
function f, it is helpful to try to reduce the number of dimensions.
Sometimes a simple technique works. We illustrate it first in the two-dimensional case, $A = A_1 \times A_2$. Fix an element y^* in A_2. (This could be the
"best" or "worst" possibility in A_2, for example 0.) Given $\mathbf{a} = (a_1, a_2)$ in A,
find $x = \pi(\mathbf{a})$ in A_1 so that $\mathbf{a}E(x, y^*)$, where E is indifference and is
defined by Eq. (5.3). Then, assuming that (A, R) is a strict weak order, for
$\mathbf{a} = (a_1, a_2)$ and $\mathbf{b} = (b_1, b_2)$, we have

$$\mathbf{a}R\mathbf{b} \Leftrightarrow [\pi(\mathbf{a}), y^*]R[\pi(\mathbf{b}), y^*].$$

The alternatives of the form (x, y^*) for fixed y^* are essentially one-dimensional, and so we have reduced our problem of calculating a utility

---

*Put another way, aRb iff a is "better than" b, aSb iff a is "at least as good as" b, and aEb
iff a and b are "equally good."
†See Dole *et al.* [1968] for an example.

function on two-dimensional alternatives to the problem of calculating a utility function over one-dimensional alternatives. Of course, this procedure only works if for every $\mathbf{a}$, there always is a $\pi(\mathbf{a})$.

A similar reduction technique works for higher dimensions. We present an example from medical decisionmaking due to Raiffa [1969]. In considering the results of alternative medical treatments, we might consider the following dimensions:

a_1 = amount of money spent for treatment, drugs, etc.

a_2 = number of days in bed with a high index of discomfort.

a_3 = number of days in bed with a medium index of discomfort.

a_4 = number of days in bed with a low index of discomfort.

$$a_5 = \begin{cases} 1 & \text{occurs,} \\ & \text{if complication } A \\ 0 & \text{does not occur.} \end{cases}$$

$$a_6 = \begin{cases} 1 & \text{occurs,} \\ & \text{if complication } B \\ 0 & \text{does not occur.} \end{cases}$$

$$a_7 = \begin{cases} 1 & \text{occurs,} \\ & \text{if complication } C \\ 0 & \text{does not occur.} \end{cases}$$

If a_1, a_4, a_5, a_6, and a_7 are kept fixed, and a_2 is changed to 0, let us ask what value of the third component will compensate for this change, i.e., for what value a_3' is

$$(a_1, a_2, a_3, a_4, a_5, a_6, a_7)$$

judged indifferent to

$$(a_1, 0, a_3', a_4, a_5, a_6, a_7).$$

That is, we think about how many days at medium level of discomfort we would trade for a given number of days at high discomfort, all other things being equal. In the same way, we find a_3'' and a_3''' so that

$$(a_1, 0, a_3', a_4, a_5, a_6, a_7) E(a_1, 0, a_3'', 0, a_5, a_6, a_7)$$

and

$$(a_1, 0, a_3'', 0, a_5, a_6, a_7) E(0, 0, a_3''', 0, a_5, a_6, a_7).$$

We might think of a_3''' as representing a certain number of days in bed with medium discomfort which corresponds to the vector (a_1, a_2, a_3, a_4). In any case, we have reduced from seven dimensions to four dimensions. In Section 7.3.1, we shall show how for this example one can reduce to two dimensions. For recent work on the use of indifference judgments to reduce dimensions and otherwise simplify the assessment of multidimensional utility functions, see MacCrimmon and Siu [1974], MacCrimmon and Wehrung [1978], or Keeney [1971, 1972].

Exercise

(Raiffa [1969]) Suppose (x_1, x_2) represents an amount of cash (say salary) incoming in two time periods, 1 and 2. Money incoming in period 1 can be reinvested to produce more money. Also, consumption now is (often considered) sweeter than consumption later. Thus, we might find a constant λ so that (x_1, x_2) is judged indifferent to $(x_1 + \lambda x_2, 0)$. The number λ can be thought of as the (subjective) *discount rate*. Find an (explicit) utility function over the set of pairs (x_1, x_2). (*Note*: The oversimplification here is that the discount rate λ is constant. In general, we will have

$$(x_1, x_2)E[x_1 + \lambda(x_1, x_2)x_2, 0],$$

where $\lambda(x_1, x_2)$ is a variable discount rate. Since we can usually get a greater return on investment for large amounts invested, $\lambda(x_1, x_2)$ might decrease for fixed x_1 as x_2 increases.)

5.3 Ordinal Utility Functions over Commodity Bundles

Suppose as in the previous section that R is a binary relation on a set of alternatives A which has a product structure. In this section, we shall usually think of R as strict preference. We shall seek conditions on (A, R) sufficient for the existence of an ordinal utility function, a real-valued function f on A which satisfies

$$(a_1, a_2, \ldots, a_n)R(b_1, b_2, \ldots, b_n) \Leftrightarrow f(a_1, a_2, \ldots, a_n) > f(b_1, b_2, \ldots, b_n).$$
$$(5.4)$$

Of course, the Birkhoff–Milgram Theorem (Theorem 3.4, Corollary 1) gives conditions on (A, R) necessary and sufficient for the existence of such a function f. But we shall want conditions that are more special to a product structure.

We state a representation theorem for the representation (5.4) for the case where A has a numerical product structure, that is, each A_i is a set of real numbers.* For simplicity, we assume that $A = Re^n$; that is, we allow

*We follow Luce and Suppes [1965, p. 259], who credit Wold and Jureen [1953] and Uzawa [1960].

all numerical values on each dimension. Essentially the same theorem holds if each component of the product is a real interval rather than all of Re.

Suppose $\mathbf{a} = (a_1, a_2, \ldots, a_n)$ and $\mathbf{b} = (b_1, b_2, \ldots, b_n)$ are two elements of A. We say that $\mathbf{a} \geq \mathbf{b}$ if $a_i \geq b_i$ for all i. We say that $\mathbf{a} > \mathbf{b}$ if $\mathbf{a} \geq \mathbf{b}$ and $\mathbf{a} \neq \mathbf{b}$, that is, $a_i \geq b_i$ for all i and $a_i > b_i$ for some i. Also, $\mathbf{a} + \mathbf{b}$ is the vector

$$(a_1 + b_1, a_2 + b_2, \ldots, a_n + b_n),$$

and if λ is a real number, then $\lambda\mathbf{a}$ is the vector

$$(\lambda a_1, \lambda a_2, \ldots, \lambda a_n).$$

The relation (A, R) is said to satisfy the *dominance condition* if, whenever $\mathbf{a} > \mathbf{b}$, then $\mathbf{a}R\mathbf{b}$. (We have already encountered a related condition in Section 1.6.) In the case of preference, not all scaling procedures lead to product structures satisfying dominance. For example, in the case of the Northeast Corridor, let us consider travel time a_1 associated with a particular transportation mode. You certainly like small a_1 better than large a_1. This can of course be remedied by using $1/$(travel time) as the value a_1. But there are more serious problems. If you live very close to your relatives—that is, if a_1 is small—you can visit often and with little expense. If a_1 is in the middle range, say 6 hours or so by car, your relatives still expect to see you often, but it is now quite an expense to make lots of trips and it is quite time-consuming, for a visit is worthwhile only if it lasts several days. If a_1 is large, say 18 hours by car, you cannot visit often, maybe just once or twice a year. But then annually, travel expenses are not so bad. In short, it is quite conceivable that you will like small a_1 best, large a_1 next best, and moderate a_1 least. Another example is due to Miller [1969]. Consider the problem of scratching an itchy portion of skin. "For a while, continued scratching is preferable to discontinued scratching; but if the scratching process is continued too long or too intensively, it is preferable to scratch more lightly or to discontinue the process altogether." Thus, for example, if a_k is "time spent scratching," then moderate a_k is preferred to small a_k, but large a_k is worst of all. As Keeney and Raiffa [1976] point out, a similar situation arises with the level of blood sugar in the body. Below a "normal" level, higher blood sugar levels are preferred. Above a "normal" level, lower levels are preferred.

There are many examples of numerical product structures that do seem to give rise to preference relations satisfying dominance. An example is where A is a set of alternative commodity bundles (market baskets): there are n products and a_i is the quantity of the ith product in your bundle. Then it seems reasonable to assume that preference satisfies the dominance condition over commodity bundles. Having 4 lamps is better than having

2, having 10 lamps is better than having 4, and having 150 lamps is better than having 10. This represents a fundamental assumption of economics: you can never be satiated with a good. Thus, the dominance condition is sometimes called *nonsatiety* or *nonsaturation*. The lexicographic ordering on the plane (Section 3.1.4) is another example of a binary relation satisfying dominance. However, even though lexicographic preference is also a strict weak order (even strict simple), we know that there is no ordinal utility function (Section 3.1.4). Thus, we shall have to add still another assumption.

To state a representation theorem, we define E on A by Eq. (5.3).

THEOREM 5.1. *Suppose* $A = Re^n$ *and* R *is a binary* (*preference*) *relation on* A. *Then there is a* (*utility*) *function* f *on* A *satisfying*

$$\mathbf{a}R\mathbf{b} \Leftrightarrow f(\mathbf{a}) > f(\mathbf{b}) \tag{5.7}$$

whenever (A, R) *satisfies the following conditions*:

(a) (A, R) *is a strict weak order.*

(b) (A, R) *satisfies the dominance condition.*

(c) (*Continuity*): *If* $\mathbf{a}R\mathbf{b}$ *and* $\mathbf{b}R\mathbf{c}$, *then there is a real number* λ *such that* $0 \leqq \lambda \leqq 1$ *and*

$$[\lambda\mathbf{a} + (1 - \lambda)\mathbf{c}]E\mathbf{b}.$$

To understand the third condition, let us observe that $\lambda\mathbf{a} + (1 - \lambda)\mathbf{c}$ is a point on the straight line joining $\mathbf{a}$ and $\mathbf{c}$. In the case of preference, the assertion is that the "indifference curve" of all elements to which element $\mathbf{b}$ is judged indifferent is continuous and hence intersects the straight line joining $\mathbf{a}$ and $\mathbf{c}$. (See Fig. 5.2.) Theorem 5.1 says that we can assume that there is an ordinal utility function so long as preferences satisfy the conditions (a), (b), and (c). We have already discussed the dominance condition. The continuity condition is in fact similar to the condition in the Birkhoff–Milgram Theorem which asserts the existence of a countable order-dense subset. However, the continuity condition has a simpler economic interpretation.

To prove Theorem 5.1, let us say that $\mathbf{a}$ is a *diagonal vector* if $a_i = a_1$, all i. We show that for all $\mathbf{a}$ there is a unique diagonal $\mathbf{a}^*$ such that $\mathbf{a}E\mathbf{a}^*$. Assuming that this is true, we shall define f as follows. If $\mathbf{a}$ is a diagonal vector, let $f(\mathbf{a}) = a_1$. If $\mathbf{a}$ is not a diagonal vector, let $f(\mathbf{a}) = f(\mathbf{a}^*)$. We show that

$$f(\mathbf{a}^*) > f(\mathbf{b}^*) \Leftrightarrow \mathbf{a}^* > \mathbf{b}^*, \tag{5.8}$$

$$\mathbf{a}^* R \mathbf{b}^* \Leftrightarrow \mathbf{a}^* > \mathbf{b}^*, \tag{5.9}$$

and

$$\mathbf{a}^* R \mathbf{b}^* \Leftrightarrow \mathbf{a}R\mathbf{b}. \tag{5.10}$$

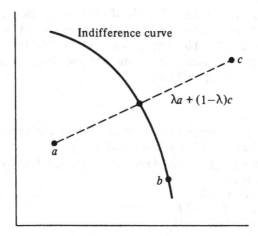

Figure 5.2. The continuity axiom.

These three conditions imply (5.7). Now,

$$f(\mathbf{a}^*) > f(\mathbf{b}^*) \Leftrightarrow (\mathbf{a}^*)_1 > (\mathbf{b}^*)_1$$
$$\Leftrightarrow \mathbf{a}^* > \mathbf{b}^*,$$

which proves (5.8). Using dominance we find that

$$\mathbf{a}^* > \mathbf{b}^* \Rightarrow \mathbf{a}^* R \mathbf{b}^*.$$

Next,

$$\sim (\mathbf{a}^* > \mathbf{b}^*) \Rightarrow [\mathbf{b}^* > \mathbf{a}^* \text{ or } \mathbf{b}^* = \mathbf{a}^*] \quad \text{(since } \mathbf{a}^*, \mathbf{b}^* \text{ are diagonal)}$$
$$\Rightarrow [\mathbf{b}^* R \mathbf{a}^* \text{ or } \mathbf{b}^* = \mathbf{a}^*] \quad \text{(by dominance)}$$
$$\Rightarrow [\mathbf{b}^* R \mathbf{a}^* \text{ or } \mathbf{b}^* E \mathbf{a}^*] \quad (R \text{ is irreflexive, since it is strict weak)}$$
$$\Rightarrow \sim \mathbf{a}^* R \mathbf{b}^* \quad \text{(since } R \text{ is strict weak).}$$

This proves (5.9). Finally, $\mathbf{a}E\mathbf{a}^*$ and $\mathbf{b}E\mathbf{b}^*$, so it is easy to show (5.10), using the properties of a strict weak order.

It is left to prove that for all $\mathbf{a}$ in Re^n there is a unique diagonal $\mathbf{a}^*$ such that $\mathbf{a}E\mathbf{a}^*$. We first prove that there is an $\mathbf{a}^*$. To prove this, let $a_m = \min_i \{a_i\}$, let $a_M = \max_i \{a_i\}$, and let $\mathbf{a}' = (a_m, a_m, \ldots, a_m)$ and $\mathbf{a}'' = (a_M, a_M, \ldots, a_M)$. If $\mathbf{a}' = \mathbf{a}$, let $\mathbf{a}^* = \mathbf{a}'$. If $\mathbf{a}'' = \mathbf{a}$, let $\mathbf{a}^* = \mathbf{a}''$. If $\mathbf{a} \neq \mathbf{a}', \mathbf{a}''$, then by dominance $\mathbf{a}'' R \mathbf{a} R \mathbf{a}'$. By continuity, there is $\lambda \in [0, 1]$ such that

$$\mathbf{a}E[\lambda \mathbf{a}' + (1 - \lambda)\mathbf{a}''].$$

Take $\mathbf{a}^*$ to be $\lambda\mathbf{a}' + (1 - \lambda)\mathbf{a}''$. Finally, if $\mathbf{a}^{**}$ is any other diagonal vector such that $\mathbf{a}E\mathbf{a}^{**}$, we have $\mathbf{a}^*E\mathbf{a}$ & $\mathbf{a}E\mathbf{a}^{**}$. Since (A, R) is a strict weak order, E is transitive (by Theorem 1.3) and $\mathbf{a}^*E\mathbf{a}^{**}$ follows. We conclude by dominance and the fact that $\mathbf{a}^*$ and $\mathbf{a}^{**}$ are diagonal that $\mathbf{a}^* = \mathbf{a}^{**}$. This completes the proof of Theorem 5.1.

Exercises

1. Show that the lexicographic ordering of the plane satisfies conditions (a) and (b) of Theorem 5.1, but not condition (c).

2. Which of the conditions of Theorem 5.1 are satisfied by the following binary relations R on Re^n:
   (a) $\mathbf{a}R\mathbf{b}$ iff $\mathbf{a} > \mathbf{b}$.
   (b) $\mathbf{a}R\mathbf{b}$ iff $\Sigma a_i > \Sigma b_i$.
   (c) $\mathbf{a}R\mathbf{b}$ iff $\Pi a_i > \Pi b_i$.
   (d) $\mathbf{a}R\mathbf{b}$ iff $a_1 > b_1$.
   (e) $\mathbf{a}R\mathbf{b}$ iff $\max a_i > \max b_i$.

3. (a) Show that under the hypotheses of Theorem 5.1, if $\mathbf{a}R\mathbf{b}$ and $\mathbf{b}R\mathbf{c}$, then there is exactly one λ, $0 \leqq \lambda \leqq 1$, so that

$$[\lambda\mathbf{a} + (1 - \lambda)\mathbf{c}]E\mathbf{b}.$$

   (b) However, show that if $\mathbf{a}S\mathbf{b}$ and $\mathbf{b}S\mathbf{c}$, where S is defined by Eq. (5.2), there may be more than one λ, $0 \leqq \lambda \leqq 1$, so that

$$[\lambda\mathbf{a} + (1 - \lambda)\mathbf{c}]E\mathbf{b}.$$

4. Which of the conditions (a) through (c) are necessary for the representation (5.7)?

5. Develop an alternative proof of Theorem 5.1 by showing that under conditions (a) through (c), (A^*, R^*) contains a countable order-dense subset, and then applying Corollary 1 of Theorem 3.4. (This method of proof is presented by Fishburn [1970, p. 32].)

6. We prefer the present proof of Theorem 5.1 because it gives a constructive proof that conditions (a) through (c) guarantee the existence of a continuous utility function, using the usual topology on Re^n and Re; for the function constructed is continuous. Show this. Cf. Exer. 22a, Section 3.1.

5.4 Conjoint Measurement

5.4.1 *Additivity*

Suppose a set of alternatives A has a product structure, i.e., A is $A_1 \times A_2 \times \cdots \times A_n$. If we want to calculate utility of elements of A, it would be easier to calculate utility on each attribute separately, and then add. Specifically, we would like to express an ordinal utility function $f: A \to Re$ as a sum of real-valued functions $f_1, f_2, \ldots, f_n$ on $A_1, A_2, \ldots, A_n$, respectively. That is, we would like

$$f(a_1, a_2, \ldots, a_n) = f_1(a_1) + f_2(a_2) + \cdots + f_n(a_n). \qquad (5.5)$$

A utility function f for which there are functions f_i satisfying (5.5) is called *additive*. More generally, we would like to calculate utility on each attribute separately and then combine in some way. That is, we would like to find real-valued functions $f_1, f_2, \ldots, f_n$ on $A_1, A_2, \ldots, A_n$, respectively, and a function $F: Re^n \to Re$ such that

$$f(a_1, a_2, \ldots, a_n) = F[f_1(a_1), f_2(a_2), \ldots, f_n(a_n)]. \qquad (5.11)$$

In this section, we consider the special case

$$F(x_1, x_2, \ldots, x_n) = x_1 + x_2 + \cdots + x_n,$$

that is, the representation (5.5). We consider other examples of composition rules or functions F in Section 5.5.

Suppose R is the binary relation of strict preference on A. Then we seek real-valued functions f_i on A_i so that for all $\mathbf{a} = (a_1, a_2, \ldots, a_n)$ and $\mathbf{b} = (b_1, b_2, \ldots, b_n)$ in A,

$$\mathbf{a} R \mathbf{b} \Leftrightarrow f_1(a_1) + f_2(a_2) + \cdots + f_n(a_n) > f_1(b_1) + f_2(b_2) + \cdots + f_n(b_n).$$
$$(5.12)$$

Technically, the representation (5.12) does not fit the general framework for fundamental measurement, as described in Section 2.1. We are not seeking a homomorphism from one relational system to another. However, we shall treat the representation (5.12) in the same spirit as the representations studied in Chapter 3, and seek (necessary and) sufficient conditions for the existence of functions f_i satisfying (5.12). The representation (5.12) is often called (*additive*) *conjoint measurement*, because different components are being measured conjointly.

Conjoint measurement has potential applications in areas other than utility. In studying response strength, let D be a set of drives and K a set of incentives. Let R be a binary relation on $D \times K$, with $(d_1, k_1) R(d_2, k_2)$ interpreted as follows: if drive d_1 is coupled with incentive k_1, then the response is stronger than if drive d_2 is coupled with incentive k_2. We seek functions $\delta: D \to Re$ and $\kappa: K \to Re$ so that for all $d_1, d_2 \in D$ and $k_1, k_2 \in K$,

$$(d_1, k_1) R(d_2, k_2) \Leftrightarrow \delta(d_1) + \kappa(k_1) > \delta(d_2) + \kappa(k_2). \qquad (5.13)$$

If we bring in habit strength, we have a third set H of habit strengths, R is a binary relation "responds more strongly than" on $D \times K \times H$, and we seek functions δ, κ, and $\mu\colon H \to Re$ such that for all $d_1, d_2 \in D$, $k_1, k_2 \in K$, and $h_1, h_2 \in H$,

$$(d_1, k_1, h_1)R(d_2, k_2, h_2) \Leftrightarrow \delta(d_1) + \kappa(k_1) + \mu(h_1) > \delta(d_2) + \kappa(k_2) + \mu(h_2).$$
(5.14)

In studies of binaural loudness in auditory perception, let $\mathcal{L}$ be a set of sounds presented to the left ear and $\mathcal{R}$ be a set of sounds presented to the right ear. Let R be a binary relation on $\mathcal{L} \times \mathcal{R}$, with $(\ell_1, r_1)R(\ell_2, r_2)$ interpreted as follows: If sounds ℓ_1 and r_1 are presented simultaneously to the left and right ear, respectively, the result is judged louder than if sounds ℓ_2 and r_2 are presented simultaneously to the left and right ear, respectively. Then, assuming the effects to the two ears are additive,* psychologists seek functions $\ell\colon\mathcal{L} \to Re$ and $r\colon\mathcal{R} \to Re$ so that for all $\ell_1, \ell_2 \in \mathcal{L}$ and $r_1, r_2 \in \mathcal{R}$,

$$(\ell_1, r_1)R(\ell_2, r_2) \Leftrightarrow \ell(\ell_1) + r(r_1) > \ell(\ell_2) + r(r_2).$$
(5.15)

In mental testing, to study the interaction between a subject and an item on a mental test, suppose S is a set of subjects and T a set of items. Let R be a binary relation on $S \times T$, with $(s_1, t_1)R(s_2, t_2)$ interpreted to mean that subject s_1 scores better on item t_1 than subject s_2 does on item t_2. It is tempting to try to assume that subjects and items are independent, and that the score of a subject depends only on his ability and the difficulty of the item, and not on how hard the item is for him personally. Then, we would like ability and difficulty functions $\alpha\colon S \to Re$ and $\delta\colon T \to Re$ so that for all $s_1, s_2 \in S$ and $t_1, t_2 \in T$,

$$(s_1, t_1)R(s_2, t_2) \Leftrightarrow \alpha(s_1) + \delta(t_1) > \alpha(s_2) + \delta(t_2).$$
(5.16)

Finally, in studying discomfort under different weather conditions, suppose T is a set of temperatures being studied and H a set of humidities. Let R be a binary relation on $T \times H$, with $(t_1, h_1)R(t_2, h_2)$ interpreted as follows: A subject is more uncomfortable at temperature t_1 and humidity h_1 than at temperature t_2 and humidity h_2. Then we seek to measure discomfort by a discomfort index or a temperature–humidity index. If the temperature and humidity effects can be separated and added, then we can find functions $\tau\colon T \to Re$ and $\gamma\colon H \to Re$ so that for all $t_1, t_2 \in T$ and $h_1, h_2 \in H$,

$$(t_1, h_1)R(t_2, h_2) \Leftrightarrow \tau(t_1) + \gamma(h_1) > \tau(t_2) + \gamma(h_2).$$
(5.17)

The number $\tau(t) + \gamma(h)$ is the discomfort index.

---

*This assumption, sometimes known as the *loudness summation hypothesis*, has been subjected to extensive discussion. See footnote ‡, p. 154 for references.

We shall seek conditions (necessary and) sufficient for the existence of functions satisfying representations like (5.12) through (5.17). Concentrating on (5.12), let us take $n = 2$ and give an example. Suppose $A_1 = \{\alpha, \beta\}$ and $A_2 = \{x, y\}$, and suppose that preference is a strict simple order in which (α, x) is strictly preferred to (α, y), which is strictly preferred to (β, x), which is strictly preferred to (β, y). Then an ordinal utility function f is given by

$$f(\alpha, x) = 3, \qquad f(\alpha, y) = 2, \qquad f(\beta, x) = 1, \qquad f(\beta, y) = 0.$$

The function f is additive, for we may take

$$f_1(\alpha) = 2, \qquad f_1(\beta) = 0, \qquad f_2(x) = 1, \qquad f_2(y) = 0.$$

Then $f(a_1, a_2) = f_1(a_1) + f_2(a_2)$. Also, of course, the functions f_1 and f_2 satisfy (5.12), for

$$f_1(\alpha) + f_2(x) > f_1(\alpha) + f_2(y) > f_1(\beta) + f_2(x) > f_1(\beta) + f_2(y).$$

Using the same A_1 and A_2, suppose (β, x) is strictly preferred to (β, y) and (α, y) is strictly preferred to (α, x). Then no additive conjoint representation exists. For if one did, then $(\beta, x)R(\beta, y)$ implies

$$f_1(\beta) + f_2(x) > f_1(\beta) + f_2(y),$$

so $f_2(x) > f_2(y)$. However, $(\alpha, y)R(\alpha, x)$ implies

$$f_1(\alpha) + f_2(y) > f_1(\alpha) + f_2(x),$$

so $f_2(y) > f_2(x)$, a contradiction.

Conditions sufficient for additive conjoint measurement were first presented by Debreu [1960]. Some of his conditions were topological in nature. Algebraic sufficient conditions in the spirit of those in Chapter 3 were first presented by Luce and Tukey [1964]. More refined conditions can be found in Krantz et al. [1971, Chapter 6]. In Section 5.4.3, we present these conditions. In Section 5.4.5, we present necessary and sufficient conditions for additive conjoint measurement in the case where each A_i is finite. Without the assumption of finiteness, presentation of such necessary and sufficient conditions is still an open problem.

Before leaving this subsection, we should remark that, in spite of historical prejudice, there is nothing magic about addition as opposed to some other operation. For, just as in the case of extensive measurement, if $f_1, f_2, \ldots, f_n$ satisfy Eq. (5.12), then $g_1 = e^{f_1}, g_2 = e^{f_2}, \ldots, g_n = e^{f_n}$, are positive and satisfy

$$\mathbf{a}R\mathbf{b} \Leftrightarrow g_1(a_1)g_2(a_2) \ldots g_n(a_n) > g_1(b_1)g_2(b_2) \ldots g_n(b_n). \qquad (5.18)$$

Conversely, if $g_1, g_2, \ldots, g_n$ are positive and satisfy Eq. (5.18), then $f_1 = \ln g_1, f_2 = \ln g_2, \ldots, f_n = \ln g_n$ satisfy Eq. (5.12). The multiplicative representation (5.18) is often more useful than the additive one. For example, in physics, momentum has a multiplicative representation, as mass times velocity.

5.4.2 Conjoint Measurement and the Balance of Trade*

As an aside, we present in this subsection an amusing application of additive conjoint measurement, which is based on an idea of David Gale.[†] Imagine that there are two countries, each of which produces just one product. In Country A, the average individual receives 4 units of this product (measured, for example, in dollars) in his lifetime, split evenly into 2 units in the first half of his life and 2 units in the second half of his life. In Country B, the average individual also receives 4 units of the product in his lifetime, but 1 in the first half and 3 in the second half. Now individuals in Country B live a "deprived" youth and a "wealthy" old age. They might be willing to trade some of the wealth they expect in their old age for more in their youth. Individuals in Country A might be willing to part with a small portion of the wealth they receive in their youth, if they are compensated in their old age, with interest, thus achieving a lifetime increase in wealth.

In general, if a represents the amount of the product obtained in the first half of an individual's life and b represents the amount obtained in the second half, a pair (a, b) represents a possible distribution.[‡] According to our argument, individuals in Country B might prefer $(4/3, 5/2)$ to $(1, 3)$ and individuals in Country A might prefer $(5/3, 5/2)$ to $(2, 2)$. Even if there is no unequal treatment of the two time periods, these preferences can be accounted for by an additive utility function, namely

$$f(a, b) = \log_{10} a + \log_{10} b.$$

For then $f(4/3, 5/2) = .523$, $f(1, 3) = .477$, $f(5/3, 5/2) = .620$, and $f(2, 2) = .602$. (Such a utility function is not unreasonable, at least if a is measured in dollars. The idea that utility of a dollars is a logarithmic function of a goes back to Daniel Bernoulli (Stevens [1959, p. 47]). We discussed this point in Section 4.3.4 and return to it in Chapter 7.)

This highly idealized example is used by Gale to argue that a regular negative balance of trade is not necessarily a bad state of affairs. The average individual in Country B trades a $(1, 3)$ distribution for a $(4/3, 5/2)$ distribution, thus having a negative balance of trade of

---

*This subsection may be omitted without loss of continuity.
†Presented to the American Association for the Advancement of Science, January 1975.
‡Compare the exercise of Section 5.2.

$1 + 3 - \frac{4}{3} - \frac{5}{2} = \frac{1}{6}$ units. If it is assumed that both populations remain fixed, Country B can have a perpetual negative trade balance, and yet be happy with it.

5.4.3 *The Luce–Tukey Theorem*

In this section, we present the Krantz *et al.* [1971] refinement of the Luce–Tukey theorem, which gives sufficient conditions for additive conjoint measurement. We shall consider the case $n = 2$, namely the representation

$$(a_1, a_2)R(b_1, b_2) \Leftrightarrow f_1(a_1) + f_2(a_2) > f_1(b_1) + f_2(b_2). \qquad (5.19)$$

We begin by stating some necessary conditions. If Eq. (5.19) holds, (A, R) must be a strict weak order. We shall want to assume that. If a and b have the same second component, then it follows from Eq. (5.19) that whether or not aRb holds does not depend on the second component. That is, for all $x, y \in A_1$ and $q, r \in A_2$,

$$(x, q)R(y, q) \Leftrightarrow (x, r)R(y, r). \qquad (5.20)$$

For, we have

$$f_1(x) + f_2(q) > f_1(y) + f_2(q) \Leftrightarrow f_1(x) > f_1(y)$$
$$\Leftrightarrow f_1(x) + f_2(r) > f_1(y) + f_2(r).$$

Similarly, for all $x, y \in A_1$ and $q, r \in A_2$,

$$(x, q)R(x, r) \Leftrightarrow (y, q)R(y, r). \qquad (5.21)$$

We say that a binary relation $(A_1 \times A_2, R)$ satisfies *independence on the first component* if (5.20) holds for all $x, y \in A_1$ and $p, q \in A_2$, *independence on the second component* if (5.21) holds for all $x, y \in A_1$ and $p, q \in A_2$, and *independence* if both of these conditions hold.

We once again define the relations S and E from R by Eqs. (5.2) and (5.3), respectively:

$$aSb \Leftrightarrow \; \sim bRa \qquad (5.2)$$

and

$$aEb \Leftrightarrow \; \sim aRb \; \& \; \sim bRa. \qquad (5.3)$$

The reader will recall that if R is strict preference, then S is weak preference and E is indifference. A binary relation $(A_1 \times A_2, R)$ satisfies the *Thomsen condition* if, for all $x, y, z \in A_1$ and $q, r, s \in A_2$,

$$(x, s)E(z, r) \ \& \ (z, q)E(y, s) \Rightarrow (x, q)E(y, r).$$

The Thomsen condition also follows from the representation (5.19). For,

$$(x, s)E(z, r) \Leftrightarrow f_1(x) + f_2(s) = f_1(z) + f_2(r)$$

and

$$(z, q)E(y, s) \Leftrightarrow f_1(z) + f_2(q) = f_1(y) + f_2(s).$$

By adding the right-hand sides and canceling, we see

$$f_1(x) + f_2(q) = f_1(y) + f_2(r),$$

so

$$(x, q)E(y, r).$$

This condition will be more clearly understood if we define a binary relation D on $A_1 \times A_2$ by

$$(x, p)D(y, q) \Leftrightarrow (x, q)E(y, p).$$

Then the Thomsen condition says that D is transitive.

We mentioned in Section 3.2 that most measurement axiomatizations have some form of Archimedean condition. The next condition we introduce is the Archimedean condition. To do this, we need to introduce some notation. A binary relation R on a product $A_1 \times A_2$ induces binary relations R_1 and R_2 on the components A_1 and A_2, respectively, as follows:

$$xR_1y \Leftrightarrow (\exists q \in A_2)[(x, q)R(y, q)] \tag{5.22}$$

and

$$qR_2r \Leftrightarrow (\exists x \in A_1)[(x, q)R(x, r)]. \tag{5.23}$$

It is easy to see that if $(A_1 \times A_2, R)$ is a strict weak order and satisfies independence, then (A_i, R_i) is a strict weak order for $i = 1, 2$. The proof is left to the reader. We may define S_i and E_i from R_i by equations analogous to (5.2) and (5.3), namely by

$$xS_iy \Leftrightarrow \sim yR_ix \tag{5.24}$$

and

$$xE_iy \Leftrightarrow \sim xR_iy \ \& \sim yR_ix. \tag{5.25}$$

If R is strict preference, then R_i is an induced strict preference relation on the ith component, S_i is weak preference on the ith component, and E_i is indifference on the ith component.

To state our Archimedean condition, we introduce the notion of a standard sequence, a sequence of equally spaced elements on one of the components. This is a notion we encountered previously in our axiomatization of difference measurement in Section 3.3. Suppose $(A_1 \times A_2, R)$ is an independent strict weak order. Let N be a (finite or infinite) set of consecutive integers (positive or negative). The sequence

$$\{x_i\colon x_i \in A_1, i \in N\}$$

is a *standard sequence* (on the first component) if there are $q, r \in A_2$ such that $\sim qE_2 r$ and for all $i, i + 1 \in N$,

$$(x_i, q)E(x_{i+1}, r).$$

The standard sequence is *strictly bounded* if there are x and y in A_1 such that $xR_1 x_i$ and $x_i R_1 y$ for all $i \in N$. Similar definitions apply on the second component. The idea of a standard sequence is that the difference between two successive elements is the same. This follows from the representation (5.19), since

$$(x_i, q)E(x_{i+1}, r) \Rightarrow f_1(x_{i+1}) - f_1(x_i) = f_2(q) - f_2(r).$$

The Archimedean condition on the real numbers can be restated to say that if we constantly add the same (nonzero) amount, then we eventually overstep any bound. Consequently, our Archimedean axiom will state that any strictly bounded standard sequence (on either component) is finite. This follows from the representation (5.19). For, suppose $\{x_i\}$ is an infinite standard sequence (on the first component). Without loss of generality take $N =$ the set of all integers. We have already seen that

$$f_1(x_{i+1}) - f_1(x_i) = f_2(q) - f_2(r).$$

Thus if $n > 0$,

$$f_1(x_n) = f_1(x_0) - n[f_2(r) - f_2(q)].$$

Since $\sim qE_2 r$, we have $f_2(r) - f_2(q) \neq 0$. If $f_2(r) - f_2(q) > 0$, fix any $x \in A_1$. By the Archimedean condition on the reals, there is a positive integer n such that

$$f_1(x_0) - n[f_2(r) - f_2(q)] < f_1(x),$$

so $xR_1 x_n$. Similarly, if $f_2(r) - f_2(q) < 0$, there is for each $x \in A_1$ a positive integer n such that $x_n R_1 x$. Thus, $\{x_i\}$ is not strictly bounded.

The next two conditions we consider are no longer necessary; that is, they do not follow from the representation (5.19). To the author's knowledge, the problem of finding conditions on $(A_1 \times A_2, R)$ that are both necessary and sufficient for the representation (5.19) has not been solved in general. It has been solved for the case where A is finite by Scott [1964]; Scott's result is stated in Section 5.4.5. A nonstandard generalization is provided by Narens [1974].

We say that the binary relation $(A_1 \times A_2, R)$ satisfies *restricted solvability* (on the first component) if, whenever $x, \bar{y}, \underline{y} \in A_1$ and $q, r \in A_2$ and

$$(\bar{y}, r)R(x, q)R(\underline{y}, r),$$

then there exists $y \in A_1$ such that $(y, r)E(x, q)$. A similar definition holds on the second component. The term solvability is usually used for a condition which says that certain equations can be solved. A *solvability* axiom on the first component would take the form: given $x \in A_1$, and q and $r \in A_2$, there is y such that $(x, q)E(y, r)$; and similarly on the second component. We only require solvability under certain restrictions, hence the terminology restricted solvability. It is easy to see that even restricted solvability is not a necessary condition for the representation (5.19). For, let $A_1 = \{1, 3\}$ and $A_2 = \{6, 7\}$, define $f(a_1, a_2) = a_1 + a_2$, and define

$$(a_1, a_2)R(b_1, b_2) \Leftrightarrow f(a_1, a_2) > f(b_1, b_2).$$

Then $(A_1 \times A_2, R)$ is trivially representable, but restricted solvability fails, since

$$(3, 6)R(1, 7)R(1, 6)$$

and there is no $y \in A_1$ such that $(y, 6)E(1, 7)$.

Given an independent strict weak order $(A_1 \times A_2, R)$, we say that the ith component $(i = 1, 2)$ is *essential* if there are $x, y \in A_i$ such that $\sim xE_i y$. We shall assume that each component is essential. This is of course not necessary. Although essentiality is simply a nontriviality assumption, it plays an important role. We shall see in Exer. 19 that if all the other assumptions hold and essentiality fails, then our representation theorem fails also.

If $(A, R) = (A_1 \times A_2, R)$ is a binary relation, we shall say that it is an *additive conjoint structure* if it satisfies the following conditions:

AXIOM C1. *(A, R) is a strict weak order.*

AXIOM C2. *(A, R) satisfies independence.*

AXIOM C3. *(A, R) satisfies the Thomsen condition.*

AXIOM C4. *Every strictly bounded standard sequence on either component is finite.*

AXIOM C5. *Restricted solvability holds (on each component).*

AXIOM C6. *Each component is essential.*

THEOREM 5.2 (Luce and Tukey). *Suppose $(A_1 \times A_2, R)$ is an additive conjoint structure. Then there exist real-valued functions f_1 on A_1 and f_2 on A_2 such that for all (a_1, a_2) and $(b_1, b_2) \in A_1 \times A_2$,*

$$(a_1, a_2)R(b_1, b_2) \Leftrightarrow f_1(a_1) + f_2(a_2) > f_1(b_1) + f_2(b_2). \qquad (5.19)$$

Moreover, if $f_1{}'$ and $f_2{}'$ are two other real-valued functions on A_1 and A_2, respectively, with the same property, then there are real numbers α, β, and γ, with $\alpha > 0$, such that

$$f_1{}' = \alpha f_1 + \beta, \quad f_2{}' = \alpha f_2 + \gamma.^*$$

This theorem is proved in Krantz et al. [1971, p. 275]. We shall omit the proof. It is instructive, however, to sketch the proof of a much simpler theorem than Theorem 5.2. We do that in the next subsection.

Before leaving this subsection, let us discuss Axioms C1 through C6 as axioms for preference among multidimensional alternatives. We have already discussed in Section 3.1 the assumption that preference is a strict weak order. The second axiom, independence, can be questioned. You might prefer ginger ale to coffee if you are having only a beverage, but prefer coffee and a doughnut to ginger ale and a doughnut. Even with market baskets, you can question independence. Suppose the first component is coffee and the second sugar. You might prefer $(1, 1)$ to $(0, 1)$ and $(0, 0)$ to $(1, 0)$ if you have a violent dislike of coffee without sugar and so would just have to find a place to dispose of it.

Keeney and Raiffa [1976] give another example. A farmer's preferences for various combinations of sunshine and rain will probably violate independence. For at one level of rain, the farmer might prefer more sunshine, whereas at another level of rain, he might prefer less.

Independence can also be violated in the mental testing situation. For example, if subject s_1 is good at arithmetic and subject s_2 is good at vocabulary, and if item t_1 is an arithmetic item and item t_2 is a vocabulary item, we might very well get $(s_1, t_1)R(s_2, t_1)$ but not $(s_1, t_2)R(s_2, t_2)$.

---

*Thus, in particular, each f_i is a (regular) interval scale. This uniqueness result does not hold for all systems $(A_1 \times A_2, R)$ representable in the form (5.19). See Exer. 24. In general, it would be interesting to know under what conditions on a representable $(A_1 \times A_2, R)$ one can obtain this uniqueness result.

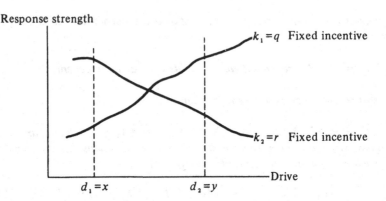

Figure 5.3. Curves of response strength versus drive for different levels of incentive. Cross-over interaction violates independence.

Suppose we plot the first variable A_1 on the horizontal axis and the strength of response (as determined by a measure preserving the relation R) on the vertical axis. Let us consider curves of response strength over varying a_1 in A_1 and fixed a_2 in A_2. If these curves cross over, then there is a violation of independence. For example, in the drive–incentive situation pictured in Fig. 5.3, there is a violation of independence, for

$$(x, r)R(x, q) \quad \text{and} \quad (y, q)R(y, r).$$

The third axiom, the Thomsen condition, is not very intuitive, and so it is hard to determine by a "thought experiment" whether or not preference on multidimensional alternatives would satisfy it. However, Levelt *et al.* [1972] have done an experimental test of a more general condition than the Thomsen condition, in the binaural loudness situation. This more general condition is called *double cancellation*, and it says that for all $x, y, z \in A_1$ and $q, r, s \in A_2$,

$$(x, s)S(z, r) \ \& \ (z, q)S(y, s) \Rightarrow (x, q)S(y, r).$$

It is easy to show that double cancellation follows from the representation (5.19) and that if (A, R) is strict weak, the Thomsen condition follows from double cancellation. In the Levelt *et al.* experiment, binaural pairs of stimuli (ℓ_1, r_1) were ordered according to loudness, assuming the independence axiom. Then double cancellation was tested on this ordering, and it was found to be satisfied by more than 98% of the cases x, y, z and q, r, s. Coombs and Komorita [1958] tested the double cancellation axiom using preferences among gambles. A more recent test of the double cancellation axiom was carried out by Wallsten [1976] in the context of information processing under uncertainty.

At first glance, the fourth axiom, the Archimedean axiom, seems reasonable, at least if A is thought of as a collection of market baskets or commodity bundles. However, if A is not a collection of market baskets, there is some question about the axiom. In Section 3.2, we presented an argument against an Archimedean axiom. The argument was that no number of lamps could compensate for a long, healthy life. This argument seems to apply here as well. Thus if a is having a long, healthy life and a' is having a short, sickly life, and if n is having n lamps, then $(a, 1)$ might be preferred to (a', n), for all n. This violates a form of the Archimedean axiom. Exercise 13 investigates this example further, and asks the reader to show that it does not violate the form of the Archimedean axiom we have introduced. However, as Exer. 14 points out, a modification of this example does violate a variant of our Archimedean axiom.

A related example is given by David Krantz.* If ℓ is being a lawyer and d is being a dishwasher, and if n is receiving a "bribe" of n dollars, then it is possible that $(\ell, 0)$ is preferred to every (d, n); a person might not want to be a dishwasher no matter what you bribed him. As usual, it is impossible to "verify" an Archimedean axiom by test, since there would need to be infinitely many tests made.

The fifth axiom, restricted solvability, probably is a reasonable one for commodity bundles. If $(\bar{y}, r)$ is preferred to (x, q), which is preferred to (y, r), then some amount y in between $\bar{y}$ and y, combined with r, should be judged equally preferable to (x, q).† For other interpretations, this axiom can be questioned. Suppose r and q are vehicle designs, and x is a power source. It is quite possible we can find a power source $\bar{y}$ for vehicle design r which makes a system preferable to the system with source x and vehicle design q, and another power source y for design r which makes the system less preferable to the system (x, q), but for there to be no power source y for design r for which (y, r) and (x, q) are judged equally preferable.

Finally, the last axiom, essentiality, is introduced to avoid trivial situations.

Experimental tests of the conjoint measurement axioms C1 through C6 have concentrated on the independence axiom and on the double cancellation axiom (hence the Thomsen condition). As Falmagne [1976] points out, most experimental tests of measurement axioms such as those for additive conjoint measurement are likely to lead to a few violations. Simply counting number of violations does not give us a good feel for fit of a model, and little statistical theory is available for making tests of fit. Also, some violations may be genuinely indicative of a systematic statistical

*Seminar on "Testability of Measurement Axioms," Center for Advanced Study in the Behavioral Sciences, Palo Alto, California, January, 1971.
†A stronger solvability axiom might not hold for commodity bundles. For example, in the coffee–sugar example discussed above, there is probably no y so that
$$(1,1)E(y,0).$$

departure from the model. What is called for, according to Falmagne, is a statistical or random analogue of a measurement theory. He develops such an analogue for conjoint measurement, and uses it to make a preliminary test of the double cancellation axiom. The test rejects additivity, but the results are too limited in scope to permit a definite conclusion. See Falmagne [1979] and Falmagne, Iverson, and Marcovici [1978] for additional results. Further work along these lines is in progress, and more is certainly needed.

The additive representation (5.19 or 5.13 through 5.17) can be tested directly. In one such test, computer programs are used to "fit" the best possible additive representation, and then statistical tests are used to see if the data is accounted for by this additive representation. Using this procedure in the binaural loudness situation, Levelt et al. [1972] discovered that an additive representation fit their data very well. Tversky [1967] used a similar procedure in testing additivity of choices under risk.

5.4.4 Conjoint Measurement under Equal Spacing

To give the reader a feel for the conjoint measurement representation, we state a simpler theorem than Theorem 5.2, and sketch a proof. Suppose $(A_1 \times A_2, R)$ is an independent strict weak order. Define binary relations J_i $(i = 1, 2)$ on A_i by

$$a_i J_i b_i \Leftrightarrow a_i R_i b_i \ \& \ (\forall c_i \in A_i)(c_i S_i a_i \ \text{ or } \ b_i S_i c_i \ \text{ but not both}),$$

where S_i is defined in Eq. (5.24). Note that $a_i J_i b_i$ holds if a_i is strictly preferred to b_i and there is no element c_i strictly in between in the strict weak order R_i. If $a_i J_i b_i$, let us call (a_i, b_i) a J_i-interval. We say that the structure $(A_1 \times A_2, R)$ is equally spaced if for all $x, y \in A_1$ and $q, r \in A_2$,

$$x J_1 y \ \& \ q J_2 r \Rightarrow (x, r) E(y, q). \tag{5.26}$$

Condition (5.26) says that the length of any J_1-interval is the same as that of any J_2-interval, and so all J_i $(i = 1, 2)$ intervals have the same length. These results follow, since $(x, r)E(y, q)$ means that

$$f_1(x) + f_2(r) = f_1(y) + f_2(q),$$

so

$$f_1(x) - f_1(y) = f_2(q) - f_2(r).$$

We say that $(A_1 \times A_2, R)$ is an *equally spaced additive conjoint structure* if it satisfies the following conditions:

AXIOM E1. *Strict weak order.*

AXIOM E2. *Independence.*

AXIOM E3. *Equal spacing.*

THEOREM 5.3. *Suppose $(A_1 \times A_2, R)$ is an equally spaced additive conjoint structure and A_1 and A_2 are finite. Then there exist real-valued functions f_1 on A_1 and f_2 on A_2 such that for all (a_1, a_2) and (b_1, b_2) belonging to $A_1 \times A_2$,*

$$(a_1, a_2)R(b_1, b_2) \Leftrightarrow f_1(a_1) + f_2(a_2) > f_1(b_1) + f_2(b_2). \tag{5.19}$$

Moreover, if f_1' and f_2' are two other functions on A_1 and A_2, respectively, with the same property, then there are real numbers α, β, and γ, with $\alpha > 0$, such that

$$f_1' = \alpha f_1 + \beta, \quad f_2' = \alpha f_2 + \gamma.$$

We sketch the proof much as do Krantz *et al.* [1971, pp. 36, 37]. Let A_1 have m elements and let A_2 have n elements. Assume first that (A_1, R_1) and (A_2, R_2) are both strict simple orders. Since A_1 and A_2 are finite, we may list their elements in increasing R_i-order. That is, we may suppose that $A_1 = \{x_1, x_2, \ldots, x_m\}$ with $x_m R_1 x_{m-1} R_1 \ldots R_1 x_1$, and that $A_2 = \{q_1, q_2, \ldots, q_n\}$ with $q_n R_2 q_{n-1} R_2 \ldots R_2 q_1$. We first observe that

$$i + j = k + \ell \Rightarrow (x_i, q_j)E(x_k, q_\ell). \tag{5.27}$$

The proof is by the equal-spacing assumption (Eq. 5.26) and mathematical induction and is left to the reader. Next, note that

$$i + j > k + \ell \Rightarrow (x_i, q_j)R(x_k, q_\ell). \tag{5.28}$$

For $k + \ell = i + u$, where $j > u$. Thus by independence $(x_i, q_j)R(x_i, q_u)$, and by (5.27), $(x_i, q_u)E(x_k, q_\ell)$. Equation (5.28) follows. Now define $f_1(x_i) = i, f_2(q_j) = j$. Equation (5.19) follows by (5.27) and (5.28). This completes the proof in the case that each R_i is strict simple. If each R_i is strict weak, use its reduction $(A_i^*, R_i^*) = (A_i/E, R_i/E)$.

To prove the uniqueness statement, one observes that if $g_1(x_i) = i$ and $g_2(q_j) = j$, then g_1, g_2 satisfy (5.19). Finally, if f_1, f_2 satisfy (5.19) and

$f_1(x_1) = \sigma_1, f_2(q_1) = \sigma_2,$ and $f_1(x_2) = \tau,$ then

$$f_1 = (\tau - \sigma_1)(g_1 - 1) + \sigma_1 \tag{5.29}$$

and

$$f_2 = (\tau - \sigma_1)(g_2 - 1) + \sigma_2. \tag{5.30}$$

Equation (5.29) follows from equal spacing, which implies that

$$f_1(x_i) = (i - 1)[f_1(x_2) - f_1(x_1)] + f_1(x_1).$$

Equation (5.30) follows similarly. Similar equations hold for any other f_1', f_2' satisfying (5.19). The uniqueness results follow.

5.4.5 Scott's Theorem

In this section, we present Scott's [1964] necessary and sufficient axioms for additive conjoint measurement in the case where each A_i is finite. We again take $n = 2$. It is convenient to state Scott's axioms in terms of the binary relation S (weak preference) defined by Eq. (5.2):

$$aSb \Leftrightarrow \sim bRa. \tag{5.2}$$

The first Scott condition is the following:

AXIOM SC1: *For all a_1, b_1 in A_1 and a_2, b_2 in A_2,*

$$(a_1, a_2)S(b_1, b_2) \quad \text{or} \quad (b_1, b_2)S(a_1, a_2).$$

This axiom follows from the stronger Luce–Tukey assumption that (A, R) is strict weak. Axiom SC1 also clearly follows from the representation (5.19).

The second Scott condition is the following.

AXIOM SC2: *Suppose $x_0, x_1, \ldots, x_{k-1}$ are in A_1 and $y_0, y_1, \ldots, y_{k-1}$ are in A_2, and suppose π and σ are permutations of $\{0, 1, \ldots, k - 1\}$. If*

$$(x_i, y_i)S(x_{\pi(i)}, y_{\sigma(i)})$$

for all $i = 1, 2, \ldots, k - 1$, then

$$(x_{\pi(0)}, y_{\sigma(0)})S(x_0, y_0).$$

Axiom SC2 is similar to Axiom SD3 in Scott's axiomatization of difference measurement (Section 3.3). To illustrate Axiom SC2, let us take $A_1 =$ $\{\alpha, \beta\}$, $A_2 = \{*, \dagger\}$, $k = 2$, $x_0 = \beta$, $x_1 = \alpha$, $y_0 = *$, and $y_1 = \dagger$. Suppose π is the identity permutation, the permutation that takes 0 into 0 and 1 into 1. Suppose σ is the permutation that takes 0 into 1 and 1 into 0. Axiom SC2 says that if $(x_1, y_1)S(x_{\pi(1)}, y_{\sigma(1)})$, then $(x_{\pi(0)}, y_{\sigma(0)})S(x_0, y_0)$. Thus, it says that $(x_1, y_1)S(x_1, y_0)$ implies $(x_0, y_1)S(x_0, y_0)$. In our example, this says that $(\alpha, \dagger)S(\alpha, *)$ implies $(\beta, \dagger)S(\beta, *)$. This is a necessary condition for the representation (5.19). (It is a special case of independence, Section 5.4.3.)

To show that Axiom SC2 follows from the representation (5.19), note that since π and σ are permutations,

$$\sum_{i=0}^{k-1} [f_1(x_i) + f_2(y_i)] = \sum_{i=0}^{k-1} f_1(x_i) + \sum_{i=0}^{k-1} f_2(y_i)$$

$$= \sum_{i=0}^{k-1} f_1(x_{\pi(i)}) + \sum_{i=0}^{k-1} f_2(y_{\sigma(i)})$$

$$= \sum_{i=0}^{k-1} [f_1(x_{\pi(i)}) + f_2(y_{\sigma(i)})],$$

for each x_j and y_j is listed once and only once in the next-to-last expression. Note also that $(x_i, y_i)S(x_{\pi(i)}, y_{\sigma(i)})$ for $i = 1, 2, \ldots, k - 1$ implies that

$$\sum_{i=1}^{k-1} [f_1(x_i) + f_2(y_i)] \geq \sum_{i=1}^{k-1} [f_1(x_{\pi(i)}) + f_2(y_{\sigma(i)})].$$

Thus,

$$f_1(x_0) + f_2(y_0) \leqq f_1(x_{\pi(0)}) + f_2(y_{\sigma(0)}),$$

so

$$(x_{\pi(0)}, y_{\sigma(0)})S(x_0, y_0).$$

THEOREM 5.4 (Scott). *Suppose A_1 and A_2 are finite sets and R is a binary relation on $A_1 \times A_2$. Then Axioms SC1 and SC2 are necessary and sufficient for the existence of real-valued functions f_1 on A_1 and f_2 on A_2 such that for all (a_1, a_2) and $(b_1, b_2) \in A_1 \times A_2$,*

$$(a_1, a_2)R(b_1, b_2) \Leftrightarrow f_1(a_1) + f_2(a_2) > f_1(b_1) + f_2(b_2). \qquad (5.19)$$

As in the case of Scott's axioms for difference measurement, the sufficiency proof in Theorem 5.4 uses a clever variant of the famous separating hyperplane theorem. We omit the proof.

Axiom SC1 seems quite reasonable for the case of preference. Axiom SC2 also seems reasonable. But it is impossible to test it empirically.

Finally, we observe that Axiom SC2 is an infinite bundle of axioms, one for each k. We do not get away with using only a finite number of k, since the x_i and y_i are not necessarily distinct. Thus, $x_0, x_1, \ldots, x_{k-1}$ might all be the same element.

Exercises

1. Suppose in a mental test, subject s_1 does better than subject s_2 on both test items, t_1 and t_2. Suppose subject s_1 does better on item t_2 than on item t_1, but s_2 does better on t_1 than on t_2. Show that the additive representation (5.16) fails. Do this
   (a) by assuming the additive representation and reaching a contradiction;
   (b) by showing that one of the necessary axioms for an additive conjoint structure is violated;
   (c) by showing that one of Scott's axioms is violated.

2. (a) Show that if $A_1 = A_2 = Re$ and R is lexicographic preference on $A = A_1 \times A_2$, then (A, R) does not satisfy additive conjoint measurement.
   (b) Which axioms for an additive conjoint structure are violated?

3. (Krantz et al. [1971, pp. 445–446]) Sidowski and Anderson [1967] asked subjects to judge the attractiveness of working at certain occupations in certain cities. The results (mean rating over subjects) are shown in Table 5.1. Suppose O is the set of occupations considered and C the set of cities, and suppose $r(a, b)$ is the rating of the alternative having occupation a in city b. Then there are functions $o: O \to Re$ and $c: C \to Re$, so that for all $a_1, b_1 \in O$ and $a_2, b_2 \in C$,

$$r(a_1, a_2) > r(b_1, b_2) \Leftrightarrow o(a_1) + c(a_2) > o(b_1) + c(b_2).$$

   (a) Verify that such functions are given by
   $o(\text{Lawyer}) = 4,\quad o(\text{Teacher}) = 10,\quad o(\text{Accountant}) = 0$
   $c(A) = 5.9,\quad c(B) = 5.4,\quad c(C) = 4.3,\quad c(D) = 3.0.$
Thus conjoint measurement (or additivity) is satisfied.
   (b) Are there such functions if $r(\text{Teacher}, A)$ is changed to 7.4?
   (c) What if $r(\text{Teacher}, A)$ is changed to 7.1?

Table 5.1. Rating $r(a, b)$ of Attractiveness of City–Occupation Combinations*
(The entry 7.3 − means less than 7.3; the entry 3.2 + means more than 3.2.)

	City			
Occupation	A	B	C	D
Lawyer	7.3	6.8	5.7	4.4
Teacher	7.3 −	6.7	5.3	3.2 +
Accountant	5.9	5.4	4.3	3.2

*Data from Sidowski and Anderson [1967].

4. (Fishburn [1970, p. 51]) Suppose $A_1 = A_2 = \{1, 2, \ldots, n\}$, and suppose

$$(a_1, a_2)R(b_1, b_2) \Leftrightarrow f(a_1, a_2) > f(b_1, b_2).$$

(a) Show that if $f(a_1, a_2) = a_1 a_2$, then $(A_1 \times A_2, R)$ has an additive conjoint representation.

(b) Show that if $f(a_1, a_2) = a_1 + a_2 + a_1 a_2$, then $(A_1 \times A_2, R)$ does not have an additive conjoint representation.

(c) For each of the following functions f, does $(A_1 \times A_2, R)$ have an additive conjoint representation?

    (i) $f(a_1, a_2) = a_1^2 + a_1 a_2$.

    (ii) $f(a_1, a_2) = \max\{a_1, a_2\}$.

    (iii) $f(a_1, a_2) = a_1 - a_2$.

    (iv) $f(a_1, a_2) = 1/a_1 a_2$.

    (v) $f(a_1, a_2) = a_1/(a_1 + a_2)$.

5. Suppose $A_1 = T$ is a set of (dry-bulb air) temperatures and $A_2 = H$ is a set of relative humidities. Then the *temperature–humidity* index (THI) is defined as

$$\text{THI } (t, h) = t - (0.55 - 0.55h)(t - 58),$$

where $t \in T$ and $h \in H$ (Conway [1963]). Suppose R is a binary relation on $T \times H$, with aRb interpreted as "is more uncomfortable under a than b." Show that if THI preserves R, then there are no functions τ and γ on T and H, respectively, so that for all $t_1, t_2 \in T$ and $h_1, h_2 \in H$,

$$(t_1, h_1)R(t_2, h_2) \Leftrightarrow \tau(t_1) + \gamma(h_1) > \tau(t_2) + \gamma(h_2). \qquad (5.17)$$

6. The THI can also be defined (Conway [1963]) as $(0.4)(td + tw) + 15$, where td is the dry-bulb air temperature and tw is the wet-bulb air temperature. Suppose D is a set of dry-bulb temperatures and W a set of wet-bulb temperatures and THI preserves the relation R of Exer. 5. Show that there are functions δ on D and ω on W so that for all $td_1, td_2 \in D$ and $tw_1, tw_2 \in W$,

$$(td_1, tw_1)R(td_2, tw_2) \Leftrightarrow \delta(td_1) + \omega(tw_1) > \delta(td_2) + \omega(tw_2).$$

7. Suppose $A_1 = A_2 = Re$ and

$$(a_1, a_2)R(b_1, b_2) \Leftrightarrow [a_1 > b_1 \text{ and } a_2 > b_2].$$

Which of the axioms for an additive conjoint structure hold?

8. For each of the following binary relations R on $Re \times Re$, check which of the axioms for an additive conjoint structure are satisfied:

    (a) $\mathbf{a}R\mathbf{b}$ iff $\max\{a_1, a_2\} > \max\{b_1, b_2\}$.

    (b) $\mathbf{a}R\mathbf{b}$ iff $a_1 a_2 > b_1 b_2$.

    (c) $\mathbf{a}R\mathbf{b}$ iff $a_1 > b_1$.

9. Does the binary relation $(A_1 \times A_2, R)$ defined in Exer. 7 satisfy Scott's Axiom SC2?

10. Check which of Scott's axioms are satisfied by the binary relations of Exer. 8 if each is considered a relation on

$$\{1, 2, \ldots, n\} \times \{1, 2, \ldots, n\}.$$

11. Suppose $A_1 = \{\alpha, \beta\}$, $A_2 = \{*, \dagger\}$, and R is the following binary relation on $A_1 \times A_2$:

$$\{\langle(\alpha, *), (\alpha, \dagger)\rangle, \langle(\alpha, *), (\beta, *)\rangle, \langle(\alpha, *), (\beta, \dagger)\rangle,$$
$$\langle(\alpha, \dagger), (\beta, \dagger)\rangle, \langle(\beta, *), (\beta, \dagger)\rangle\}.$$

Is $(A_1 \times A_2, R)$ an equally spaced additive conjoint structure?

12. Suppose $A_1 = A_2 = \{0, 1\}$ and R is lexicographic preference on $A = A_1 \times A_2$. Show that (A, R) has an additive conjoint representation (5.19), but it is not an equally spaced additive conjoint structure.

13. The next two exercises investigate an example used to argue against the Archimedean axiom for additive conjoint measurement. Suppose $A_1 = \{a, a'\}$ and $A_2 = N$. Suppose

$$(a, i)R(a, j) \Leftrightarrow i > j,$$
$$(a', i)R(a', j) \Leftrightarrow i > j,$$
$$(a, i)R(a', j), \quad \text{all } i, j.$$

(a) Show directly that every strictly bounded standard sequence on either component is finite.

(b) Verify (a) indirectly by showing that $(A_1 \times A_2, R)$ is representable in the form (5.19).

14. Suppose $A_1 = \{a, a', b\}$ and $A_2 = N$. Suppose on $\{a, a'\} \times N$, R is defined as in Exer. 13. Suppose moreover that

$$(b, i)R(b, j) \Leftrightarrow i > j,$$
$$(a, i)R(b, i + 100), \quad \text{all } i,$$
$$(b, i)R(a', j), \quad \text{all } i, j,$$

and otherwise R is defined so as to make a strict weak order. Think of b as the alternative "having a long, healthy life with one sickness."

(a) Show that every strictly bounded standard sequence on either component is finite.

(b) Show that $(A_1 \times A_2, R)$ is not representable in the form (5.19).

(c) Which of the axioms for additive conjoint measurement fail?

(d) Show that the following statement is true, and hence a variant of the Archimedean axiom fails. The sequence

$$1, 101, 201, \ldots$$

forms a standard sequence on the second component and the sequence

$$(a', 1), (a', 101), (a', 201), \ldots$$

is strictly bounded and infinite.

15. Suppose f_1 and f_2 define an additive conjoint representation.
    (a) Show that the following statements are meaningful:
        (i) $f(a_1, a_2) > f(b_1, b_2)$.
        (ii) $f(a_1, a_2)$ is constant.
    (b) Show that the following statement is not meaningful:

$$f(a_1, a_2) = 2f(b_1, b_2).$$

16. (Krantz et al. [1971]) Let

$$A_1 = A_2 = \{2^n: n \text{ is a positive integer}\}.$$

Let

$$(a_1, a_2)R(b_1, b_2) \Leftrightarrow a_1 + a_2 > b_1 + b_2.$$

Show that $(A_1 \times A_2, R)$ satisfies Axioms C1 through C4 and C6 of additive conjoint measurement, but not Axiom C5. (In particular, this shows that Axioms C1 through C4 are not sufficient for conjoint measurement.)

17. Let A_1 and A_2 be as in Exer. 16, and let R on $A_1 \times A_2$ be the lexicographic ordering. Show that $(A_1 \times A_2, R)$ satisfies Axioms C1 through C4 and C6 of additive conjoint measurement, but not Axiom C5.

18. Let A_1 and A_2 be as in Exer. 16, and let

$$(a_1, a_2)R(b_1, b_2) \Leftrightarrow [a_1 \geq b_1 \ \& \ a_2 \geq b_2 \ \& \ (a_1 > b_1 \text{ or } a_2 > b_2)];$$

that is, $\mathbf{a}R\mathbf{b} \Leftrightarrow \mathbf{a} > \mathbf{b}$. Show that $(A_1 \times A_2, R)$ satisfies Axioms C2, C4, C5, and C6 of additive conjoint measurement, but not Axioms C1 and C3.

19. This exercise and the next one are intended to investigate the role of essentiality (Axiom C6) in the axioms for an additive conjoint structure. Suppose $A_1 = Re \times Re$, $A_2 = \{0\}$, T is the lexicographic ordering of A_1, and

$$(\langle a, b \rangle, 0)R(\langle c, d \rangle, 0) \Leftrightarrow \langle a, b \rangle T \langle c, d \rangle.$$

    (a) Show that $(A_1 \times A_2, R)$ satisfies Axioms C1 through C5, but not C6.

    (b) Moreover, show that the representation (5.19) fails to hold.

20. Suppose we modify the example of the previous exercise to make $A_1 = Re \times Re$, $A_2 = \{0, 1\}$, and T the lexicographic ordering of A_1. Suppose

$$\langle a, b \rangle J \langle c, d \rangle \Leftrightarrow \sim \langle a, b \rangle T \langle c, d \rangle \ \& \ \sim \langle c, d \rangle T \langle a, b \rangle.$$

Let

$$(\langle a, b \rangle, i) R(\langle c, d \rangle, j) \Leftrightarrow [\langle a, b \rangle T \langle c, d \rangle \text{ or } (\langle a, b \rangle J \langle c, d \rangle \& i > j)].$$

    (a) Show that the representation (5.19) fails to hold.
    (b) Show that essentiality, Axiom C6, now holds.
    (c) Which of the axioms C1 through C5 fail and which hold?

    21. (Krantz *et al.* [1971]) Show that the axioms for an additive conjoint structure are independent.

    22. Prove the following assertions made in the proof of Theorem 5.3:
    (a) Equation (5.27) holds.
    (b) Equations (5.29) and (5.30) hold.
    (c) The uniqueness results follow from (5.29) and (5.30) for both f_1, f_2 and f_1', f_2'.

    23. (Krantz *et al.* [1971]) Suppose that $(A_1 \times A_2, R)$ satisfies Axioms C1 and C4 of additive conjoint measurement and also the following conditions:

    (i) (Unrestricted) solvability on both components.
    (ii) Double cancellation.
    (iii) At least one component is essential.
Show that $(A_1 \times A_2, R)$ is an additive conjoint structure.

    24. Suppose $|A_1| = 1$ and f_1 and f_2 satisfy Eq. (5.19). Show that if f_1' and f_2' also satisfy Eq. (5.19), the uniqueness result of Theorem 5.2 does not hold. In particular, show that all monotone increasing transformations of f_2 are admissible.

    25. (Krantz *et al.* [1971, Section 10.7.2], Luce [1965], Marley [1968], Roberts [1974]) Suppose A is a set of physical objects in motion, E is kinetic energy, m is mass, and v is velocity. Then $E = \frac{1}{2}mv^2$. Now $m = f_1$ and $v = f_2$ can be looked at as extensive measures (Section 3.2) by introducing a relation $\succ_1$ and an operation o_1 on $A_1 = A$ and similar $\succ_2$ and o_2 on $A_2 = A$. Moreover, if $h_1 = \frac{1}{2}m$ and $h_2 = v^2$, then $g_1 = \log h_1$ and $g_2 = \log h_2$ can be looked at as conjoint measures on $A_1 \times A_2$ for an appropriate relation R. These conjoint measures and extensive measures are related: there are constants $\gamma_1, \gamma_2, \alpha_1 > 0, \alpha_2 > 0, \alpha_1/\alpha_2$ rational, so that

$$h_1 = \gamma_1 f_1^{\alpha_1}, \quad h_2 = \gamma_2 f_2^{\alpha_2}. \tag{5.31}$$

In some sense, therefore, the conjoint measures are closely related to the extensive measures, and the two ways of measuring kinetic energy are consistent. This exercise explores conditions under which this result holds in general, and sketches out a proof that these conditions work.

    Suppose $(A_1 \times A_2, R)$ is an additive conjoint structure, and suppose $(A_i, \succ_i, o_i)$ is an extensive structure (Section 3.2) for $i = 1, 2$. We say that a *law of exchange* holds if there are positive integers m and n such that for all positive integers i and j and for all a in A_1 and u in A_2,

$$(i^m a, j^n u) E(j^m a, i^n u).$$

For $i = 1, 2$, suppose $g_i : A_i \to Re$ and $f_i : A_i \to Re$ are such that the following conditions hold:

(i) $(a, b)R(c, d) \Leftrightarrow g_1(a) + g_2(b) > g_1(c) + g_2(d)$.

(ii) $a \succ_i b \Leftrightarrow f_i(a) > f_i(b)$.

(iii) $f_i(a \circ_i b) = f_i(a) + f_i(b)$.

Choose f_i so that $f_i(a) = 1$ for some a. (This can always be done.) Let $h_i = e^{g_i}$. Then

(iv) $(a, b)R(c, d) \Leftrightarrow h_1(a)h_2(b) > h_1(c)h_2(d)$.

(a) Define ϕ_i on the range of f_i by $\phi_i(f_i(a)) = h_i(a)$. Show that ϕ_i is strictly increasing.

(b) Suppose (5.31) holds for some $\gamma_1, \gamma_2, \alpha_1 > 0, \alpha_2 > 0$, with α_1 / α_2 rational. Show that a law of exchange holds for some m, n.

(c) Prove that if a law of exchange holds with positive integers m and n, then for all a in A_1 and u in A_2, and positive integers i and j,

$$h_1(i^m a)/h_1(j^m a) = h_2(i^n u)/h_2(j^n u). \qquad (5.32)$$

(d) If i and j are positive integers, let $h(i, j)$ denote the common ratio in (5.32). Show that if i, j, k, ℓ are positive integers, then if $i/j = k/\ell$, it follows that $h(i, j) = h(k, \ell)$.

(e) Show from the above that we may write $h(i, j)$ as $h(i/j)$.

(f) Observe that h is a function from the positive rationals to the reals. Moreover, h is strictly increasing, and for r and s positive rationals, $h(rs) = h(r)h(s)$.

(g) It is easy to extend h to all positive reals by letting

$$h(x) = \sup\{h(y) : 0 < y \leq x, \ y \text{ positive rational}\}.$$

Show that $h(x) = x^\beta$, some $\beta > 0$. [The proof uses Cauchy's fourth equation (Section 4.2) and the observation (Exer. 12 of Section 4.2) that our solution to this holds for monotone increasing functions.]

(h) If $f_1(a) = \lambda^m$, some a in A_1, define $\psi_1(\lambda) = \phi_1(\lambda^m)$. Show that if i is a positive integer, and λ is in the domain of ψ_1, then so is $i\lambda$.

(i) Setting $j = 1$ in (5.32) and using the result of (g), show that $h_1(i^m a) = i^\beta h_1(a)$.

(j) Show from the results of (i) and (h) that $\psi_1(i\lambda) = i^\beta \psi_1(\lambda)$.

(k) By assumption, 1 is in the range of f_1, and so $\psi_1(1)$ is defined. Show that $\psi_1(\lambda) = \psi_1(1)\lambda^\beta$. Conclude that $\phi_1(\lambda) = \psi_1(\lambda^{1/m}) = \psi_1(1)\lambda^{\beta/m} = \gamma_1 \lambda^{\alpha_1}$, where $\alpha_1 = \beta/m$. A similar argument shows that $\phi_2(\lambda) = \gamma_2 \lambda^{\alpha_2}$, $\alpha_2 = \beta/n$. Thus, (5.31) follows with α_1/α_2 rational—in fact, equal to n/m.

(Note: An alternative approach to the relation between conjoint and extensive measurement is based on distributive laws. See Luce [1978] and Narens and Luce [1976] for a development.)

26. Discuss density as a fundamental scale. (In Chapter 2 we discuss it as a derived scale.) In particular, if D is "denser than," discuss axioms for the existence of functions M and V such that

$$aDb \Leftrightarrow M(a)V(b) < M(b)V(a).$$

27. In Section 2.6, Exer. 12, we mentioned an experiment designed to rank-order a collection of stereo speakers. Consider this experiment in the light of the conditions for conjoint measurement discussed in this section. For example, consider what is being assumed about the parameters in order to sum scores over dimensions.

5.5 Nonadditive Representations

5.5.1 Decomposability and Polynomial Conjoint Measurement

The measurement representations (5.12) and (5.19) are based on two principles: We can decompose utilities (or other scales) into utilities on individual dimensions, and we can then add the results. The idea of decomposability has great applicability even in situations where additivity does not apply. In general, if A has a product structure i.e., if A equals $A_1 \times A_2 \times \cdots \times A_n$, and R is a binary relation on A, we say that (A, R) is *decomposable* if there are real-valued functions $f_1, f_2, \ldots, f_n$ on $A_1, A_2, \ldots, A_n$, respectively, and a function

$$F:[f_1(A_1) \times f_2(A_2) \times \cdots \times f_n(A_n)] \to Re$$

such that for all $\mathbf{a} = (a_1, a_2, \ldots, a_n)$ and $\mathbf{b} = (b_1, b_2, \ldots, b_n)$ in A,

$$\mathbf{a}R\mathbf{b} \Leftrightarrow F[f_1(a_1), f_2(a_2), \ldots, f_n(a_n)] > F[f_1(b_1), f_2(b_2), \ldots, f_n(b_n)].$$

$$(5.33)$$

The function F is called a *composition rule*. It is frequently assumed that F is one-to-one or strictly increasing in each variable. The representation (5.33) applies in very general situations. Of course, the additive representation arises if F satisfies

$$F(x_1, x_2, \ldots, x_n) = x_1 + x_2 + \cdots + x_n. \qquad (5.34)$$

We have already seen (Section 5.4.1) that the representation (5.33) with F given in (5.34) is closely related to the representation (5.33) with F given as a product,

$$F(x_1, x_2, \ldots, x_n) = x_1 x_2 \ldots x_n. \qquad (5.35)$$

The product composition rule arises in the response strength situation with three factors. For example, Hull [1952] has argued that the appropriate representation here is

$$(d_1, k_1, h_1)R(d_2, k_2, h_2) \Leftrightarrow \delta(d_1)\kappa(k_1)\mu(h_1) > \delta(d_2)\kappa(k_2)\mu(h_2). \quad (5.36)$$

On the other hand, Spence [1956] has argued that the more appropriate representation is

$$(d_1, k_1, h_1)R(d_2, k_2, h_2) \Leftrightarrow [\delta(d_1) + \kappa(k_1)]\,\mu(h_1) > [\delta(d_2) + \kappa(k_2)]\,\mu(h_2).$$
$$(5.37)$$

The composition rule in Spence's model is the distributive rule

$$F(x_1, x_2, x_3) = (x_1 + x_2)x_3.$$

This distributive rule, in turn, also arises in the study of perceived risk (Coombs and Huang [1970]), where x_1 is a function of expected regret, x_2 of expected value, and x_3 of the number of plays in a gamble. See Krantz and Tversky [1971] and Krantz et al. [1971] for examples of other composition rules.

We first state some necessary and sufficient conditions for decomposability with the function F being one-to-one in each variable and then investigate briefly some necessary conditions for various composition rules. By the Birkhoff–Milgram Theorem (Corollary 1 to Theorem 3.4), decomposability implies that (A, R) is a strict weak order and (A^*, R^*) has a countable order-dense subset. Moreover, since F is one-to-one in each variable,

$$F(x_1, x_2, \ldots, x_{i-1}, y_i, x_{i+1}, \ldots, x_n) = F(x_1, x_2, \ldots, x_{i-1}, y_i', x_{i+1}, \ldots, x_n)$$

implies that $y_i = y_i'$. Thus, decomposability with such an F implies that for all i and all a_i, a_i' in A_i and all b_j, c_j in $A_j, j \neq i$, we have

$$(b_1, b_2, \ldots, b_{i-1}, a_i, b_{i+1}, \ldots, b_n)E(b_1, b_2, \ldots, b_{i-1}, a_i', b_{i+1}, \ldots, b_n)$$
$$\Leftrightarrow \quad (5.38)$$
$$(c_1, c_2, \ldots, c_{i-1}, a_i, c_{i+1}, \ldots, c_n)E(c_1, c_2, \ldots, c_{i-1}, a_i', c_{i+1}, \ldots, c_n),$$

where E is the indifference relation defined from R by Eq. (5.3). If condition (5.38) holds, we say (A, R) satisfies *substitutability*. The following theorem is proved in Krantz et al. [1971].

THEOREM 5.5. *The product structure (A, R) is decomposable with a function F which is one-to-one in each variable if and only if (A, R) is a strict weak order, (A^*, R^*) has a countable order-dense subset, and (A, R) satisfies substitutability.*

Proof. Necessity has already been shown. To show sufficiency, note that by the Birkhoff–Milgram Theorem (Corollary 1 to Theorem 3.4), there is a real-valued function f on A such that for $\mathbf{a}, \mathbf{b}$ in A,

$$\mathbf{a}R\mathbf{b} \Leftrightarrow f(\mathbf{a}) > f(\mathbf{b}).$$

Fixing x_j in A_j, $j = 1, 2, \ldots, n$, we define f_i on A_i by

$$f_i(a_i) = f(x_1, x_2, \ldots, x_{i-1}, a_i, x_{i+1}, \ldots, x_n).$$

Finally, let $F:[f_1(A_1) \times f_2(A_2) \times \cdots \times f_n(A_n)] \to Re$ be defined as

$$F[f_1(a_1), f_2(a_2), \ldots, f_n(a_n)] = f(a_1, a_2, \ldots, a_n).$$

It is easy to verify that F is well-defined and is one-to-one in each variable. ∎

The composition rules that have been of most interest in the measurement literature are those where F is a polynomial, a function of the form

$$F(x_1, x_2, \ldots, x_n) = \sum_k \alpha_k x_1^{\beta_{1k}} x_2^{\beta_{2k}} \ldots x_n^{\beta_{nk}},$$

where α_k is a real number and the β_{ik} are nonnegative integers. In case F is a polynomial, the representation (5.33) is sometimes called *polynomial conjoint measurement*. Krantz et al. [1971, Chapter 7] summarize some conditions either necessary or sufficient for a variety of polynomial representations. However, there is still much work to be done in this area. It is sometimes of sufficient practical use to specify necessary conditions for a polynomial representation. For example, in the response strength application, if one has a choice between the Hullian and Spencian models, one has a choice between the two polynomial composition rules (5.36) and (5.37). As Krantz and Tversky [1971] point out, if one can derive necessary conditions for each and show that one of these necessary conditions is violated, this eliminates one of the possible models. For example, assuming that δ, κ, and μ all take on only positive values, one necessary condition for the representation (5.36) is the following independence condition (known as a *joint independence condition*):

$$(d_1, k_1, h_1)R(d_2, k_1, h_2) \Rightarrow (d_1, k_2, h_1)R(d_2, k_2, h_2). \tag{5.39}$$

However, this condition is not necessary for the representation (5.37). For, as Krantz and Tversky [1971] point out, suppose we take $\delta(d_1) = 1$, $\delta(d_2) = 2$, $\kappa(k_1) = 4$, $\kappa(k_2) = 2$, $\mu(h_1) = 5$, and $\mu(h_2) = 4$. Then if R is *defined* by (5.37), we have a violation of (5.39):

$$(d_1, k_1, h_1)R(d_2, k_1, h_2) \quad \text{since} \quad (1 + 4)5 > (2 + 4)4,$$

$$\sim (d_1, k_2, h_1)R(d_2, k_2, h_2) \quad \text{since} \quad (1 + 2)5 \not> (2 + 2)4.$$

For a discussion of similar conditions useful in the testing of alternative polynomial conjoint measurement models, see Exer. 5 below and see Krantz [1968], Krantz and Tversky [1971], or Krantz *et al.* [1971, Chapter 7].

5.5.2 Nondecomposable Representations

Even if decomposability is not satisfied, one can usefully study numerical representations on product structures. For example, in case R is preference, Raiffa [1969] and Fishburn [1974] seek real-valued functions f_1 and λ_1 on A_1 and f_2 and λ_2 on A_2 such that for all a_1, b_1 in A_1 and a_2, b_2 in A_2,

$$(a_1, a_2)R(b_1, b_2) \Leftrightarrow f_1(a_1) + f_2(a_2) + \lambda_1(a_1)\lambda_2(a_2) > \atop f_1(b_1) + f_2(b_2) + \lambda_1(b_1)\lambda_2(b_2). \tag{5.40}$$

The term $\lambda_1(a_1)\lambda_2(a_2)$ represents an interaction effect. The representation (5.40) is called the *quasi-additive representation*, and we shall encounter a variant of it in Section 7.3.3, where we shall give some sufficient conditions for the existence of a quasi-additive utility function and mention a variety of applications. See also Farquhar [1977] for a discussion.

Tversky [1969] studies the representation where there are real-valued functions $f_1, f_2, \ldots, f_n$ on $A_1, A_2, \ldots, A_n$, respectively, and increasing, continuous functions $\phi_1, \phi_2, \ldots, \phi_n$ from Re to Re, so that for all i,

$$\phi_i(-\delta) = -\phi_i(\delta),$$

and so that

$$\mathbf{a}R\mathbf{b} \Leftrightarrow \sum_{i=1}^{n} \phi_i[f_i(a_i) - f_i(b_i)] > 0. \tag{5.41}$$

The representation (5.41) is called the *additive difference model*. See Wallsten [1976] for a recent application and Beals, Krantz, and Tversky [1968] for an analogous representation if indifference is used in place of preference. If each ϕ_i is a linear function, $\phi_i(\delta_i) = t_i\delta_i$ for some positive t_i, then the additive difference model implies additive conjoint measurement. For

$$\sum_{i=1}^{n} \phi_i[f_i(a_i) - f_i(b_i)] = \sum_{i=1}^{n} [t_if_i(a_i) - t_if_i(b_i)],$$

and we have

$$\mathbf{a}R\mathbf{b} \Leftrightarrow \Sigma(t_i f_i)(a_i) > \Sigma(t_i f_i)(b_i).$$

The representation (5.41) can be generalized to the representation

$$\mathbf{a}R\mathbf{b} \Leftrightarrow \phi\{[f_1(a_1) - f_1(b_1)], [f_2(a_2) - f_2(b_2)], \ldots, [f_n(a_n) - f_n(b_n)]\} > 0,$$

or the representation

$$\mathbf{a}R\mathbf{b} \Leftrightarrow F\left[\sum_{i=1}^{n} \phi_i(a_i, b_i)\right] > 0.$$

Such representations are called by Krantz [1968] *absolute difference models*. They have been studied rather extensively by Pfanzagl [1959].

Another interesting class of representations of current interest is the general class known as fractional hypercube representations, which are discussed by Farquhar [1974, 1975, 1976]. Farquhar develops techniques for deriving sufficient conditions for a wide variety of these representations. Farquhar [1977] surveys a number of other representations of current interest.

Exercises

1. Suppose there is a function $f: A \rightarrow Re$ such that for all $\mathbf{a}, \mathbf{b} \in A$,

$$\mathbf{a}R\mathbf{b} \Leftrightarrow f(\mathbf{a}) > f(\mathbf{b}).$$

(a) Show that if $f(a_1, a_2) = a_1 + a_2 + a_1 a_2$, then (A, R) is decomposable.

(b) For each of the following functions f, determine if (A, R) is decomposable:

  (i) $f(a_1, a_2) = a_1^2 + a_1 a_2$.
  (ii) $f(a_1, a_2) = \max\{a_1, a_2\}$.
  (iii) $f(a_1, a_2) = a_1 - a_2$.
  (iv) $f(a_1, a_2) = 1/a_1 a_2$.
  (v) $f(a_1, a_2) = a_1/(a_1 + a_2)$.

2. Observe that the function THI(t, h) of Exer. 5, Section 5.4, defines a quasi-additive representation that is also a polynomial representation.

3. Show that in both the Hullian and Spencian models, if the functions δ, κ, and μ all take on only positive values, then the following independence condition is satisfied:

$$(d_1, k_1, h_1)R(d_2, k_1, h_1) \Leftrightarrow (d_1, k_2, h_2)R(d_2, k_2, h_2).$$

4. Show that F as defined in the proof of Theorem 5.5 is well-defined and is one-to-one in each variable.

5. (Krantz and Tversky [1971]) (a) The *simple polynomials* are defined as follows: Any monomial x_i is a simple polynomial. If two simple polynomials have no variables in common, their sum and product are simple polynomials. Show that, up to permutation of labels, there are only four simple polynomials $F(x_1, x_2, x_3)$ of three variables:

$x_1 + x_2 + x_3$ (additive),
$(x_1 + x_2)x_3$ (distributive),
$x_1x_2 + x_3$ (dual distributive),
$x_1x_2x_3$ (multiplicative).

(b) In the following, we shall assume that $A = A_1 \times A_2 \times A_3$ and that (5.33) holds with all $f_i(x_i) > 0$. We shall investigate various necessary conditions which can be used to differentiate among the simple polynomial composition rules of part (a). For a more complete discussion, see Krantz [1968], Krantz and Tversky [1971], or Krantz *et al.* [1971, Chapter 7]. We say that A_1 is *independent* (of A_2 and A_3) if, for all $a, b \in A_1$, $p, q \in A_2$, and $u, v \in A_3$,

$$(a, p, u)R(b, p, u) \Leftrightarrow (a, q, v)R(b, q, v).$$

(The rule in Exer. 3 exemplifies this.) Independence of A_2 and of A_3 is defined similarly. These notions are generalizations of the independence notions that arose in Section 5.4.3. Show that for any of the F's of part (a), independence holds for each A_i.

(c) Again generalizing a notion of Section 5.4.3, we say that *double cancellation* holds for A_1 and A_2 if

$$[(a, q, u)S(b, r, u) \ \& \ (b, p, u)S(c, q, u)] \Rightarrow (a, p, u)S(c, r, u),$$

where S is defined from R by Eq. (5.2). A similar definition holds for any A_i and A_j. Show that if F is any of the polynomials of part (a), then double cancellation holds for any A_i and A_j.

(d) We say that A_1 and A_2 satisfy *joint independence* (from A_3) if

$$(a, p, u)S(b, q, u) \Leftrightarrow (a, p, v)S(b, q, v).$$

Similar definitions hold for joint independence of A_i and A_j (from A_k). Show that joint independence for all pairs i and j implies independence for all single factors.

(e) Show that if F is any of the polynomials of part (a), then some pair of factors is jointly independent (from the third).

(f) Show that if F is the additive polynomial of part (a), then joint independence holds for each pair.

(g) We say that *distributive cancellation* holds if, whenever the conditions $(a, p, u)S(c, r, v)$, $(b, q, u)S(d, s, v)$, and $(d, r, v)S(b, p, u)$ all hold, then $(a, q, u)S(c, s, v)$. Show that if F is the distributive polynomial of part (a), then distributive cancellation holds.

(h) Show that distributive cancellation also follows if F is the additive polynomial of part (a).

(i) However, show that distributive cancellation may fail if F is the dual distributive polynomial of part (a).

6. Show that if the quasi-additive representation holds, then (A, R) is a strict weak order.

7. Show that if the additive difference model holds, then (A, R) does not have to be a strict weak order; in particular, R does not have to be transitive. Thus, the additive difference model is a measurement model for preferences which may not be transitive. (Tversky [1969] shows that if $n \geq 3$, and the additive difference model holds, then (A, R) is transitive if and only if all the functions ϕ_i are linear, that is, there are real numbers t_i so that $\phi_i(\delta) = t_i\delta$, all δ.)

8. Huang [1975] and Kirk [1977] consider the following representation, which they call *nonsimple distributive conjoint measurement*:

$$\mathbf{aRb} \Leftrightarrow \sum_{i=1}^{n-1} g_i(a_1) f_{i+1}(a_{i+1}) > \sum_{i=1}^{n-1} g_i(b_1) f_{i+1}(b_{i+1}).$$

Which of the axioms for additive conjoint measurement are necessary for this representation?

5.6 Joint Scales of Individuals and Alternatives

In this section, we turn to a quite different kind of product structure, that where the set of individuals making judgments is one dimension and the set of alternatives or objects about which judgments are made is a second dimension. This kind of situation arose in the mental testing situation, and we shall see a variety of other situations in which it arises. We present this material mostly to illustrate problems of measurement quite different from the preservation of ordinal preference data.

5.6.1 Guttman Scales

Suppose S is a set of individuals whose reactions or experiences are being studied and E is a set of reactions or experiences. Let aRb mean that individual a had (or experienced) reaction or experience b. Then R defines a binary relation on $S \cup E$; in particular, R is a subset of $S \times E$. In a classical experiment, Stouffer *et al.* [1950] studied fear symptoms of United States soldiers during World War II. The set S here is the group of soldiers being studied, and the set E consists of certain fear reactions such as violent pounding of the heart, shaking or trembling all over, or losing control of the bowels. The experimenters found that it was possible to order the reactions such that if a soldier experienced a reaction, he (tended to) experience all reactions coming before it in the order. Thus, it was possible to simultaneously order the individuals and reactions in such a way that individual a had reaction b if and only if a followed b in the ordering. This *joint ordering* of individuals and reactions suggests that there is a natural ordering of the fear reactions, from least to most severe. In terms of a representation, we can think of finding two real-valued functions, s on S and e on E, such that for all a in S and b in E,

$$aRb \Leftrightarrow s(a) > e(b). \tag{5.42}$$

The two functions s and e satisfying Eq. (5.42) define what is called a *Guttman scale*, after Louis Guttman [1944]. Notice that we are not seeking a homomorphism from one relational system into another. However, obtaining functions s and e satisfying Eq. (5.42) can be thought of as a measurement problem, and we can ask for a representation theorem.

In general, given a triple (S, E, R), with R a subset of $S \times E$, we ask whether (S, E, R) possesses a Guttman scale. This question arises in a variety of contexts other than the one we have discussed. For example, if S is a set of individuals and E is a set of statements, the relation aRb can be taken to mean that individual a agrees with statement b. Then the existence of a Guttman scale implies that the statements have a certain natural ordering vis-à-vis the subjects. If S is a set of individuals and E a set of test items, and aRb means that individual a answers test item b correctly, then a Guttman scale leads to a natural joint ordering of items as to difficulty vis-à-vis the level of skill of individuals, with a subject answering an item correctly if and only if his skill level is above the level of difficulty of the item.

It is clear that for a Guttman scale to exist, it is not possible to have

$$aRb \quad \text{and} \quad \sim a'Rb \quad \text{while} \quad a'Rb' \quad \text{and} \quad \sim aRb'. \tag{5.43}$$

If condition (5.43) fails for all $a, a' \in S$ and $b, b' \in E$, we say that (S, E, R) is *consistent*.

THEOREM 5.6. *If S and E are finite sets and $R \subseteq S \times E$, then (S, E, R) possesses a Guttman scale if and only if (S, E, R) is consistent.*

Proof. Omitted. See, for example, Ducamp and Falmagne [1969].

We shall generalize the notion of Guttman scale in Exers. 32 and 33, Section 6.1. For recent results on existence of Guttman scales, see Leibowitz [1978].

5.6.2 *Unfolding*

Let us again consider the situation where S is a set of subjects and E is a set of experiences, reactions, statements, or alternatives—let us think of statements to be concrete. We imagine that the statements can be measured (for example, on a scale of degree of conservatism)—let $e(b)$ be the measure of statement b. We imagine that each individual a associates an ideal value or degree on the scale (for example ideal degree of conservatism)—let $s(a)$ be this ideal degree. Then we can certainly imagine the individual a as preferring statement x to statement y, or agreeing more with statement x than with statement y, if and only if statement x is closer to a's ideal than is statement y, that is, if and only if

$$|s(a) - e(x)| < |s(a) - e(y)|.$$

The functions s and e are sometimes said to define a *joint scale* of individuals and statements. Suppose we let $xR_a y$ mean that individual a agrees more with statement x than with statement y. Then we have for all a in S and x, y in E,

$$xR_a y \Leftrightarrow |s(a) - e(x)| < |s(a) - e(y)|. \tag{5.44}$$

For each a, it follows from (5.44) that the binary relation R_a defines a strict weak order: an ordinal utility representation can be obtained by "folding" the real line at $s(a)$.

Conversely, suppose each individual a gives us his preferences among statements, with $xR_a y$ having the interpretation that a agrees more with x than with y. We ask: Are there real-valued functions s on S and e on E satisfying Eq. (5.44)? If so, we say that s and e define a *Coombs scale*, after Clyde Coombs [1950]. As we observed above, in order for a Coombs scale to exist, the individual preference relations R_a must all be strict weak orders. The Coombs scale can be thought of as a joint "unfolding" of the individual preference orderings.

The representation (5.44) can be generalized in a natural way. Namely, this representation assumes that we can measure distance $|s(a) - e(x)|$, but uses a very specific distance measure. Other metrics can be used in place of this. In the most general situation, we can think of a metric space (X, d) and functions $s: S \to X$ and $e: E \to X$ such that for all a in S and x, y in E,

$$xR_a y \Leftrightarrow d[s(a), e(x)] < d[s(a), e(y)]. \tag{5.45}$$

In particular, if (X, d) is a higher-dimensional Euclidean space, we speak of *multidimensional unfolding*. Multidimensional unfolding has been studied by Bennett and Hays [1960] and Suppes and Zinnes [1963]. See also Coombs [1964, Chapter 7].

Not much progress has been made on an axiom system for the representation (5.44), let alone its generalizations. We shall present some results for (5.44) in the exercises.

It should be noted that neither Guttman scales nor Coombs scales are usually obtainable if the representation is asked to hold "exactly." The representations can usually be obtained only "approximately" at best, and the emphasis in the literature is often to assume that the representation holds and to find the "best-fitting" functions s and e.

Exercises

1. A triple (S, E, R) with $R \subseteq S \times E$ can be represented by a matrix of 0's and 1's whose rows correspond to elements of S and whose columns correspond to elements of E, with the i, j entry equal to 1 if and only if iRj. What property of this matrix after a permutation of rows and columns corresponds to existence of a Guttman scale?

2. Suppose $S = \{1, 2, 3\}$ and $E = \{\alpha, \beta, \gamma\}$.
   (a) Show that if $R = \{(1, \alpha), (2, \beta), (3, \gamma)\}$, there is no Guttman scale.
   (b) Check if there is a Guttman scale in the following cases:
   (i) $R = \{(1, \alpha), (2, \alpha), (3, \alpha)\}$.
   (ii) $R = \{(1, \alpha), (1, \beta), (2, \alpha), (3, \alpha), (3, \beta)\}$.

3. Suppose that $S = \{1, 2, 3\}$, $E = \{\alpha, \beta, \gamma\}$, and R_1, R_2, R_3 are the following rankings (strict simple orders):
   R_1: α over β over γ,
   R_2: β over γ over α,
   R_3: γ over α over β.
Show that there is no Coombs scale.

4. If S and E are as in Exer. 3, determine whether there are Coombs scales in the following cases:
   (a) R_1: α over β over γ,
       R_2: β over γ over α,
       R_3: β over γ over α.
   (b) R_1: α over β over γ,
       R_2: γ over β over α,
       R_3: α over γ over β.

5. (Ducamp and Falmagne [1969], Ore [1962]) Imagine a set S of patients and a set E of symptoms. Let aRb hold if and only if patient a has symptom b. The problem is to assign to each patient and to each symptom a disease such that a patient has all the symptoms of his disease, and only these symptoms. Put another way, find functions $s: S \rightarrow Re$ and $e: E \rightarrow Re$ such that for all $a \in S$ and $b \in E$,

$$aRb \Leftrightarrow s(a) = e(b). \qquad (5.46)$$

Show that if S and E are finite and $R \subseteq S \times E$, then there are functions s and e satisfying Eq. (5.46) if and only if for all $a, a' \in S$ and $b, b' \in E$,

$$(aRb \ \& \ a'Rb \ \& \ a'Rb') \Rightarrow aRb'.$$

6. Suppose R is a binary relation on $A \times A$. Krantz et al. [1971, Section 4.1.2] give sufficient conditions for the existence of a function $f: A \rightarrow Re$

such that for all $a, x, y \in A$,

$$(a, x)R(a, y) \Leftrightarrow |f(a) - f(x)| > |f(a) - f(y)|. \qquad (5.47)$$

This representation is related to the representation (5.44) by taking $S = E = A$ and defining R_a on $E = A$ by

$$xR_a y \Leftrightarrow (a, y)R(a, x).$$

Krantz *et al.* [1971, Section 4.10] also study the following related representation:

$$(a, b)R(c, d) \Leftrightarrow |f(a) - f(b)| > |f(c) - f(d)|. \qquad (5.48)$$

The relation R can be derived on the basis of judged proximity, or on the basis of the response "a and b are further apart than c and d." The first representation theorem for the representation (5.48) was given by Hölder [1901]. More modern axiomatizations were first given by Suppes and Winet [1955] and by Tversky and Krantz [1970]. If (X, d) is a metric space, and f is a function from A into X, then the following representation generalizes (5.48):

$$(a, b)R(c, d) \Leftrightarrow d(a, b) > d(c, d). \qquad (5.49)$$

The representation (5.49), when used with higher dimensional Euclidean metrics, is at the foundation of *multidimensional scaling* in psychology. It has been studied theoretically by Beals, Krantz, and Tversky [1968]. Beginning with the work of Shepard [1962a, b] and Kruskal [1964a, b], a large number of computer programs have been developed to fit data to representations of the form (5.49).

Exercises 6 through 8 study the representation (5.47).

(a) Show that the following condition of negative transitivity is necessary for the representation (5.47):

$$[\sim (b, d)R(a, d) \ \& \sim (c, d)R(b, d)] \Rightarrow \sim (c, d)R(a, d).$$

(b) Consider the necessity of the following conditions as well:

(i) $[\sim (b, c)R(a, c) \ \& \ \sim (a, b)R(c, b)] \Rightarrow \sim (b, a)R(c, a)$.

(ii) $[\sim (b, d)R(a, d) \ \& \ \sim (c, b)R(d, b) \ \& \ \sim (a, c)R(b, c)] \Rightarrow \sim (c, a)R(d, a)$.

7. Given $(A \times A, R)$, we define a ternary relation of betweenness B on A as follows:

$$B(a, b, c) \Leftrightarrow [\sim (b, c)R(a, c) \ \& \sim (b, a)R(c, a)].$$

Which of the following conditions on the betweenness relation, assumed by Krantz *et al.*, are necessary for the representation (5.47)?

(a) If $b \neq c$, then $B(a, b, c)$ and $B(b, c, d)$ imply $B(a, b, c)$, $B(a, c, d)$, $B(a, b, d)$, and $B(b, c, d)$.

(b) If $B(a, b, c)$ and $B(a, c, d)$, then $B(a, b, d)$ and $B(b, c, d)$.

8. If R is a binary relation on $A \times A$ and $a, b, c \in A$, $a \neq b$, we say c is a *midpoint* of a and b if $\sim (c, a)R(c, b)$ and $\sim (c, b)R(c, a)$. Show that it is possible for there to be a function f satisfying Eq. (5.47), but for there to be some $a \neq b$ in A such that there is no midpoint c of a and b.

References

Baker, K. A., Fishburn, P. C., and Roberts, F. S., "Partial Orders of Dimension 2," *Networks*, 2 (1971), 11-28.

Beals, R., Krantz, D. H., and Tversky, A., "Foundations of Multidimensional Scaling," *Psychol. Rev.*, 75 (1968), 127-142.

Bennett, J. F., and Hays, W. L., "Multidimensional Unfolding: Determining the Dimensionality of Ranked Preference Data," *Psychometrika*, 25 (1960), 27-43.

Cochrane, M., and Zeleny, M. (eds.), *Multiple Criteria Decision Making*, University of South Carolina Press, Columbia, South Carolina, 1973.

Conway, H. M. (ed.), *The Weather Handbook*, Conway, Atlanta, Georgia, 1963.

Coombs, C. H., "Psychological Scaling without a Unit of Measurement," *Psychol. Rev.*, 57 (1950), 145-158.

Coombs, C. H., *A Theory of Data*, Wiley, New York, 1964.

Coombs, C. H., and Huang, L., "Polynomial Psychophysics of Risk," *J. Math. Psychol.*, 7 (1970), 317-338.

Coombs, C. H., and Komorita, S. S., "Measuring Utility of Money through Decisions," *Amer. J. Psychol.*, 71 (1958), 383-389.

Debreu, G., "Topological Methods in Cardinal Utility Theory," in K. J. Arrow, S. Karlin, and P. Suppes (eds.), *Mathematical Methods in the Social Sciences*, Stanford University Press, Stanford, California, 1960, pp. 16-26.

Dole, S. H., Campbell, H. G., Dreyfuss, D., Gosch, W. D., Harris, E. D., Lewis, D. E., Parker, T. M., Ranftl, J. W., and String, J., "Methodologies for Analyzing the Comparative Effectiveness and Costs of Alternative Space Plans," Memorandum RM-5656-NASA, The RAND Corporation, Santa Monica, California, August 1968.

Ducamp, A., and Falmagne, J. C., "Composite Measurement," *J. Math. Psychol.*, 6 (1969), 359-390.

Falmagne, J. C., "Random Conjoint Measurement and Loudness Summation," *Psychol. Rev.*, 83 (1976), 65-79.

Falmagne, J. C., "On a Class of Probabilistic Conjoint Measurement Models: Some Diagnostic Properties," *J. Math. Psychol.*, 19 (1979), 73-88.

Falmagne, J. C., Iverson, G., and Marcovici, S., "Binaural Loudness Summation: Probabilistic Theory and Data," manuscript, Department of Psychology, New York University, New York, New York, 1978.

Farquhar, P. H., "Fractional Hypercube Decompositions of Multiattribute Utility Functions," Tech. Rept. 222, Department of Operations Research, Cornell University, Ithaca, New York, August 1974.

Farquhar, P. H., "A Fractional Hypercube Decomposition Theorem for Multiattribute Utility Functions," *Operations Research*, 23 (1975), 941-967.

Farquhar, P. H., "Pyramid and Semicube Decompositions of Multiattribute Utility Functions," *Operations Research*, 24 (1976), 256-271.

Farquhar, P. H., "A Survey of Multiattribute Utility Theory and Applications," in M. Starr and M. Zeleny (eds.), *TIMS/North Holland Studies in the Management Sciences: Multiple Criteria Decisionmaking*, 6 (1977), 59-89.

Fishburn, P. C., *Utility Theory for Decisionmaking*, Wiley, New York, 1970.

Fishburn, P. C., "Von Neumann–Morgenstern Utility Functions on Two Attributes," *Operations Research*, **22** (1974), 35–45.

Guttman, L., "A Basis for Scaling Qualitative Data," *Amer. Sociol. Rev.*, **9** (1944), 139–150.

Hölder, O., "Die Axiome der Quantität und die Lehre vom Mass," *Ber. Verh. Kgl. Sächsis., Ges. Wiss. Leipzig, Math.-Phys. Klasse*, **53** (1901), 1–64.

Huang, L. C., "A Nonsimple Conjoint Measurement Model," *J. Math. Psychol.*, **12** (1975), 437–448.

Hull, C. L., *A Behavior System*, Yale University Press, New Haven, Connecticut, 1952.

Keeney, R. L., "Utility Independence and Preferences for Multiattributed Consequences," *Operations Research*, **19** (1971), 875–893.

Keeney, R. L., "Utility Functions for Multiattribute Consequences," *Management Sci.*, **18** (1972), 276–287.

Keeney, R. L., and Raiffa, H., *Decisions with Multiple Objectives: Preferences and Value Tradeoffs*, Wiley, New York, 1976.

Kirk, D. B., "Nonsimple Distributive Conjoint Measurement," Michigan Mathematical Psychology Program, Tech. Rept. MMPP 77-3, Department of Psychology, University of Michigan, Ann Arbor, Michigan, 1977.

Krantz, D. H., "A Survey of Measurement Theory," in G. B. Dantzig and A. F. Veinott, Jr. (eds), *Mathematics of the Decision Sciences, Part 2*, Lectures in Applied Mathematics, Vol. 12, American Mathematical Society, Providence, Rhode Island, 1968, pp. 314–350.

Krantz, D. H., and Tversky, A., "Conjoint Measurement Analysis of Composition Rules in Psychology," *Psychol. Rev.*, **78** (1971), 151–169.

Krantz, D. H., Luce, R. D., Suppes, P., and Tversky, A., *Foundations of Measurement*, Vol. I, Academic Press, New York, 1971.

Kruskal, J. B., "Multidimensional Scaling by Optimizing Goodness of Fit to a Nonmetric Hypothesis," *Psychometrika*, **29** (1964a), 1–28.

Kruskal, J. B., "Nonmetric Multidimensional Scaling: A Numerical Method," *Psychometrika*, **29** (1964b), 115–130.

Kryter, K. D., *The Effects of Noise on Man*, Academic Press, New York, 1970.

Leibowitz, R., "Interval Counts and Threshold Numbers of Graphs," Doctoral Dissertation, Department of Mathematics, Rutgers University, New Brunswick, New Jersey, 1978.

Levelt, W. J. M., Riemersma, J. B., and Bunt, A. A., "Binaural Additivity of Loudness," *Brit. J. Math. Stat. Psychol.*, **25** (1972), 51–68.

Luce, R. D., "A 'Fundamental' Axiomatization of Multiplicative Power Relations among Three Variables," *Phil. Sci.*, **32** (1965), 301–309.

Luce, R. D., "Dimensionally Invariant Numerical Laws Correspond to Meaningful Qualitative Relations," *Phil. Sci.*, *45 (1978), 1–16.*

Luce, R. D., and Suppes, P., "Preference, Utility, and Subjective Probability," in R. D. Luce, R. R. Bush, and E. Galanter (eds.), *Handbook of Mathematical Psychology*, Vol. 3, Wiley, New York, 1965, pp. 249–410.

Luce, R. D., and Tukey, J. W., "Simultaneous Conjoint Measurement: A New Type of Fundamental Measurement," *J. Math. Psychol.*, **1** (1964), 1–27.

MacCrimmon, K. R., and Siu, J. K., "Making Trade-offs," *Decision Sci.*, **5** (1974), 680–704.

MacCrimmon, K. R., and Wehrung, D. A., "Trade-off Analysis: Indifference and Preferred Proportion," in D. E. Bell (ed.), *Conflicting Objectives in Decisions*, Wiley, New York, 1978.

Manheim, M. L., and Hall, F. L., "Abstract Representation of Goals," Paper P-67-24, Department of Civil Engineering, Massachusetts Institute of Technology, Cambridge, Massachusetts, January 1968.

Marley, A. A. J., "An Alternative 'Fundamental' Axiomatization of Multiplicative Power Relations among Three Variables," *Phil. Sci.*, **35** (1968), 185–186.

Miller, J. R., "Assessing Alternate Transportation Systems," Memorandum RM-5865-DOT, The RAND Corporation, Santa Monica, California, 1969.

Narens, L., "Minimal Conditions for Additive Conjoint Measurement and Qualitative Probability," *J. Math. Psychol.*, **11** (1974), 404–430.

Narens, L, and Luce, R. D., "The Algebra of Measurement," *J. Pure Appl. Algebra*, **8** (1976), 197–233.

Ore, O., *Theory of Graphs*, American Mathematical Society Colloquium Publications, Vol. XXXVIII, Providence, Rhode Island, 1962.

Pfanzagl, J., "Die Axiomatischen Gründlagen einer Allgemeinen Theorie des Messens," *Schriftenreihen Statist. Instit. Univ. Wien, Neue Folge*, Nr. 1, Physica Verlag, Würzburg, 1959.

Raiffa, H., "Preferences for Multi-attributed Consequences," Memorandum RM-5868-DOT, The RAND Corporation, Santa Monica, California, 1969.

Roberts, F. S., "What If Utility Functions Do Not Exist?", *Theory and Decision*, **3** (1972), 126–139.

Roberts, F. S., "Laws of Exchange and Their Applications," *SIAM J. Appl. Math.*, **26** (1974), 260–284.

Scott, D., "Measurement Models and Linear Inequalities," *J. Math. Psychol.*, **1** (1964), 233–247.

Shepard, R. N., "The Analysis of Proximities: Multidimensional Scaling with an Unknown Distance Function. I," *Psychometrika*, **27** (1962a), 125–140.

Shepard, R. N., "The Analysis of Proximities: Multidimensional Scaling with an Unknown Distance Function. II," *Psychometrika*, **27** (1962b), 219–246.

Sidowski, J. B., and Anderson, N. H., "Judgments of City-Occupation Combinations," *Psychon. Sci.*, **7** (1967), 279–280.

Spence, W. K., *Behavior Theory and Conditioning*, Yale University Press, New Haven, Connecticut, 1956.

Stevens, S. S., "Measurement, Psychophysics, and Utility," in C. W. Churchman and P. Ratoosh (eds.), *Measurement: Definitions and Theories*, Wiley, New York, 1959, pp. 18–63.

Stouffer, S. A., Guttman, L., Suchman, E. A., Lazarsfeld, P. F., Star, S. A., and Clausen, J. A., *Measurement and Prediction*, Princeton University Press, Princeton, New Jersey, 1950.

Suppes, P., and Winet, M., "An Axiomatization of Utility Based on the Notion of Utility Differences," *Management Sci.*, **1** (1955), 259–270.

Suppes, P., and Zinnes, J. L., "Basic Measurement Theory, " in R. D. Luce, R. R. Bush, and E. Galanter (eds.), *Handbook of Mathematical Psychology*, Vol. I, Wiley, New York, 1963, pp. 1–76.

Tversky, A., "Additivity, Utility, and Subjective Probability," *J. Math. Psychol.*, **4** (1967), 175–201.

Tversky, A., "Intransitivity of Preferences," *Psychol. Rev.*, **76** (1969), 31–48.

Tversky, A., and Krantz, D. H., "The Dimensional Representation and the Metric Structure of Similarity Data," *J. Math. Psychol.*, **7** (1970), 572–596.

Uzawa, H., "Preference in Rational Choice in the Theory of Consumption," in K. J. Arrow, S. Karlin, and P. Suppes (eds.), *Mathematical Methods in the Social Sciences*, Stanford University Press, Stanford, California, 1960, pp. 129–148.

Wallsten, T. S., "Using Conjoint-Measurement Models To Investigate a Theory about Probabilistic Information Processing," *J. Math. Psychol.*, **14** (1976), 144–185.

Wold, H., and Jureen, L., *Demand Analysis, A Study in Econometrics*, Wiley, New York, 1953.

CHAPTER 6

Nontransitive Indifference, Probabilistic Consistency, and Measurement without Numbers

6.1 Semiorders and Interval Orders

6.1.1 *Nontransitivity of Indifference*

In Section 3.1, we gave examples where R is preference and there is no homomorphism from (A, R) into $(Re, >)$. In this chapter, we give some additional examples and then we ask whether or not measurement is still possible if such a homomorphism does not exist. We are led in the process to consider several unorthodox examples of measurement, including measurement without numbers, and measurement that starts with probabilities or proportions instead of relations. The results have applications beyond preference, in particular to measurement of psychophysical quantities such as loudness, which we studied in Chapter 4.

If R is (strict) preference on a set A, then indifference corresponds to the binary relation I on A defined by

$$aIb \Leftrightarrow \sim aRb \,\&\, \sim bRa. \qquad (6.1)$$

That is, you are indifferent between a and b if and only if you prefer neither a to b nor b to a*. Suppose there is an ordinal utility function f on A, that is, a function $f : A \to Re$ satisfying

$$aRb \Leftrightarrow f(a) > f(b). \qquad (6.2)$$

---

*In previous chapters, we used the notation E for this relation. Here we use I, for reasons to be explained below.

ENCYCLOPEDIA OF MATHEMATICS and Its Applications, Gian-Carlo Rota (ed.). Vol. 7: Fred S. Roberts, Measurement Theory

If f exists, then

$$aIb \Leftrightarrow f(a) = f(b). \tag{6.3}$$

Equation (6.3) implies that I is transitive, for aIb and bIc imply $f(a) = f(b)$ and $f(b) = f(c)$, whence $f(a) = f(c)$, so aIc.

The economist Armstrong [1939, 1948, 1950, 1951] was one of the first to argue that indifference is not necessarily transitive.† (Menger [1951] claims that attacks on the transitivity of indifference go back as far as Poincaré in the nineteenth century.) Luce [1956] suggests as one argument against the transitivity of indifference the following. Most people would prefer a cup of coffee with one spoon of sugar to a cup with five spoons. But if sugar were added to the first cup at the rate of 1/100 of a gram, they would almost certainly be indifferent between successive cups. If indifference were transitive, they would have to be indifferent between the cup with one spoon and the cup with five spoons. Similarly if preference between air environments is determined on the basis of eye irritation, then you probably prefer an air environment with .05 parts per million (ppm) of ozone to one with .5 ppm. But you remain indifferent if ozone is added in amounts of 10^{-10} ppm at a time. To give a related example, in the judgments that one sound is louder than another, we might easily find three sounds a, b, and c such that a and b as well as b and c are judged equally loud, because they are sufficiently close, while a and c are sufficiently far apart so we can recognize one as louder. Thus, the attack on transitivity of the relation I extends beyond the case of preference.

A considerably different example is the following. Suppose you are indifferent between two alternative plans for government support of the arts, plans a and b, where plan a would allocate a budget of 200 million dollars to a federal Institute for the Arts and plan b would allocate 200 million dollars to various state institutes. It seems likely that you would still be indifferent between plan a and plan b', which would allocate 200 million and one dollars to the state institutes. For probably if you have any preference when budgets are so close it will be based on a choice of a particular approach to support of the arts (federal versus state). On the other hand, if you want to see the government spend money on the arts, you would certainly prefer b' to b, which violates transitivity of indifference.

A famous related example in utility theory, due to Armstrong [1939], is the following. Suppose a boy is indifferent between receiving as a gift a pony or a bicycle. He will undoubtedly prefer the bicycle if a bell is added to the bicycle without the bell. But he is still likely to be indifferent

---

†Hence, indifference is not an equivalence relation, which is one reason why we choose to use I rather than E for indifference in this chapter.

between the bicycle with bell and the pony. Hence, indifference is not transitive.

Other arguments against the transitivity of indifference and many references to the literature of this issue can be found in Fishburn [1970a] and in Krantz et al. [to appear].*

The problem of nontransitivity of indifference led Luce [1956] to slightly modify the demands in the measurement of preference. Motivated by examples like the first two, and the notion of threshold in psychophysics, he suggested that we seek a real valued function f on A so that for all a, $b \in A$, a is preferred to b if and only if $f(a)$ is not only larger than $f(b)$ but "sufficiently larger" so that we can tell a and b apart. To formalize this representation problem, we fix a positive number δ, the threshold, and ask for conditions on the relational system (A, R) necessary and sufficient for the existence of a real-valued function f on A such that for all a, $b \in A$,

$$aRb \Leftrightarrow f(a) > f(b) + \delta. \tag{6.4}$$

This representation is obviously of interest for judgments of relative loudness as well as for judgments of preference. To formulate this representation in the measurement-theoretic terms of Section 2.1, we define a binary relation $>_\delta$ on Re by

$$x >_\delta y \Leftrightarrow x > y + \delta.$$

Then we ask for a homomorphism from (A, R) into $(Re, >_\delta)$.†

We shall restrict our discussion to the case where A is a finite set, in which case conditions on (A, R) necessary and sufficient for the representation (6.4) can be explicitly stated. We should note that this representation, which is designed to account for examples like the cups of coffee, the comparison of air environments, and the comparison of loudness, does not account for examples like the alternative budgets and the pony–bicycle. For example, if a' is the plan of budgeting 200 million and one dollars to the federal government, you probably prefer a' to a, but are

---

*Kramer [1968] argues that the nontransitivity of indifference may be due to the organism's limited capacity. Computer scientists have made analogous statements, to the effect that the relation of equality between numbers is nontransitive for a computer due to round off error, resulting from limited memory, speed, or available time. (See Hamming [1965] and Rothstein [1965].) These points were made by Professor Jacob Marschak in a Western Management Science Colloquium at U.C.L.A. in 1971. The limited capacity argument might explain the preferences for cups of coffee and for air environments, but it is not clear that it explains the alternative budgets or the pony–bicycle example.

†Other representations for preference using the notion of threshold are considered in Fishburn [1970a, b], Krantz et al. [to appear], Luce [1956], and Roberts [1969a, 1971b]. In these treatments, the threshold is often allowed to vary from place to place rather than remaining constant as it does here.

indifferent between a' and b and between a' and b'. Thus, your preference relation R on the set of plans $\{a, a', b, b'\}$ is probably the relation $\{(a', a), (b', b)\}$. But this relation cannot be represented in the form (6.4). For if it could, then we would have

$$f(a') > f(a) + \delta \geqq f(b') > f(b) + \delta,$$

whence $a'Rb$. We return to the alternative budgets example in Section 6.2.3, where we discuss the condition of strong stochastic transitivity.

6.1.2 The Scott–Suppes Theorem

Conditions on (A, R) necessary and sufficient for the representation (6.4) are embodied in the concept of a semiorder, a concept introduced by Luce [1956]. Our definition of semiorder is formulated following that of Scott and Suppes [1958]. The binary relation (A, R) is called a *semiorder* if, for all $a, b, c, d \in A$, the following axioms are satisfied:

AXIOM S1. $\sim aRa$.

AXIOM S2. aRb & $cRd \Rightarrow [aRd$ or $cRb]$.

AXIOM S3. aRb & $bRc \Rightarrow [aRd$ or $dRc]$.

To explain the axioms, and to see that they follow from the representation (6.4), let us first note that Axiom S1 says that (A, R) is irreflexive, which follows since $f(a)$ can never be larger than $f(a) + \delta$. To see that Axiom S3 holds, suppose aRb and bRc. Then $f(a)$, $f(b)$, and $f(c)$ have positions like those in Fig. 6.1. Now $f(d) \geqq f(b)$ implies dRc, and $f(d) \leqq f(b)$ implies aRd. To see that Axiom S2 holds, consider two cases: $f(a) \geqq f(c)$ and $f(c) \geqq f(a)$. In the first case, we have

$$f(a) \geqq f(c) > f(d) + \delta,$$

so aRd. In the second case, we have

$$f(c) \geqq f(a) > f(b) + \delta,$$

so cRb.

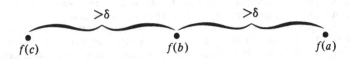

Figure 6.1. Axiom S3 of the definition of a semiorder is a necessary condition.

In the budgetary example where $A = \{a, a', b, b'\}$ and $R = \{(a', a), (b', b)\}$, we see clearly that Axiom S2 is violated so that the preference relation is not a semiorder.

It should be remarked in passing that every semiorder is a strict partial order (Section 1.6). To prove this, we note that transitivity follows from Axiom S1 by taking $d = c$ in Axiom S3.

We shall prove the following theorem, due to Scott and Suppes [1958].

THEOREM 6.1 (Scott and Suppes). *Suppose R is a binary relation on a finite set A and δ is a positive number. Then (A, R) is a semiorder if and only if there is a real-valued function f on A such that for all $a, b \in A$,*

$$aRb \Leftrightarrow f(a) > f(b) + \delta. \tag{6.4}$$

COROLLARY. *If a binary relation R on a finite set A is representable in the form (6.4) for some positive number δ, then it is representable in the form (6.4) for any positive number δ. In particular, it is representable in the form*

$$aRb \Leftrightarrow f(a) > f(b) + 1. \tag{6.5}$$

Proof of Corollary. If f satisfies (6.4), then $(\delta'/\delta)f$ satisfies (6.4) with δ' in place of δ.

We defer a proof of Theorem 6.1 until Section 6.1.7. Note that finiteness is not needed for the necessity of the semiorder axioms, only for their sufficiency.

Before closing this section, we give an example. Let

$$A = \{w, x, y, z, \alpha, \beta, \gamma\}$$

and let

$$R = \{(w, x), (w, y), (w, z), (w, \alpha), (w, \beta), (w, \gamma), (x, \alpha),$$
$$(x, \beta), (x, \gamma), (y, \beta), (y, \gamma), (z, \gamma), (\alpha, \gamma)\}. \tag{6.6}$$

Then (A, R) is not a strict weak order, for $\sim xRz$ and $\sim zR\alpha$, but $xR\alpha$, which violates negative transitivity. Thus, there is no homomorphism from (A, R) into $(Re, >)$. But there is a function f on A satisfying Eq. (6.4), for (A, R) is a semiorder. Axiom S1 for semiorders is straightforward. Axioms S2 and S3 are tedious to check by hand. For example, we note that $xR\alpha$ and $zR\gamma$. Thus, Axiom S2 requires that $xR\gamma$ or $zR\alpha$ holds. The former is the case. Similar checks must be made for many cases to verify Axiom S2. Similarly, we note that $xR\alpha$ and $\alpha R\gamma$. Using $d = z$, we note that, by Axiom S3, either xRz or $zR\gamma$ must hold. We have $zR\gamma$. Similar checks must be made case by case to verify Axiom S3.

An easier way to check that (A, R) is a semiorder is to find a function f satisfying Eq. (6.4). If $\delta = 1$, such a function is given for our example by

$$f(w) = 5, \quad f(x) = 3, \quad f(y) = 2.7, \quad f(z) = 2.5,$$
$$f(\alpha) = 1.9, \quad f(\beta) = 1.6, \quad f(\gamma) = .8.$$

It is left to the reader to check that f satisfies Eq. (6.4).

6.1.3 Uniqueness

The representation (6.4), that is, the representation

$$\mathfrak{A} = (A, R) \to \mathfrak{B} = (Re, >_\delta),$$

is our first example of an irregular representation. To see this, let $A = \{x, y, z\}$ and let $R = \{(x, z), (x, y)\}$. Then two functions satisfying Eq. (6.4) with $\delta = 1$ are given by

$$f(x) = 2, \quad f(y) = 0, \quad f(z) = 0 \tag{6.7}$$

and

$$g(x) = 2, \quad g(y) = .1, \quad g(z) = 0. \tag{6.8}$$

By Theorem 2.1, $(\mathfrak{A}, \mathfrak{B}, f)$ is not a regular scale. (We have already encountered this example in Section 2.2.) A uniqueness theorem that specifies a class of admissible transformations would not be helpful here, since this class could differ from homomorphism to homomorphism. Thus the theory of scale type discussed in Section 2.3 does not apply to all semiorders.* The theory of meaningfulness does apply, however, if we use the more general definition that a statement is meaningful if its truth or falsity is unchanged when scales in the statement are replaced by other acceptable scales. In this sense, in the case of a homomorphism f from (A, R) into $(Re, >_\delta)$, the statement

$$f(a) > f(b)$$

is not meaningful. For if a is y and b is z, this is not true for the homomorphism f of Eq. (6.7), but is true for the homomorphism g of Eq.

---

*However, if $\mathfrak{A}$ is a semiorder, then according to the discussion of Section 2.2.2, there is an isomorphism F from the reduction $\mathfrak{A}^*$ of $\mathfrak{A}$ into $\mathfrak{B} = (Re, >_\delta)$. (Hence, $\mathfrak{A}^*$ is a semiorder.) The Corollary to Theorem 2.1 implies that $(\mathfrak{A}^*, \mathfrak{B}, F)$ is regular. The class of admissible transformations of such a system $(\mathfrak{A}^*, \mathfrak{B}, F)$ does not (usually) take any of the standard forms (such as similarity transformations), and its characterization in general is an open problem. However, Manders [1977] has recently obtained some interesting results on this problem.

(6.8). It is meaningful, however, to assert that

$$f(a) > f(b) + \delta,$$

that is, to assert that $f(a)$ is "sufficiently larger" than $f(b)$. We return to the uniqueness question for semiorders in Section 6.1.6.

6.1.4 *Interval Orders and Measurement without Numbers*

For another insight into the representation (6.4), let us consider the interval

$$J(a) = [f(a) - \delta/2, f(a) + \delta/2].$$

If J and J' are two real intervals, we shall say that

$$J \succ J' \quad \text{iff} \quad a > b \quad \text{for all} \quad a \in J \quad \text{and} \quad b \in J'.$$

If $J \succ J'$, we say that J *strictly follows* J'. If f satisfies (6.4), then

$$aRb \Leftrightarrow J(a) \succ J(b). \tag{6.9}$$

We may think of $J(a)$ as a range of fuzziness about a, or a range of possible values. For example, if we are estimating the monetary value of a particular product (as is done in one popular television program), $J(a)$ could be a range of estimates. The model of behavior embodied in Eq. (6.9) says that we prefer a to b if and only if we are sure that every possible value of a is larger than every possible value of b. If (A, R) is a semiorder, then all the intervals $J(a)$ have the same length. But it is interesting to think of the possibility of letting them have different lengths. Certainly in the case of estimating the monetary values of different products, we want to allow the ranges of values to be of different lengths for different products. We now ask: When does there exist an assignment to each a in A of an interval $J(a)$ so that for all a, b in A, (6.9) is satisfied? That is, under what circumstances is a person acting (at least) as if he satisfies the model (6.9)? If we take a more general point of view than we did in Section 2.1, then the assignment of intervals satisfying Eq. (6.9) is as legitimate a form of measurement as the assignment of numbers satisfying the representation

$$aRb \Leftrightarrow f(a) > f(b).$$

For, one of the goals of measurement is to reflect empirical relations by well-known relations on mathematical objects. Having translated an empirical relational system into what we shall loosely call a mathematical relational system, we can apply the whole collection of mathematical tools at our disposal to better understand the mathematical system and hence

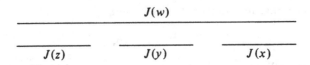

Figure 6.2. An interval representation for the interval order (A, R), where $A = \{x, y, z, w\}$ and $R = \{(x, y), (y, z), (x, z)\}$. Intervals are displaced vertically for ease of comparison.

the empirical one. In particular, we can apply our mathematical tools to help in decisionmaking. In this broad sense, assignment of vectors, sets, intervals, geometric objects, etc., is a perfectly legitimate form of measurement if a representation theorem stating a homomorphism from an empirical relational system to a mathematical relational system can be proved. A similar point of view is expressed in Krantz [1968] and in Coombs, Raiffa, and Thrall [1954].

Having expressed this point of view, let us state a representation theorem for the representation (6.9). A binary relation (A, R) is called an *interval order* if it satisfies the first two axioms in the definition of a semiorder. Clearly, every semiorder is an interval order. But it is not too hard to give an example of an interval order that is not a semiorder. Let $A = \{x, y, z, w\}$ and define R on A by

$$R = \{(x, y), (y, z), (x, z)\}.$$

An interval representation satisfying Eq. (6.9) for (A, R) is shown in Fig. 6.2. But (A, R) is not a semiorder, since xRy and yRz, but $\sim xRw$ and $\sim wRz$.

We now have the following representation theorem.

THEOREM 6.2 (Fishburn [1970 b,c]). *Suppose (A, R) is a binary relation on a finite set A. Then (A, R) is an interval order if and only if there is an assignment of an interval $J(a)$ to each a in A so that for all a, b in A,*

$$aRb \Leftrightarrow J(a) > J(b). \tag{6.9}$$

To the best of the author's knowledge, there has not been much work done on the uniqueness of this representation. However, Greenough and Bogart [1979] define the length of an interval order and show that an interval order (without duplicated holdings*) of length n has a unique representation as a collection of intervals using no more than n points as end points. W.T. Trotter (personal communication) has also obtained some recent results. Uniqueness questions whose answer would be of particular interest for applications revolve around the question of how overlapping

---

*See Greenough and Bogart for a definition.

intervals overlap—in particular, when must one be contained inside another?

Let us comment briefly on why these kinds of questions are of interest. In many problems in the social sciences, we wish to put some objects into serial order. For example, in political science we wish to list candidates ranging from liberal to conservative. In psychology, we wish to place individuals in order of stages of development. In archaeology, we wish to place artifacts in chronological order. The general problem of seriation has close connections with interval orders (and with the interval graphs studied in Exers. 25, 26, and 30). Let us discuss the seriation problem in the context of sequence dating in archaeology. For references on this subject, see Kendall [1963, 1969a,b, 1971a,b,c]. Each (type of) artifact in a collection of interest was in use over a certain period (interval) of time. Suppose we know for two artifacts, a and b, whether or not the time period of a strictly followed that of b. By Theorem 6.2, an assignment of a time interval $J(a)$ to each artifact a which preserves the observed relation of strict following exists if and only if this observed relation defines an interval order. If such an assignment exists, we can still get relationships among the time intervals wrong. For example, we might in reality have the time interval for x beginning after that for y and ending before that for y. However, an assignment J satisfying Eq. (6.9) might not have this property. It is only required that $J(x)$ and $J(y)$ overlap. It would be helpful to know under what circumstances two different interval assignments J and J' satisfying Eq. (6.9) can have the property that $J(x)$ is contained in $J(y)$ while $J'(x)$ is not contained in $J'(y)$, and under what circumstances this cannot happen. For a further discussion of seriation problems and their connection with interval orders and interval graphs, see Coombs and Smith [1973], Hubert [1974], Roberts [1976, Section 3.4] or Roberts [1978b, Sections 3.3 and 4.2].

6.1.5 Compatibility between a Weak Order and a Semiorder

Returning to the Scott–Suppes Theorem, we shall show that the theorem is false without the assumption that A is finite. Then, we shall see how to generalize the theorem. Our results will be useful in proving the Scott–Suppes Theorem, and they will be applied in our discussion of probabilistic consistency in Section 6.2.

To show that the Scott–Suppes Theorem is false without the assumption that A is finite, let N be the set of positive integers and let α be any element not in N. Take A to be $N \cup \{\alpha\}$ and define R on A by

$$\left\{\begin{array}{ll} aRb \Leftrightarrow a > b + 1 & \text{for } a, b \in N, \\ \alpha Ra & \text{for all } a \in N. \end{array}\right\} \tag{6.10}$$

It is not hard to verify that (A, R) is a semiorder. But there is no function $f: A \to Re$ satisfying Eq. (6.4). For suppose such a function f exists. Note that since $2R0$, $f(2) > f(0) + \delta$. By induction, $f(2n) > f(0) + n\delta$. Now $\alpha R 2n$ for all n, so $f(\alpha) > f(2n) + \delta > f(0) + (n + 1)\delta$. Thus, $f(\alpha)$ is larger than every positive number, which is impossible.

To see how to generalize the Scott–Suppes Theorem, let us suppose for the moment that a function f satisfying Eq. (6.4) exists. Then we define a binary relation W on A by

$$aWb \Leftrightarrow f(a) \geq f(b). \tag{6.11}$$

By Corollary 2 to Theorem 3.4, (A, W) is a weak order. It corresponds to the weak order on the reals "weakly to the right of" and gives us the relative order of the f values, if not the specific values. Now before we have calculated the function f, we do not know what W is. But we shall be able to uncover an appropriate W by defining it explicitly in terms of R. Namely, we take

$$aWb \Leftrightarrow (\forall c)[(bRc \Rightarrow aRc) \& (cRa \Rightarrow cRb)]. \tag{6.12}$$

If W is the relation of Eq. (6.11), then certainly it satisfies the implication $\Rightarrow$ of Eq. (6.12). For

$$aWb \ \& \ bRc \Rightarrow [f(a) \geq f(b)] \ \& \ [f(b) > f(c) + \delta]$$
$$\Rightarrow f(a) > f(c) + \delta$$
$$\Rightarrow aRc.$$

Similarly,

$$aWb \ \& \ cRa \Rightarrow cRb.$$

However, W does not necessarily satisfy the implication $\Leftarrow$ of Eq. (6.12).

To illustrate Eq. (6.12), let us consider the example $A = \{w, x, y, z, \alpha, \beta, \gamma\}$ with R defined by Eq. (6.6). Then xWz holds. For $zRc \Rightarrow xRc$. (There is only one case to check: $zR\gamma$ holds and also $xR\gamma$.) Similarly, nothing is in the relation R to x, so of course $cRx \Rightarrow cRz$. Similarly, xWx, xWy, $zW\alpha$, etc. W is the weak order which ranks w largest, then x, then y, then z, then α, then β, and then γ. W here is a simple order. In general, W is only a weak order, as we shall prove shortly. To give a second example, if $A = N \cup \{\alpha\}$ and R is defined by Eq. (6.10), then the weak order W is given by the order $>$ on N with $\alpha W a$ for all a in N. W is again simple.

LEMMA 6.3. *If (A, R) is a semiorder and W is defined by Eq. (6.12), then (A, W) is a weak order.*

Proof. (A, W) is a weak order if and only if it is transitive and strongly complete. To verify transitivity, suppose xWy and yWz. To show xWz, choose c in A and show

$$zRc \Rightarrow xRc$$

and

$$cRx \Rightarrow cRz.$$

If zRc, then since yWz, yRc follows. Since xWy, yRc implies xRc. Thus, $zRc \Rightarrow xRc$. A similar proof shows that $cRx \Rightarrow cRz$. So far, the semiorder axioms have not been used. Proof of strong completeness proceeds by cases and uses the semiorder axioms. Details are left to the reader. ∎

The binary relation W defined by Eq. (6.12) is called the *weak order associated with R*. Where several semiorders R exist, it will be convenient to denote the associated weak orders by $W(R)$. The definition of $W(R)$ is originally due to Luce [1956], and in the present form is due to Scott and Suppes [1958].

If (A, R) is a semiorder, then as before define a binary relation I on A by

$$aIb \Leftrightarrow \sim aRb \;\&\; \sim bRa. \tag{6.1}$$

If R is preference, then I is indifference. As we have observed, if R is a semiorder, I is not necessarily an equivalence relation as it was when R was a strict weak order; for I may not be transitive.

LEMMA 6.4. *If (A, R) is a semiorder and (A, W) is its associated weak order, then for all $a, b, c \in A$,*

$$aRb \Rightarrow aWb \tag{6.13}$$

and

$$aWbWc \;\&\; aIc \Rightarrow aIb \;\&\; bIc. \tag{6.14}$$

Condition (6.14) is known as the *weak mapping rule*, and was introduced by Goodman [1951] in the following equivalent form:

$$aWbWcWd \;\&\; aId \Rightarrow bIc. \tag{6.15}$$

Equation (6.15) says that intervals of preference cannot be contained within intervals of indifference. That is, we cannot prefer b to c or c to b and at the same time be indifferent between a and d, if a is weakly to the right of b, which is weakly to the right of c, which is weakly to the right of d. Similarly, alternatives that are not within threshold cannot be contained within alternatives that are within threshold. (Proof of the equivalence of (6.14) and (6.15) is left to the reader.) The weak mapping rule has recently found applications to seriation problems in archaeology, psychology, and political science. See Hubert [1974] and Roberts [1979].

Proof of Lemma 6.4. Note that (A, R) is transitive—we already observed that this follows from the third semiorder axiom. Suppose that aRb. Then $bRc \Rightarrow aRc$ and $cRa \Rightarrow cRb$, since R is transitive. This proves (6.13). To prove (6.14), suppose $aWbWc$ and aIc. To show aIb, suppose aIb is false. Then by Eq. (6.1), either aRb or bRa. If aRb, then bWc implies aRc, which contradicts aIc. If bRa, then since aWb, bRa implies aRa, using $c = a$. This contradicts the first semiorder axiom. A similar proof establishes bIc. ∎

We say that a binary relation (A, R) and a weak order (A, W) are *compatible* if for all $a, b, c \in A$, Eqs. (6.13) and (6.14) hold. Thus, we have seen that every semiorder has a compatible weak order. The converse is also true, if (A, R) is asymmetric, and will be useful in various applications.

THEOREM 6.5 (Roberts [1971b]). *Let (A, R) be an asymmetric binary relation. Then there is some weak order on A compatible with (A, R) if and only if (A, R) is a semiorder.*

Proof. The semiorder axioms can be verified directly from asymmetry and Eqs. (6.13) and (6.14). The details are left to the reader.

If (A, R) is a semiorder, if $f:A \to Re$ satisfies Eq. (6.4), and if W is defined on A by Eq. (6.11), then (A, W) is a weak order on A compatible with (A, R). Proof is straightforward. It is not necessarily the case that W satisfies Eq. (6.12). However, we shall observe in Theorem 6.6 that the weak order defined by (6.12) and W are "essentially" the same.

6.1.6 *Uniqueness Revisited*

There is no uniqueness theorem for the representation (6.4) which defines the class of admissible transformations of the function f. However, there is a uniqueness theorem in a different sense. Recall that in the case of strict weak orders (A, R), the relation I of indifference defined by Eq. (6.1) was an equivalence relation (Theorem 1.3). For semiorders (A, R) it is not: nontransitivity of I was the motivation for the concept of semiorder. We introduce an equivalence relation E on A by

$$aEb \Leftrightarrow (\forall c \in A)(aIc \Leftrightarrow bIc). \tag{6.16}$$

Two alternatives a and b are in the relation E if and only if they are (considered) indifferent to exactly the same alternatives c. The relation E turns out to be exactly the perfect substitutes relation defined from $\mathfrak{A} = (A, R)$ by the method of Section 1.8, as is easy to prove. In our example of Eq. (6.6), we do not have aEb for any $a \neq b$. For example, $\sim xEy$, since $\sim xIa$, but yIa. To give an example that has some equivalent elements, let $A = \{x, y, z\}$ and $R = \{(x, z), (y, z)\}$. Then xEy, since xIx, xIy, yIx, and yIy. In fact, in this example, if $u \neq v$, then uEv iff $\{u, v\} = \{x, y\}$.

Suppose W is a weak order on A compatible with the semiorder (A, R). Proceeding by the method of Section 1.8, we now define $A^* = A/E$ to be the collection of equivalence classes under E and define $W^* = W/E$ on A^* by

$$a^* W^* b^* \Leftrightarrow aWb. \tag{6.17}$$

As usual with this procedure, W^* is well-defined. Moreover, it is a simple order on A^*, since (A, W) is a weak order. If (A, W) is already simple, then every equivalence class has only one element, and (A^*, W^*) is just the same as (A, W). We define R^* on A^* by

$$a^* R^* b^* \Leftrightarrow aRb. \tag{6.18}$$

The reader can easily verify that (A^*, R^*) is well-defined, that it is a semiorder, and that

$$a^* W^* b^* \Leftrightarrow a^* W(R^*) b^*, \tag{6.19}$$

where $W(R^*)$ is the weak order associated with R^* by means of Eq. (6.12).

Suppose next that W and W' are two weak orders on A compatible with (A, R). If W^* and W'^* are defined from W and W', respectively, by Eq. (6.17), then Eq. (6.19) implies that $W^* = W'^*$. It is now not hard to show that the only freedom allowed in the weak orders (A, W) or (A, W') is to vary the ordering of points within equivalence classes under the relation E; this ordering within equivalence classes can be varied arbitrarily. To illustrate, if $A = \{x, y, z\}$ and $R = \{(x, z), (y, z)\}$, then xEy, and three weak orders on A compatible with R are

$$W = \{(x, x), (y, y), (z, z), (x, y), (x, z), (y, z)\},$$
$$W' = \{(x, x), (y, y), (z, z), (y, x), (x, z), (y, z)\},$$

and

$$W'' = \{(x, x), (y, y), (z, z), (x, y), (y, x), (x, z), (y, z)\}.$$

To summarize:

THEOREM 6.6 (Roberts [1971b]). *Let (A, R) be a semiorder and suppose (A, W) and (A, W') are two weak orders on A.*
  (a) *If both (A, W) and (A, W') are compatible with R, then if $\sim aEb$, we have $aWb \Leftrightarrow aW'b$.*
  (b) *If (A, W) is compatible with R and if, whenever $\sim aEb$, we have $aWb \Leftrightarrow aW'b$, then (A, W') is compatible with R.*

COROLLARY 1. *If (A, R) is a semiorder, then $W(R^*)$ defined from (A^*, R^*) by Eq. (6.12) is the unique simple order compatible with (A^*, R^*).*

COROLLARY 2. *If (A, R) is a semiorder and (A, W) is a compatible weak order on A, then (A, W) is obtained from $(A^*, W(R^*))$ by ordering elements within equivalence classes.*

COROLLARY 3. *If (A, R) is a semiorder, then it has a compatible simple order W.*

Proof. Obtain W from $W(R^*)$ by using a simple order within each equivalence class.

6.1.7 *Proof of the Scott–Suppes Theorem*

We are now ready to present a proof of the Scott–Suppes Theorem. We have already shown in Section 6.1.2 that the semiorder axioms follow from the representation (6.4). Let us now prove the converse. This proof is constructive.[*] By the Corollary to Theorem 6.1, we may set $\delta = 1$. Let (A, R) be a semiorder, with A finite. By Corollary 3 to Theorem 6.6, there is a simple order W on A compatible with R. We shall prove by induction on $n = |A|$ that there is a function $f : A \to Re$ satisfying the following three conditions for all $a, b \in A$:

$$aRb \Leftrightarrow f(a) > f(b) + 1, \tag{6.20}$$

$$aWb \Leftrightarrow f(a) \geq f(b), \tag{6.21}$$

$$|f(a) - f(b)| \neq 1. \tag{6.22}$$

---

[*]The author thanks a referee for suggesting the idea of this proof, which is different from the original Scott–Suppes proof. For an alternative proof, see Rabinovitch [1977].

If $|A| = 1$, the result is clear. Thus, let us assume the result for $|A| = n - 1$ and show it for $|A| = n$.

By virtue of Corollary 2 of Theorem 3.1, we may list the elements of A as $a_1, a_2, \ldots, a_n$ in such a way that $a_i W a_j$ iff $i \geq j$.

LEMMA 6.7. $a_i R a_j \Leftrightarrow i > j$ and $\sim a_i I a_j$.   (6.23)

Proof. The implication $\Rightarrow$ follows from the definition of I and the compatibility between W and R, which implies that $a_i W a_j$. To prove $\Leftarrow$, note that since W is simple, $i > j \Rightarrow \sim a_j W a_i$. Thus, $\sim a_j R a_i$. ∎

Let $A' = A - \{a_n\}$ and let R' and W' be the respective restrictions of R and W to A'. It is clear that (A', R') is a semiorder and (A', W') is a compatible simple order. Thus, by the induction hypothesis, there is a function $f : A' \to Re$ satisfying conditions (6.20) through (6.22) for all a, b in A'. We shall extend f to A by defining $f(a_n)$. This is defined by considering three cases.

Case 1. $a_n R a_{n-1}$.

In this case, compatibility implies that $a_n R a_i$ for all $i < n$. Define $f(a_n)$ to be $f(a_{n-1}) + 2$. It is clear that the extended function f satisfies Eqs. (6.20) through (6.22).

Case 2. $\sim a_n R a_1$.

In this case, by Lemma 6.7, $a_n I a_i$ for all $i < n$. Let

$$\lambda = f(a_{n-1}) - f(a_1).$$

By condition (6.22), $\lambda \neq 1$, and by compatibility, $a_{n-1} I a_1$, so $\lambda < 1$. Let

$$f(a_n) = f(a_1) + \frac{\lambda + 1}{2}.$$

Then clearly

$$|f(a_n) - f(a_1)| < 1$$

and

$$f(a_n) > f(a_{n-1}).$$

It is easy to see that (6.20) through (6.22) now hold on A.

Case 3. $a_n R a_1$ and $\sim a_n R a_{n-1}$.
It is clear from compatibility and Lemma 6.7 that there is an i such that

$$a_n R a_j \quad \text{for} \quad j \leqq i,$$

and

$$\sim a_n R a_j \quad \text{for} \quad j \geqq i + 1.$$

Thus, to obtain a function f satisfying Eqs. (6.20) through (6.22), we want to define $f(a_n)$ so that

$$f(a_n) > f(a_i) + 1, \tag{6.24}$$

$$f(a_n) < f(a_{i+1}) + 1, \tag{6.25}$$

$$f(a_n) > f(a_{n-1}). \tag{6.26}$$

Equation (6.24) comes from requirement (6.20), Eq. (6.25) from requirements (6.20) and (6.22), and Eq. (6.26) from requirement (6.21). Now since $\sim a_n R a_{i+1}$ and since $i + 1 \leqq n - 1 < n$, we have $a_{i+1} I a_{n-1}$ and so by (6.20) and (6.22) on A',

$$f(a_{n-1}) < f(a_{i+1}) + 1.$$

Also, we have

$$f(a_i) + 1 < f(a_{i+1}) + 1.$$

We simply choose $f(a_n)$ to be any number bigger than $f(a_i) + 1$ and bigger than $f(a_{n-1})$, but smaller than $f(a_{i+1}) + 1$. To make a specific choice, let

$$\lambda' = \max\{ f(a_{n-1}), f(a_i) + 1 \},$$

and let $f(a_n)$ be halfway between $f(a_{i+1}) + 1$ and λ'; that is, let

$$f(a_n) = \frac{f(a_{i+1}) + 1 + \lambda'}{2}.$$

This guarantees that conditions (6.24), (6.25), and (6.26) hold and hence that f satisfies (6.20) through (6.22). This completes the proof of the Scott–Suppes Theorem.

The inductive proof given above clearly gives rise to a stepwise construction of a function f. We first define $f(a_1)$, then $f(a_2)$, etc. To illustrate, let

$$A = \{w, x, y, z, \alpha, \beta, \gamma\},$$

and let R be defined on A by Eq. (6.6). Then, as we have noted, W can be taken to be the simple order $a_7 = w$, $a_6 = x$, $a_5 = y$, $a_4 = z$, $a_3 = \alpha$, $a_2 = \beta$, $a_1 = \gamma$.

Step 1. Arbitrarily define $f(a_1) = f(\gamma)$ to be 0.
Step 2. To define $f(a_2) = f(\beta)$, note that Case 2 applies, and we have

$$\lambda = f(a_{n-1}) - f(a_1) = f(a_1) - f(a_1) = 0.$$

Thus,

$$f(\beta) = f(a_2) = f(a_1) + \frac{\lambda + 1}{2} = f(\gamma) + \frac{1}{2} = \frac{1}{2}.$$

Step 3. To define $f(a_3) = f(\alpha)$, note that $\alpha R \gamma$ and $\sim \alpha R \beta$, so Case 3 applies with $i = 1$. Then

$$\lambda' = \max\{ f(a_2), f(a_1) + 1 \} = \max\{ \tfrac{1}{2}, 1 \} = 1.$$

Hence,

$$f(\alpha) = f(a_3) = \frac{f(a_2) + 1 + \lambda'}{2} = \frac{f(\beta) + 1 + 1}{2} = \frac{5}{4}.$$

Step 4. To define $f(a_4) = f(z)$, note that $z R \gamma$ but $\sim z R \beta$, so again Case 3 applies, with $i = 1$. Then

$$\lambda' = \max\{ f(a_3), f(a_1) + 1 \} = \frac{5}{4}$$

and

$$f(z) = f(a_4) = \frac{f(a_2) + 1 + \lambda'}{2} = \frac{11}{8}.$$

Step 5. To define $f(a_5) = f(y)$, note that Case 3 again applies, with $i = 2$. Thus,

$$\lambda' = \frac{3}{2}$$

and

$$f(y) = f(a_5) = \frac{f(a_3) + 1 + \lambda'}{2} = \frac{15}{8}.$$

Step 6. To define $f(a_6) = f(x)$, note that Case 3 again applies, with $i = 3$. Then

$$\lambda' = \frac{9}{4}$$

and

$$f(x) = f(a_6) = \frac{f(a_4) + 1 + \lambda'}{2} = \frac{37}{16}.$$

Step 7. To define $f(a_7) = f(w)$, note that Case 1 applies, and so

$$f(w) = f(a_7) = f(a_6) + 2 = \frac{69}{16}.$$

To sum up, the function f is defined by Table 6.1.

Table 6.1

u	γ	β	α	z	y	x	w
$f(u)$	0	8/16	20/16	22/16	30/16	37/16	69/16

6.1.8 *Semiordered Versions of Other Measurement Representations*

Many of the representations studied earlier in this volume have analogues if we use the same idea that motivated the semiorder representation (6.4). For example, it is natural to study the representation

$$aRb \Leftrightarrow f(a) > f(b) + \delta \tag{6.4}$$

and

$$f(a \circ b) = f(a) + f(b), \tag{6.27}$$

a modification of extensive measurement (Section 3.2). A variant of the representation (6.4), (6.27) has been studied by Adams [1965], Krantz [1967], Luce [1973], and Krantz *et al.* [to appear]. These authors find it awkward to deal with an operation $\circ$; they use a set X and comparisons between finite subsets of X, with union $\cup$ playing the role of $\circ$, though not quite analogously. In place of (6.27), they require that

$$f(a \cup b) = f(a) + f(b),$$

provided $a \cap b = \varnothing$.

Luce [1973] and Krantz *et al.* [to appear] study a semiordered version of conjoint measurement (Section 5.4), namely, the representation

$$(a_1, a_2)R(b_1, b_2) \Leftrightarrow f_1(a_1) + f_2(a_2) > f_1(b_1) + f_2(b_2) + \delta.$$

The representation of subjective probability measurement, which we study in Chapter 8, also has a semiordered analogue. This has been studied by Fishburn [1969], Domotor [1969], Stelzer [1967], Domotor and Stelzer [1971], Luce [1973], and Krantz *et al.* [to appear]. As of this writing, no one has succeeded in axiomatizing a semiordered version of the expected utility representations we shall study in Chapter 7.

Any standard relation on the reals has an analogue if equality is replaced by "closeness" or tolerance, as in the semiorder example. Some examples of such relations, for example ε-betweenness, are studied in Roberts [1973]. The geometry imposed on the reals by such relations is called *tolerance geometry*. It has potential applications in visual perception (Roberts [1970]; see also Poston [1971] and Zeeman [1962]). A systematic development of tolerance geometry has not been carried out.

Exercises

1. Raiffa [1968, p. 79] gives the following example. A man is indifferent between a paid vacation in Mexico (M) and a paid vacation in Hawaii (H). If he is offered a $1 bonus to go to M, he might still be indifferent between H and M + $1. Similar reasoning suggests the following indifferences:

$$HI(M + \$1)$$
$$(M + \$1)I(H + \$2)$$
$$(H + \$2)I(M + \$3)$$
$$\cdots$$
$$(M + \$9999)I(H + \$10,000).$$

Transitivity of indifference would imply

$$HI(H + \$10,000).$$

[It would also imply that $HI(H + \$2)$.] Can this example be explained using the threshold or semiorder model? (Incidentally, Raiffa argues that an individual confronted with this example might very well reconsider his preferences and indifferences.)

2. (a) Suppose $A = \{a, b, c, d\}$ and $R = \{(a, d)\}$. Show that (A, R) is a semiorder.

   (b) Which of the following binary relations (A, R) are semiorders?
     (i) $A = Re$, aRb iff $a < b - 1$.
     (ii) $A = \{a, b, c, d\}$, $R = \{(a, b), (b, c), (c, d)\}$.

Table 6.2. Preferences Among Cars
(The i, j entry is 1 iff i is preferred to j.)

	Buick	Cadillac	Toyota	Volkswagen
Buick	0	0	1	0
Cadillac	1	0	1	0
Toyota	0	0	0	0
Volkswagen	0	0	0	0

(iii) $A = N$, $R = >$
(iv) A = all subsets of $\{1, 2, 3\}$, $R = \subsetneqq$.

3. (a) Suppose $A = \{a, b, c, d, e\}$ and $R = \{(a, b), (b, c), (a, c)\}$. Show that (A, R) is not a semiorder, but is an interval order.

(b) Which of the examples in Exer. 2b are interval orders?

4. Show that if an individual's preferences are defined as in Table 6.2, then there is no function f satisfying Eq. (6.4). However, there is such a function if Volkswagen is preferred to Toyota.

5. (a) In an experiment of Estes (reported in Atkinson, Bower, and Crothers [1965, pp. 146–150] and Restle and Greeno [1970, p. 241]), each member of a group of subjects was asked a set of questions about which of two distinguished personalities he would rather meet and talk to. Table 6.3 shows the combined group preference. Does the group preference define a semiorder?

(b) In asking judges to compare the taste or flavor of vanilla puddings, Davidson and Bradley [1969] obtained the group data of Table 6.4. Does the group preference for vanilla puddings define a semiorder?

(c) Consider the judgments of relative importance among objectives for a library system in Dallas shown in Table 1.8 of Chapter 1. If R is "more important than" on the set $A = \{a, b, c, d, e, f\}$, is (A, R) a semiorder?

(d) Repeat part (c) for the judgments of relative importance among goals for a state environmental agency in Ohio shown in Table 1.9 of Chapter 1.

Table 6.3. Group Preference for Meeting with Distinguished Individuals*
(Entry i, j is 1 if more than 75% of the group members preferred meeting and talking to i more than to j.)

	Dwight Eisenhower	Winston Churchill	Dag Hamerskjold	William Faulkner
Dwight Eisenhower	0	0	1	1
Winston Churchill	0	0	1	1
Dag Hamerskjold	0	0	0	0
William Faulkner	0	0	0	0

*Data from an experiment of Estes (Atkinson, Bower, and Crothers [1965, pp. 146–150]). See also Table 6.9.

Table 6.4. Group Preference for Vanilla Puddings* (Entry i,j is 1 if more than 60% of the group members preferred the taste of pudding i to pudding j.)

	1	2	3	4	5
1	0	0	1	1	0
2	1	0	0	0	0
3	0	1	0	0	0
4	0	0	1	0	0
5	0	0	0	1	0

*Data from Davidson and Bradley [1969]. See also Table 6.10.

6. Suppose A is the set of sounds $\{x, y, u, v, w\}$ and p_{ab} as defined in Exer. 6, Section 3.1 gives the proportion of times a subject says that a is louder than b. Suppose aRb holds if and only if $p_{ab} \geq .75$. Is (A, R) a semiorder?

7. Is every strict weak order a semiorder?

8. Let (A, R) be the semiorder of Exer. 2a. Show that there is a homomorphism $f:(A, R) \to (Re, >_\delta)$, with $\delta = 1$, which is irregular.

9. Is every strict partial order a semiorder?

10. Show that every interval order is a strict partial order.

11. Is every strict partial order an interval order?

12. (a) Suppose (A, R) is the semiorder of Exer. 2a.
   (i) Show that the weak order $W(R)$ associated with (A, R) is given by a ranked first, then b, c tied, and then d.
   (ii) Show that all possible compatible simple orders are, from first element to last: a, b, c, d, and a, c, b, d.
   (iii) Find all possible compatible weak orders.
   (b) For the following semiorders, find all possible compatible weak orders:
   (i) $A = \{a, b, c, d, e\}$,
   $R = \{(a, b), (a, d), (a, e), (b, e), (c, d), (c, e)\}$.
   (ii) $A = \{a, b, c, d, e\}$,
   $R = \{(a, c), (a, d), (a, e), (b, e), (c, e)\}$.
   (iii) $A = \{a, b, c, d, e\}$,
   $R = \{(a, e), (a, d), (c, d), (b, d), (e, d)\}$.
   (c) For each semiorder of part (b), pick a compatible simple order and find the function f defined by the constructive method of Section 6.1.7.

13. Let $A = \{1, 2\} \times N$ and let R on A be the lexicographic ordering, that is,

$$(a, s)R(b, t) \Leftrightarrow a > b \text{ or } (a = b \& s > t).$$

Show that (A, R) is a semiorder, but there is no function $f:A \to Re$ satisfying Eq. (6.4).

14. Show that if (A, R) is a semiorder and W is defined by Eq. (6.12), then (A, W) is strongly complete.

15. Suppose (A, W) is a weak order and I is a binary relation on A. Show that (6.14) and (6.15) are equivalent.

16. Suppose (A, R) is a semiorder and E is the equivalence relation of Eq. (6.16). Show that E is the same as the perfect substitutes relation defined by the method of Section 1.8.

17. Suppose (A, R) is a semiorder, E is the equivalence relation of Eq. (6.16), and suppose $aEb \Rightarrow a = b$. Show that every compatible weak order is a simple order.

18. Suppose (A, R) is a semiorder, $A^* = A/E$, and we define

$$a^* R^* b^* \Leftrightarrow aRb.$$

Suppose there is a real-valued function F on A^* such that for all a^*, $b^* \in A^*$,

$$a^* R^* b^* \Leftrightarrow F(a^*) > F(b^*) + 1.$$

Show that if $f(a)$ is defined to be $F(a^*)$, then f satisfies Eq. (6.5).

19. In conjoint measurement, suppose we replace the representation of Eq. (5.19) with the representation

$$(a_1, a_2)R(b_1, b_2) \Leftrightarrow f_1(a_1) + f_2(a_2) > f_1(b_1) + f_2(b_2) + \delta,$$

where δ is a positive constant. Show that it still follows that every strictly bounded standard sequence (Section 5.4.3) is finite.*

20. (Roberts [1969a]) Suppose A is finite, (A, R) is a semiorder, and I is defined on A by Eq. (6.1). Prove that there is an x in A with the following property: whenever xIa and xIb, then aIb.

21. Since every semiorder is a strict partial order, one can talk about the dimension of the semiorder in the sense of Section 1.6, Exers. 12 through 18. Figure 6.3 shows the Hasse diagram of a semiorder. Verify that its dimension is 3. (Rabinovitch [1978] shows that every semiorder on a finite set has dimension at most 3. However, Bogart, Rabinovitch, and Trotter [1976] show that interval orders on finite sets can have arbitrarily large dimension.)

22. Exercises 22 through 30 deal with the case where we start with judgments of indifference only. Show that there is a function f satisfying Eq. (6.3) if and only if (A, I) is an equivalence relation.

23. If we start with judgments of indifference, the representation corresponding to Eq. (6.4) is

$$aIb \Leftrightarrow |f(a) - f(b)| \leq \delta. \qquad (6.28)$$

*The reader should take care to note that E as defined in Chapter 5 corresponds to I as defined in this chapter, and not to E as defined here.

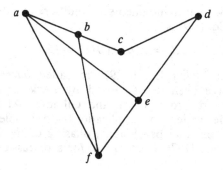

Figure 6.3. The Hasse diagram of a semiorder with dimension 3.

Show that a binary relation satisfying Eq. (6.28) is not necessarily an equivalence relation. Binary relations satisfying (6.28) are called indifference graphs and are characterized in Roberts [1969a].* See Roberts [1978b, 1979] for recent applications of indifference graphs.

24. (a) Show that the following binary relation defines an indifference graph:

$$A = \{x, y, u, v\},$$
$$I = \{(x, x), (y, y), (u, u), (v, v), (x, y), (y, x), (y, u), (u, y), (u, v), (v, u)\}.$$

(b) Which of the following binary relations are indifference graphs?

(i) $A = \{x, y, u, v\}$,
$I = \{(x, x), (y, y), (u, u), (v, v), (x, u), (u, x)\}$.

(ii) $A = \{x, y, u\}$,
$I = \{(x, x), (y, y), (u, u), (x, y), (y, x), (y, u), (u, y),$
$(u, x), (x, u)\}$.

(iii) $A = \{x, y, u, v\}$,
$I = \{(x, x), (y, y), (u, u), (v, v), (x, y), (y, x), (y, u), (u, y),$
$(u, v), (v, u), (v, x), (x, v)\}$.

(iv) $A = \{x, y, u, v\}$,
$I = \{(x, x), (y, y), (u, u), (v, v), (x, y), (y, x), (x, u),$
$(u, x), (x, v), (v, x)\}$.

*The representation (6.28) generalizes to arbitrary metric spaces (X, d). We seek a function $f{:}A \to X$ such that for all $a, b \in A$,

$$aIb \Leftrightarrow d[f(x), f(y)] \leqq \delta.$$

Not much work has been done on this representation to date, and in particular there is no known characterization of relations (A, I) for which such a representation exists with (X, d) n-dimensional Euclidean space, $n > 1$. For some partial results, see Roberts [1969b].

25. If we start with just judgments of indifference, the representation corresponding to Eq. (6.9) is

$$aIb \Leftrightarrow J(a) \cap J(b) \neq \varnothing. \tag{6.29}$$

Binary relations satisfying Eq. (6.29) are called *interval graphs*.* The interval graphs have been characterized by Lekkerkerker and Boland [1962], Fulkerson and Gross [1965], and Gilmore and Hoffman [1964]. They have a wide variety of applications—to problems of genetics, archaeology, developmental psychology, phasing of traffic lights, ecosystems, etc. See Roberts [1976, Section 3.4] for a discussion of a variety of applications.

(a) Show that every indifference graph is an interval graph.

(b) Show that the converse is false.

26. Which of the binary relations of Exers. 24a and 24b define interval graphs?

27. Suppose (A, R) is a binary relation. The *complement* of (A, R) is the binary relation (A, R^c), where

$$R^c = \{(a, b) \in A \times A : (a, b) \notin R\}.$$

The complement of an indifference graph is not reflexive and hence could not be an indifference graph. However, show that if all pairs (x, x) for x in A are added to the complement of an indifference graph, the resulting binary relation might be an indifference graph.

28. Suppose (A, K) is a symmetric binary relation. An *orientation* of (A, K) is an antisymmetric binary relation P on A such that $P \cup P^{-1} = K$. (For definitions of union $\cup$ and converse $^{-1}$, see Exers. 3 and 1 of Section 1.2.) The binary relation P picks one and only one of the pairs (x, y) and (y, x) whenever these pairs are both in the relation K and $x \neq y$. To give an example, if $A = \{x, y, z\}$ and $K = \{(x, y), (y, x), (y, z), (z, y)\}$, then one orientation P of (A, K) is given by $P = \{(x, y), (y, z)\}$. If (A, I) is an indifference graph, then one way of defining an orientation of the complement of (A, I) (Exer. 27) is to use the function f satisfying Eq. (6.28) and define

$$xPy \Leftrightarrow f(x) > f(y) + \delta. \tag{6.30}$$

---

*The representation (6.29) has a higher dimensional analogue just as does the representation (6.28). Instead of intervals on the line, we think of boxes in n-space, n-dimensional rectangles with sides parallel to the coordinate axes. We ask for an assignment of a box $B(a)$ to each a in A such that for all a, b in A,

$$aIb \Leftrightarrow B(a) \cap B(b) \neq \varnothing.$$

This representation seems particularly hard to characterize for higher dimensions, even two. It seems to have a variety of applications, including some very interesting ecological ones. For references, see Roberts [1969b, 1976, Section 3.5, 1978a], Cohen [1978], Gabai [1976], and Trotter [1977].

(a) Find such an orientation P for the indifference graph of Exer. 24a.

(b) Verify that P defined by Eq. (6.30) always defines an orientation of the complement of (A, I).

29. (a) Show that if (A, I) is an indifference graph, then its complement has an orientation that is transitive. (A symmetric binary relation that has a transitive orientation is called a *comparability graph*. Comparability graphs have been characterized by Ghouila-Houri [1962] and Gilmore and Hoffman [1964]).

(b) Show that if A is finite and (A, I) is symmetric, then (A, I) is an indifference graph if and only if its complement has an orientation which is a semiorder.

30. (a) Show that if (A, I) is an interval graph, then its complement is a comparability graph (Exer. 29).

(b) Show that if (A, I) is symmetric, then (A, I) is an interval graph if and only if its complement has an orientation which is an interval order.

31. A long-standing problem in the dimension theory of strict partial orders has been the problem of characterizing strict partial orders of a given dimension. (Dimension is defined in Exers. 12 through 18 of Section 1.6.). On the basis of a result of Dushnik and Miller [1941], Baker, Fishburn and Roberts [1971] show that a strict partial order has dimension 1 or 2 if and only if its symmetric complement (Exer. 12, Section 1.3) is a comparability graph. Use this criterion to determine whether or not the strict partial orders whose Hasse diagrams are given in Fig. 1.8 have dimension less than or equal to 2.

32. (Ducamp and Falmagne [1969]). In Section 5.6.1, we talked about a set S of individuals and a set E of reactions or experiences or statements or test items. We talked about a binary relation R on $A = (S \cup E)$ with $R \subseteq S \times E$ and aRb interpreted to mean that individual a agrees with statement b, or answers test item b correctly, etc. Let us distinguish two types of positive replies by individuals. For example, they either agree strongly with a statement, or agree, not necessarily strongly. Or they answer a test question correctly without difficulty, or they answer it correctly, perhaps with difficulty. Let R and T be binary relations on A representing these two levels of reactions, with $R \subseteq S \times E$ and $T \subseteq S \times E$. Corresponding to the semiorder idea, let us ask for functions $s: S \to Re$ and $e: E \to Re$ and real numbers δ and η, with $\delta > \eta$, such that for all $a \in S$ and $b \in E$,

$$aRb \Leftrightarrow s(a) > e(b) + \delta \qquad (6.31)$$

and

$$aTb \Leftrightarrow s(a) > e(b) + \eta. \qquad (6.32)$$

The functions s and e generalize the notion of a Guttman scale which we defined in Section 5.6.1.

(a) Give proof or counterexample: We can find a representation
(6.31), (6.32) with some $\delta > \eta$ if and only if we can find one with $\delta = 1$,
$\eta = 0$.

(b) Give proof or counterexample: If we can find a representation
with some $\delta > \eta$, then we can find one with any $\delta > \eta$.

33. Ducamp and Falmagne [1969] define a *bisemiorder* to be a quadruple (S, E, R, T) such that R and T are binary relations on $(S \cup E)$, both $R \subseteq S \times E$ and $T \subseteq S \times E$, and such that the following axioms hold for all $a, a' \in S$ and $b, b' \in E$:

AXIOM BS1. *If aRb, then aTb.*

AXIOM BS2. *If aRb, not $a'Rb$, and $a'Rb'$, then aRb'.*

AXIOM BS3. *If aTb, not $a'Tb$, and $a'Tb'$, then aTb'.*

AXIOM BS4. *If aTb, not $a'Tb$, and $a'Rb'$, then aRb'.*

AXIOM BS5. *If aRb, not $a'Tb$, and $a'Tb'$, then aRb'.*

Show that all these axioms are necessary conditions for the representation
(6.31), (6.32). (Ducamp and Falmagne also prove their sufficiency. For a
more recent proof, see Ducamp [1978].)

6.2 The Theory of Probabilistic Consistency

6.2.1 *Pair Comparison Systems*

In Section 3.1, we discussed two possible interpretations for axioms like
those that define a strict weak order or a semiorder. These either could be
considered conditions of rationality, which say something about a rational
person's judgments, or they could be considered testable conditions, subject to experimental verification. If we look at these conditions in the latter
sense, then we see that individuals are often inconsistent in their judgments. For example, a subject might at one time say he prefers a to b and
not b to a, and at another time, maybe even just a bit later in the same
session, and under circumstances that seem unchanged, say he prefers b to
a and not a to b. (It is not clear that it is possible to present judgments
under exactly identical conditions, since prior judgments can in principal
influence later judgments. However, we are trying to explain behavior, and
often individuals behave as if they are making inconsistent judgments
under seemingly identical conditions.) Similarly, a subject may at one
point say sound a is louder than sound b and b is not louder than a, and
soon thereafter say sound b is louder than sound a and a is not louder than
b. If these observations are correct, then all the measurement techniques
we have described so far break down. For "preferred to," "louder than,"

and similar judgments often do not define *relations*, the starting point of our theories.

If subjects were totally inconsistent in their judgments, then there would be no hope of developing theories of their behavior. But in many cases, there is a pattern to the inconsistencies, so a better word than "inconsistencies" would be "variations." Although a subject may not be "absolutely consistent," he may be "probabilistically consistent," to paraphrase the terminology of Block and Marschak [1959]. We shall try to make precise some notions of probabilistic consistency in this section, and relate them to measurement. We shall formulate the theory in terms of preference, though we shall keep in mind other applications, in particular to psychophysics.

Suppose A is a set. Let $p(a, b) = p_{ab}$ be the frequency with which a is preferred to b, that is, the proportion of times a subject says he prefers a to b. Obviously, similar definitions apply if judgments are of relative loudness, relative importance, and so on. The numbers p_{ab} define a function $p: A \times A \to [0, 1]$. We shall assume that each time the subject makes a judgment, the conditions are "identical." (As we have previously observed, this assumption makes it questionable whether we can ever obtain the data p_{ab}. For in an experiment, once a choice has been made between a and b, this choice is likely to affect the choice between a and b later on, even though efforts are made to mask the fact that the choice has previously been made.)

The system (A, p) is called a *pair comparison system*. We assume that each time an individual makes a judgment between a and b, he is forced to say he prefers one to the other. Thus, for all $a \neq b$ in A, we have

$$p_{ab} + p_{ba} = 1. \tag{6.33}$$

For convenience, we also assume that (6.33) holds when $a = b$, that is, that $p_{aa} = \frac{1}{2}$. (It might be more natural to assume that p_{aa} is undefined.) If the pair comparison system (A, p) satisfies (6.33) for all a, b in A, we call (A, p) a *forced choice pair comparison system*.

Forced choice pair comparison systems also arise in group decisionmaking. We assume that each of a group of individuals (experts) is consistent in his judgments (of preference); that is, his judgments of preference define a relation. Then we take p_{ab} to be the proportion of individuals who prefer a to b.

6.2.2 Probabilistic Utility Models

6.2.2.1 The Weak Utility Model

In trying to make precise the notion of probabilistic consistency, we shall introduce several representations that arise if we try to perform measurement using data from a forced choice pair comparison system.

When we think of the underlying judgment as one of preference, p_{ab} is in some sense a measure of strength of preference. The simplest idea is to try to find a function $f: A \to Re$ so that for all a, b in A,

$$p_{ab} > p_{ba} \Leftrightarrow f(a) > f(b). \tag{6.34}$$

If (6.34) holds, then $f(a) > f(b)$ if and only if a is preferred to b a majority of times. In a weak sense, f can be taken as a utility function. Following Luce and Suppes [1965], we shall say that a forced choice pair comparison system (A, p) satisfies the *weak utility model* if there is a real-valued function f on A satisfying Eq. (6.34). The weak utility model is a model of probabilistic consistency.

The representation (6.34) is in a sense derived measurement, for we start with one scale p and derive a second scale f from p. Note that f is not defined explicitly in terms of p, but rather satisfies a condition $C(p, f)$, to use the notation of Section 2.5. A representation theorem for the weak utility model may be derived from Corollary 2 to Theorem 3.4.* Let us define a binary relation W on A by

$$aWb \Leftrightarrow p_{ab} \geqq p_{ba}. \tag{6.35}$$

THEOREM 6.8. *A forced choice pair comparison system (A, p) satisfies the weak utility model if and only if (A, W) defined in Eq. (6.35) is a weak order and (A^*, W^*) has a countable order-dense subset.*

It is also easy to prove a uniqueness theorem for the weak utility model. (The reader should refer to the definitions in Section 2.5 before reading this theorem.) Note that p_{ab} is an absolute scale. Thus, the narrow and wide senses of uniqueness for f coincide.

THEOREM 6.9. *If (A, p) is a forced choice pair comparison system and the function $f: A \to Re$ satisfies the weak utility model, then f defines a (regular) ordinal scale in both the narrow and wide senses.*

Proof. It is necessary to prove that if f satisfies (6.34) and $\phi: f(A) \to Re$ is a monotone increasing transformation, then $\phi \circ f$ satisfies (6.34); and to prove that if $g: A \to Re$ also satisfies (6.34), then there is a monotone increasing $\phi: f(A) \to Re$ so that $g = \phi \circ f$. These results follow from the uniqueness theorem for the representation $(A, W) \to (Re, \geqq)$ in Corollary 2 to Theorem 3.4. ∎

---

*A similar theorem may be derived from Corollary 1 to Theorem 3.4. But we shall use this version below.

6.2.2.2 The Strong Utility Model

Sometimes it is desirable to obtain a stronger scale than an ordinal scale. Our next measurement model accomplishes this. We say that a forced choice pair comparison system (A, p) satisfies the *strong utility model* if there is a real-valued function f on A so that for all a, b, c, d in A,

$$p_{ab} > p_{cd} \Leftrightarrow f(a) - f(b) > f(c) - f(d). \qquad (6.36)$$

The idea behind this model is that the probability of making a choice depends on the difference in scale values. The strong utility model implies the weak utility model, since

$$p_{ab} > p_{ba} \Leftrightarrow f(a) - f(b) > f(b) - f(a)$$
$$\Leftrightarrow 2f(a) > 2f(b)$$
$$\Leftrightarrow f(a) > f(b).$$

Let us give an example of a pair comparison system that satisfies the weak utility model but fails to satisfy the strong utility model. Let $A = \{x, y, z\}$ and define p_{ab} by the following matrix, whose a, b entry gives p_{ab}:

$$
(p_{ab}) = \begin{array}{c} \\ x \\ y \\ z \end{array}
\begin{pmatrix}
\overset{\displaystyle x}{} & \overset{\displaystyle y}{} & \overset{\displaystyle z}{} \\
\frac{1}{2} & \frac{2}{3} & \frac{3}{4} \\
\frac{1}{3} & \frac{1}{2} & \frac{1}{2} \\
\frac{1}{4} & \frac{1}{2} & \frac{1}{2}
\end{pmatrix}. \qquad (6.37)
$$

It is easy to see that the weak utility model holds. However, if f is a function satisfying Eq. (6.36), $p_{yz} = p_{zy}$ implies that $f(y) = f(z)$. Thus $f(x) - f(y) = f(x) - f(z)$, which implies $p_{xy} = p_{xz}$, a contradiction.

It is of interest to mention several arguments to the effect that such failures of the strong utility model can occur for preferences or judgments of relative loudness. One such argument is based on Davidson and Marschak [1959, p. 237]. In preferences among different monetary amounts, we expect $p_{\$5000, \$0} = 1$ and $p_{\$1, \$0} = 1$, so $p_{\$5000, \$0} = p_{\$1, \$0}$, and therefore $f(\$5000) = f(\$1)$. Hence, $p_{\$5000, \$2}$ should equal $p_{\$1, \$2}$, which is certainly nonsense. Similar problems occur often when $p_{ab} = 1$.

A second argument suggesting that the strong utility model fails is closely related to the Davidson–Marschak argument. In psychophysical experiments, the frequency of judgments of preference, or "louder than," and so on is essentially 1 if choices are far enough apart. Thus, for example, we can have three sounds, with a much louder (more intense) than b, and b much louder than c, and hence $p_{ab} = p_{bc} = p_{ac} = 1$. If f

satisfies Eq. (6.36), then $f(a) = f(b) = f(c)$. But now suppose d is close in loudness (intensity) to a, but not to b and c. Then $p_{da} \approx \frac{1}{2}$ while $p_{db} = 1$, which is impossible. Thus, the strong utility model fails.

In Section 6.1, we discussed several examples that were arguments against the transitivity of indifference. Two of the examples, the one about support of the arts and the one about the pony and the bicycle, can also be used as arguments against the strong utility model. For, suppose plan a would allocate a budget of 200 million dollars to a federal Institute for the Arts, plan b would allocate 200 million dollars to state institutes, and plan b' would allocate 200 million and one dollars to state institutes. If you are indifferent between a and b, p_{ab} is $\frac{1}{2}$, so $f(a) - f(b)$ is 0. Now similarly $p_{ab'}$ is $\frac{1}{2}$, so $f(a) - f(b')$ is 0. However, $p_{b'b}$ is 1, so $f(b') - f(b)$ is large. This is an impossibility. (This is really an argument against the weak utility model.) The argument in the pony–bicycle example is similar. (If p_{ab} and $p_{a'b'}$ are only approximately but not exactly $\frac{1}{2}$, the argument does not work.) All these examples involve some probabilities that are 0 or 1. Some authors have chosen to require that Eq. (6.36) hold only for $p_{ab} \neq 0, 1$.

To give an example that violates the strong utility model and has no $p_{ab} = 0$ or 1, consider the data of Table 6.5. Suppose there is a function f satisfying Eq. (6.36). Since

$$p_{\text{Beethoven, Brahms}} > p_{\text{Beethoven, Wagner}},$$

we have

$$f(\text{Beethoven}) - f(\text{Brahms}) > f(\text{Beethoven}) - f(\text{Wagner}),$$

so

$$f(\text{Wagner}) > f(\text{Brahms}).$$

This implies

$$f(\text{Mozart}) - f(\text{Brahms}) > f(\text{Mozart}) - f(\text{Wagner}),$$

Table 6.5. Preferences for Composers*

(The a, b entry shows the proportion of orchestra members interviewed who preferred composer a over composer b.)

	Beethoven	Brahms	Mozart	Wagner
Beethoven	.50[†]	.67	.79	.64
Brahms	.33	.50[†]	.60	.54
Mozart	.21	.40	.50[†]	.51
Wagner	.36	.46	.49	.50[†]

*Data from Folgmann [1933].
[†]By assumption.

or

$$p_{\text{Mozart, Brahms}} > p_{\text{Mozart, Wagner}},$$

which is false.

As with the weak utility model, sufficient conditions for the strong utility model to hold can be obtained from previously known theorems by introducing a relation, in this case the quaternary relation D on A defined by

$$abDcd \Leftrightarrow p_{ab} > p_{cd}. \tag{6.38}$$

As in difference measurement, $abDcd$ is interpreted to mean that a is preferred to b at least as much as c is preferred to d.

THEOREM 6.10. *If the quaternary relation (A, D) defined by Eq. (6.38) is an algebraic difference structure (Section 3.3.1), then the forced choice pair comparison system (A, p) satisfies the strong utility model. Moreover, if (A, D) is an algebraic difference structure and the function $f: A \rightarrow Re$ satisfies the strong utility model, then f defines a (regular) interval scale in both the narrow and wide senses.*

Proof. Use the representation and uniqueness theorems of Section 3.3.

6.2.2.3 The Fechnerian Utility Model

A variant of the strong utility model is the *Fechnerian utility model*. A forced choice pair comparison system (A, p) satisfies this model if there is a real-valued function f on A and a (strictly) monotone increasing function $\phi: Re \rightarrow Re$ so that for all a, b in A,

$$p_{ab} = \phi[f(a) - f(b)]. \tag{6.39}$$

The idea, as in the strong utility model, is that the greater the difference between $f(a)$ and $f(b)$, the greater the frequency with which a is preferred to b. The Fechnerian model is sometimes called the strong utility model, but we shall distinguish the two. This model appears in classical psychophysics, where it arises in the attempt to relate a physical magnitude to a psychological one. The model implies that if $p_{ab} = p_{cd}$, then $f(a) - f(b) = f(c) - f(d)$. Thus, pairs of stimuli that are equally often confused are equally far apart. This property was a critical assumption in Fechner's derivation of the psychophysical function as a logarithmic one—see our discussion of Fechner's Law in Section 4.1. (For a more detailed discussion of Fechner's Law, see Luce and Galanter [1963, Section 2] or Falmagne [1974].)

Often in variations on this general Fechnerian utility model, the function ϕ is required to be a cumulative distribution function. Then $\phi[f(a) - f(b)]$ is the probability that $f(a)$ is larger than $f(b)$. This general idea goes back to Thurstone [1927a, b], who sought scale values $f(a)$ so that for all a, b,

$$p_{ab} = \int_{-\infty}^{f(a)-f(b)} N(x) \, dx, \tag{6.40}$$

where $N(x)$ is a normal distribution with mean 0 and standard deviation 1. A more recent variant uses ϕ as the logistic distribution $\phi(x) = \dfrac{1}{1 + e^{-x}}$ and seeks scale values $f(a)$ so that

$$p_{ab} = \frac{1}{1 + e^{-[f(a)-f(b)]}}. \tag{6.41}$$

This model is due to Guilford [1954, p. 144] and Luce [1959]. More generally, one could seek to assign a random variable $F(a)$ to each a in A so that

$$p_{ab} = Pr[F(a) \geqq F(b)]. \tag{6.42}$$

Equation (6.42) is called the *random utility model*. The interpretation is that the utilities are no longer assumed to stay fixed, but are determined by some probabilistic procedure. See Luce and Suppes [1965] or Krantz *et al.* [to appear] for a discussion.

The Fechnerian utility model clearly implies the strong utility model, for

$$p_{ab} > p_{cd} \Leftrightarrow \phi[f(a) - f(b)] > \phi[f(c) - f(d)]$$
$$\Leftrightarrow f(a) - f(b) > f(c) - f(d).$$

Thus, the examples we have given which violate the strong utility model, also violate the Fechnerian model.

Finally, let us ask if the strong utility model is equivalent to the Fechnerian utility model. If A is finite, this is true. For, suppose f satisfies the strong utility model, that is, Eq. (6.36). Let

$$B = f(A) \dotminus f(A) = \{x - y : \; x = f(a), y = f(b), \quad \text{for some } a, b \in A\}. \tag{6.43}$$

Define $\phi: B \to Re$ by $\phi(x - y) = p_{ab}$, where $x = f(a)$ and $y = f(b)$. Then ϕ is well-defined and monotone increasing on B. Moreover, for all a, b in A, Eq. (6.39) holds. (This argument works even if A is not finite.) If A is finite, B is finite, and so it is easy to extend ϕ to a (strictly) monotone increasing function ϕ on Re, by taking ϕ to be a linear function between two successive elements of B. Thus, ϕ, f satisfy the Fechnerian utility model. However, if A is not finite, this argument does not work. The function ϕ may not be extendable to a (strictly) monotone increasing function on all of Re.*

Some authors (e.g., Pfanzagl [1968, p. 173]) have chosen to define (or study) the variant of the Fechnerian utility model where ϕ is only required to be defined on $B = f(A) \dot{-} f(A)$. This model is equivalent to the strong utility model. Whether or not the version we have stated is equivalent seems to be an open question.

*To see this, let

$$A = \left\{ \tfrac{1}{4} \right\} \cup \left[\tfrac{3}{4}, 1 \right).$$

Let

$$C = A \dot{-} A = \{ a - b : a, b \in A \}.$$

Then

$$C = \left(-\tfrac{3}{4}, -\tfrac{1}{2} \right] \cup \left(-\tfrac{1}{4}, 0 \right] \cup \left[0, \tfrac{1}{4} \right) \cup \left[\tfrac{1}{2}, \tfrac{3}{4} \right).$$

Let $\psi: C \to Re$ be defined by

$$\psi(x) = \begin{cases} x + \tfrac{1}{2} & \text{if } x \in \left(-\tfrac{1}{4}, \tfrac{1}{4} \right), \\ x + \tfrac{1}{4} & \text{if } x \geq \tfrac{1}{2}, \\ x + \tfrac{3}{4} & \text{if } x \leq -\tfrac{1}{2}. \end{cases}$$

Define $p: A \times A \to [0, 1]$ by

$$p_{ab} = \psi[a - b].$$

Then we have

$$p_{ab} > p_{cd} \Leftrightarrow \psi[a - b] > \psi[c - d]$$

$$\Leftrightarrow a - b > c - d.$$

It follows that (A, p) satisfies the strong utility model. Take $f(x) = x$. Let $B = f(A) - f(A)$ be as in Eq. (6.43), and define $\phi: B \to Re$ by $\phi(x - y) = p_{ab}$ if $x = f(a)$ and $y = f(b)$. Then $B = C$ and on B, $\phi = \psi$. Now ϕ cannot be extended to a monotone increasing ϕ on all of Re. For $\phi(\tfrac{1}{2}) = \tfrac{3}{4}$, so $\phi(\tfrac{3}{8})$ would have to be less than $\tfrac{3}{4}$. However, $\sup\{\phi(x): x \in B \ \& \ x < \tfrac{1}{4}\} = \tfrac{3}{4}$.

This argument does not mean that there are no functions ϕ and f satisfying (6.39), but only that this method does not lead to such functions.

6.2.2.4 The Strict Utility Model

Our last measurement model has simple representation and uniqueness theorems associated with it. A natural requirement on a utility function f is that it satisfy

$$p_{ab} = \frac{f(a)}{f(a) + f(b)}. \tag{6.44}$$

Again following Luce and Suppes [1965], we say that a pair comparison system (A, p) satisfies the *strict utility model* if there is a real-valued function f on A satisfying Eq. (6.44). The strict utility model has been studied by many authors. The model was used by Zermelo [1929] to measure the playing power $f(a)$ of a chess player a. It also has applications to measurement of response strength in psychology, as we shall mention below. See Luce and Suppes [1965, p. 335] and Krantz *et al.* [to appear] for additional references. The pair comparison system of Eq. (6.37) does not satisfy the strict utility model. If there is a function f on A satisfying Eq. (6.44), then $p_{yz} = \frac{1}{2}$ implies

$$\frac{f(y)}{f(y) + f(z)} = \frac{1}{2},$$

so $f(y) = f(z)$. Thus,

$$\frac{f(x)}{f(x) + f(y)} = \frac{f(x)}{f(x) + f(z)},$$

so $p_{xy} = p_{xz}$, which is a contradiction.

The strict utility model never applies if any p_{ab} is 1. For then $f(b) = 0$. But then $p_{bb} = \dfrac{f(b)}{f(b) + f(b)}$ is undefined, instead of $\frac{1}{2}$. Requiring (6.44) to hold only for $a \neq b$ is not a satisfactory solution to this problem. The strict utility model still rarely applies if any $p_{ab} = 1$. For then $f(b) = 0$. Given *any other* c, $p_{cb} = f(c)/f(c)$, which is either 1 or undefined (if $f(c) = 0$). Thus, we shall assume that for all a, b in A, $p_{ab} \neq 0, 1$. If $p_{ab} \neq 0, 1$, then the "discrimination" between a and b is imperfect. If this is the case for all a, b, we call the pair comparison system (A, p) *imperfect*.

Luce [1959] and Luce and Suppes [1965] point out that if (A, p) is imperfect, the strict utility model implies the Fechnerian and strong utility models. For suppose f satisfies (6.44). If $f(a) < 0$, any a, then $f(b) < 0$, all *b.* Otherwise

$$\frac{f(b)}{f(b) + f(a)}$$

is either negative or greater than 1, and both contradict $p_{ab} \in (0, 1)$. Let $g(a) = -f(a)$. Then $g(a) > 0$, all a, and g also satisfies the strict utility model. Thus, in any case, we may assume that $f(a) \geqq 0$, all a. But $f(a) = 0$ implies that p_{aa} is undefined instead of $\frac{1}{2}$. Hence, we have $f(a) > 0$, all a. We can now define $f' = \ln f$. Let ϕ be the logistic distribution function $\phi(\lambda) = 1/(1 + e^{-\lambda})$. Then

$$p_{ab} = \frac{f(a)}{f(a) + f(b)}$$

$$= \frac{1}{1 + f(b)/f(a)}$$

$$= \frac{1}{1 + \exp\{-[f'(a) - f'(b)]\}}$$

$$= \phi[f'(a) - f'(b)],$$

so f' and ϕ satisfy the Fechnerian utility model. The strong utility model follows.

It is not hard to show by example that the Fechnerian utility model can hold for (A, p) while the strict utility model fails.

We now present a representation theorem for the strict utility model. We say that an imperfect forced choice pair comparison system (A, p) satisfies the *product rule* if for all a, b, c in A,

$$p_{ab}p_{bc}p_{ca} = p_{ac}p_{cb}p_{ba}. \tag{6.45}$$

The following theorem is proved in Suppes and Zinnes [1963], and we follow their proof. (The results also appear in Luce [1959], though not in one place.)

THEOREM 6.11 *An imperfect forced choice pair comparison system (A, p) satisfies the strict utility model if and only if it satisfies the product rule.*

Proof. Suppose the product rule holds. Let e be any element of A. Then define $f(a)$ by

$$f(a) = p_{ae}/p_{ea}.$$

(Since (A, p) is imperfect, we may divide by P_{ea}.) We show that f satisfies

(6.44). For, given a and b, we have

$$\frac{f(a)}{f(a) + f(b)} = \frac{p_{ae}/p_{ea}}{p_{ae}/p_{ea} + p_{be}/p_{eb}}$$

$$= \frac{p_{ae}/p_{ea}}{p_{ae}/p_{ea} + (p_{ba}/p_{ab})(p_{ae}/p_{ea})}$$

$$= \frac{1}{1 + (p_{ba}/p_{ab})}$$

$$= \frac{p_{ab}}{p_{ab} + p_{ba}}$$

$$= p_{ab} \qquad \text{(by forced choice)}.$$

The proof of the converse is left to the reader. ∎

Remark: In psychology, imperfect forced choice pair comparison systems satisfying the product rule are sometimes called *Bradley–Terry–Luce* (*or BTL*) *systems*, after Bradley and Terry [1952], Bradley [1954a,b, 1955], and Luce [1959]. (Cf. Suppes and Zinnes [1963].) Here, p_{ab} is interpreted as the proportion of times a subject judges stimulus a to be greater in some sense than stimulus b, and the function f is interpreted as a measure of response strength.

We next state a uniqueness theorem for the strict utility model, again following Suppes and Zinnes [1963]. If a similarity transformation may be either positive or negative (but not zero), let us refer to a *generalized ratio scale* rather than a ratio scale.

THEOREM 6.12. *If (A, p) is an imperfect forced choice pair comparison system and the function $f: A \to Re$ satisfies the strict utility model, then f defines a (regular) generalized ratio scale in both the narrow and wide senses. If f is required to be positive, then f defines a (regular) ratio scale in both the narrow and wide senses.*

Proof. Note that (A, p) is an absolute scale, so the narrow and wide senses of uniqueness for derived scales coincide. Clearly, if $\alpha \neq 0$ and $f' = \alpha f$, then f' also satisfies the strict utility model. Suppose next that f' is any other function satisfying this model. Fix e in A. Note that $f(e) \neq 0$, $f'(e) \neq 0$. This follows, since $p_{ee} = \frac{1}{2}$. Let a be any element in A. Then, since $f(e) \neq 0$ and $f'(e) \neq 0$,

$$\frac{f(a)}{f(e)} = \frac{p_{ae}}{p_{ea}} = \frac{f'(a)}{f'(e)}.$$

It follows that

$$f'(a) = \alpha f(a),$$

where α is $f'(e)/f(e)$, which is not zero. If f and f' are required to be positive, then α is positive.   ■

The four measurement models presented in Section 6.2.2 are all models of probabilistic consistency. They are successively more restrictive, in the sense that the strict utility model implies the Fechnerian utility model, which implies the strong utility model, which implies the weak utility model. Moreover, the weak utility model does not imply the strong utility model, and the Fechnerian utility model does not imply the strict utility model. As we have remarked, however, it is not known if the strong utility model implies the Fechnerian utility model.

In the next section, we shall discuss some related conditions and also several additional situations in which these various models fail or are satisfied.

6.2.3 Transitivity Conditions

The product rule of Eq. (6.45) is a useful condition in that it can be subjected to test given the data about frequency of preference. Luce and Suppes [1965] call such a condition an *observable property*. A nonobservable property is a condition like Eq. (6.44), which is stated in terms of unknown functions. In this section, we shall state three observable conditions that are closely related to the utility models of the previous section. Each of these can also be thought of as a model or condition of probabilistic consistency.

We say that a forced choice pair comparison system (A, p) satisfies *Weak Stochastic Transitivity* (*WST*) if for all a, b, c in A,

$$p_{ab} \geq \tfrac{1}{2} \ \& \ p_{bc} \geq \tfrac{1}{2} \Rightarrow p_{ac} \geq \tfrac{1}{2}. \tag{6.46}$$

In words, whenever at least $\frac{1}{2}$ the time a is preferred to b (judged louder than b), and at least $\frac{1}{2}$ the time b is preferred to c (judged louder than c), then at least $\frac{1}{2}$ the time, a is preferred to c (judged louder than c). This is a probabilistic kind of transitivity and might be more appropriately called weak probabilistic transitivity—the word stochastic is used for probabilistic processes that take place over time, and there is no notion of time involved here.

(A, p) is said to satisfy *Moderate Stochastic Transitivity* (*MST*) if for all a, b, c in A,

$$p_{ab} \geq \tfrac{1}{2} \ \& \ p_{bc} \geq \tfrac{1}{2} \Rightarrow p_{ac} \geq \min \{ p_{ab}, p_{bc} \}. \tag{6.47}$$

This condition is a stronger form of transitivity. It says that if a is preferred to b at least $\tfrac{1}{2}$ the time and b is preferred to c at least $\tfrac{1}{2}$ the time, then a is preferred to c at least as large a proportion of times as the minimum of p_{ab} and p_{bc}. Naturally, if (A, p) satisfies MST, it must satisfy WST. The converse is false, as is easy to show by example.

(A, p) is said to satisfy *Strong Stochastic Transitivity* (*SST*) if for all a, b, c in A,

$$p_{ab} \geq \tfrac{1}{2} \ \& \ p_{bc} \geq \tfrac{1}{2} \Rightarrow p_{ac} \geq \max \{ p_{ab}, p_{bc} \}. \tag{6.48}$$

Of course, SST implies MST. The converse is false. A survey of many other conditions of stochastic transitivity can be found in Fishburn [1973].

All three of the observable stochastic transitivity conditions, WST, MST, and SST, follow from the strong utility model (and hence from the Fechnerian and strict utility models). For we have the following result.

THEOREM 6.13. *For forced choice pair comparison systems, the strong utility model implies SST.*

Proof. Suppose $p_{ab} \geq \tfrac{1}{2}$ and $p_{bc} \geq \tfrac{1}{2}$. By forced choice, $p_{cc} = \tfrac{1}{2}$, so $p_{ab} \geq p_{cc}$. Thus

$$f(a) - f(b) \geq f(c) - f(c)$$
$$f(a) - f(c) \geq f(b) - f(c)$$
$$p_{ac} \geq p_{bc}.$$

Similarly, $p_{bc} \geq p_{aa}$ implies $p_{ac} \geq p_{ab}$. ∎

The converse of Theorem 6.13 is false. Take $A = \{x, y, u, v\}$ and define p as follows:

$$(p_{ab}) = \begin{array}{c} \\ x \\ y \\ u \\ v \end{array} \begin{pmatrix} \tfrac{1}{2} & 1 & 1 & 1 \\ 0 & \tfrac{1}{2} & 1 & 1 \\ 0 & 0 & \tfrac{1}{2} & \tfrac{1}{2} \\ 0 & 0 & \tfrac{1}{2} & \tfrac{1}{2} \end{pmatrix} \begin{array}{c} x \quad y \quad u \quad v \end{array}.$$

Then SST is satisfied, but the strong utility model fails. To see the latter, note that $p_{xu} = p_{yu}$, so $f(x) = f(y)$, and $p_{xy} = p_{xu}$, so $f(y) = f(u)$. It follows that $f(x) = f(u)$ and that $p_{xv} = p_{uv}$, which is false.

In the next subsection, we shall show that if A is finite, then WST is equivalent to the weak utility model. Thus, we can sum up as follows:

COROLLARY. *If A is finite, then we have the following string of implications and equivalences: Strict utility model $\Rightarrow$ Fechnerian utility model $\Rightarrow$ strong utility model $\Rightarrow$ SST $\Rightarrow$ MST $\Rightarrow$ WST $\Leftrightarrow$ weak utility model.*

In this Corollary, none of the implications are equivalences, with the possible exception of that from the Fechnerian utility model to the strong utility model. We do not know whether that is an equivalence. If A is not finite, then it is easy to show that WST does not imply the weak utility model. Indeed, even SST does not imply the weak utility model. See Exer. 6 for an example.

There have been many arguments (and experimental results) questioning conditions like SST, and hence questioning some of the utility models. We have mentioned some of the arguments against the utility models in the previous section. Here, let us mention arguments against the transitivity conditions. Suppose you are given a choice between a trip to Paris and a trip to Rome, and you have no clear preference. In fact, your frequency of choosing Paris over Rome is about $\frac{1}{2}$, say just a bit more than $\frac{1}{2}$. Thus,

$$p_{\text{Paris, Rome}} \approx \tfrac{1}{2}.$$

A travel agent offers you a "package tour" of Paris, with one dollar of spending money. Certainly you prefer this to Paris alone. Thus,

$$p_{\text{Paris}+\$1, \text{ Paris}} = 1.$$

By SST,

$$p_{\text{Paris}+\$1, \text{ Rome}} \geqq 1.$$

That is, you always prefer Paris + \$1 to Rome. On the other hand, it seems likely that you would still be pretty much undecided between Paris + \$1 and Rome, and so

$$p_{\text{Paris}+\$1, \text{ Rome}} \approx \tfrac{1}{2}.$$

(This is very similar to the example involving support of the arts and to the pony–bicycle example.)

Experimental evidence showing SST to be violated has been given by Tversky [1969] and Tversky and Russo [1969]. Tversky and Russo presented subjects with a pair of rectangles and asked them to judge which rectangle was larger. There were two kinds of rectangles, short fat ones and long thin ones. Subjects tend to be good at comparing areas of figures similar in shape, and not so good for figures of different shapes. Suppose a and b are short fat rectangles, close in area, with a somewhat larger. Suppose c is long and thin, close in area to b. In experiments, p_{ab}, the frequency with which the area of a is judged to be larger than the area of b, tends to be almost 1, while p_{bc} and p_{ac} tend to be approximately $\frac{1}{2}$. Thus, SST is violated.

In an experiment performed by Tversky [1969], subjects were given choices among alternatives that basically had three dimensions (cf. Chapter 5). The instructions suggested that if two alternatives were close (a term not precisely defined) on the first component, then choice should be made on the basis of the second and third components. Otherwise, choices should be made on the basis of the first component. A typical experimental situation involved hypothetical deans of admission, who were asked to judge candidates first on the basis of intellectual ability and then, if they were close on this basis, on the basis of emotional stability and social facility. In such a situation, suppose a and b are close on the first component, with a substantially better than b on the second and third components. Suppose b and c are also close on the first component, with b substantially better on the second and third components. Finally, suppose c is substantially enough better than a on the first component. Then people tend to choose a over b a majority of times, b over c a majority of times, and c over a a majority of times. Here, even WST is violated. To give a numerical example, suppose a is the triple $(2, 6, 8)$, b the triple $(3, 4, 4)$, c the triple $(4, 2, 2)$, and "close" means at most 1 apart. Then a is rated over b most of the time, and similarly b over c, and c over a.*

Another violation of SST was obtained in a less contrived experiment performed by Coombs [1959]. The stimuli in this experiment were 12 gray chips, labeled $A, B, C, \ldots L$, and subjects were asked to compare chips as to which was a most representative gray.† The data is shown in Table 6.6. Note that $p_{GD} = .86$ and $p_{DE} = .51$, but $p_{GE} = .79$, violating SST. Also, $p_{IJ} = .98$, $p_{JC} = .57$, and $p_{IC} = .80$. There are many other violations of SST.

---

*The Paris–Rome example, the rectangle example, and the college entrance example all are multidimensional in nature. In the first, there is a city dimension and a dollar dimension; in the second, a shape and a size dimension; and in the third, three dimensions. The transitivity conditions and the utility models we have been studying are most appropriate for "one-dimensional" sets of stimuli. However, the next example shows that even with data that is seemingly one-dimensional, there can be violations. In any case, the examples and experimental results we have mentioned suggest that a multidimensional theory of probabilistic consistency would be important to develop.

†Actually, four chips at a time were presented, and rank-ordered as to representativeness, and then pair comparisons were derived from these rank orderings.

Table 6.6. Preferences for Chips*

(The i, j entry is the relative frequency p_{ij} that i was judged more representative gray than j, by a single subject. In this and the following tables, the missing entries may be obtained from the forced choice assumption.)

	G	F	H	I	D	E	J	C	B	K	L	A
G		.53	.63	.66	.86	.79	.97	.92	.94	.97	.99	.99
F			.62	.62	.87	.87	.90	.96	.99	.93	.99	1.00
H				.54	.63	.68	.94	.80	.89	.97	.99	.97
I					.64	.61	.98	.80	.82	1.00	1.00	.99
D						.51	.60	.93	.99	.87	.96	.99
E							.71	.94	.99	.84	.96	.99
J								.57	.63	.94	1.00	.91
C									.93	.68	.93	1.00
B										.58	.84	1.00
K											1.00	.73
L												.53
A												

*Data from Coombs [1959, p. 229, Table 4].

However, in an experiment by Davidson and Marschak [1959], fifteen of seventeen subjects seemed to be satisfying SST, at least in a statistical sense. And Griswold and Luce [1962] and McLaughlin and Luce [1965] also found support for SST.

In spite of contrary examples, the stochastic transitivity conditions are still widely accepted. Indeed, most experimental data seems to confirm at least WST and, usually, MST as well. See Luce and Suppes [1965] for a discussion of such data. In the next subsection, we study the stochastic transitivity conditions in more detail.

6.2.4 *Homogeneous Families of Semiorders*

We have already seen that some of our results about relations can be translated into results about pair comparison systems. For example, if the binary relation W is defined on A by

$$aWb \Leftrightarrow p_{ab} \geq p_{ba}, \tag{6.35}$$

then Theorem 6.8 gives a representation theorem for the weak utility model in terms of (A, W). In this subsection, we shall prove similar theorems.

THEOREM 6.14. *If A is finite and (A, p) is a forced choice pair comparison system, then the following statements are equivalent:*
(a) *(A, p) satisfies the weak utility model.*
(b) *(A, p) satisfies WST.*
(c) *If W is defined from p by Eq. (6.35), then (A, W) is a weak order.*

Proof. We have already observed in Theorem 6.8 that (a) and (c) are equivalent. Now (a) implies (b), for if $p_{ab} \geq \frac{1}{2}$ and $p_{bc} \geq \frac{1}{2}$, then $f(a) \geq f(b)$ and $f(b) \geq f(c)$, so $f(a) \geq f(c)$, so $p_{ac} \geq \frac{1}{2}$. Finally, we show that (b) implies (c). To show that (A, W) is weak, note that strong completeness follows, since either $p_{ab} \geq p_{ba}$ or $p_{ba} \geq p_{ab}$. Transitivity follows, since

$$aWbWc \Rightarrow p_{ab} \geq p_{ba} \And p_{bc} \geq p_{cb}$$

$$\Rightarrow p_{ab} \geq \tfrac{1}{2} \And p_{bc} \geq \tfrac{1}{2}$$

$$\Rightarrow p_{ac} \geq \tfrac{1}{2}$$

$$\Rightarrow p_{ac} \geq p_{ca}$$

$$\Rightarrow aWc.$$

∎

As we have remarked earlier, this theorem is false without the hypothesis of finiteness. See Exer. 6.

Often in experiments with loudness judgments, a is taken to be louder than b (or beyond threshold) if it is judged louder a sufficiently large percentage of the time, for example $\frac{3}{4}$ of the time. A similar idea is useful for preference. Corresponding to any number $\lambda \in [\frac{1}{2}, 1)$, we can define a binary relation R_λ on A as follows:

$$aR_\lambda b \Leftrightarrow p_{ab} > \lambda.$$

Thus, $aR_\lambda b$ if and only if a is preferred to b a fraction of the time greater than λ. The relation R_λ was introduced by Luce [1958, 1959], who proved that under certain conditions each (A, R_λ), $\lambda \in [\frac{1}{2}, 1)$, is a semiorder. It is easy to prove that this conclusion follows from SST (strong stochastic transitivity). The proof uses the results of Section 6.1.5, specifically Theorem 6.5, which says that since (A, R_λ) is asymmetric, it is a semiorder if it is compatible with some weak order. The candidate for the compatible weak order is the relation W of Eq. (6.35). W is a weak order, since SST implies WST, which by Theorem 6.14 implies that W is weak. To show compatibility, we must show that

$$aR_\lambda b \Rightarrow aWb \tag{6.49}$$

and

$$aWbWc \And aI_\lambda c \Rightarrow [aI_\lambda b \text{ and } bI_\lambda c], \tag{6.50}$$

where

$$aI_\lambda b \Leftrightarrow \sim aR_\lambda b \And \sim bR_\lambda a. \tag{6.51}$$

Equation (6.49) follows, since from $\lambda \geqq \frac{1}{2}$, we have

$$aR_\lambda b \Rightarrow p_{ab} > \lambda$$
$$\Rightarrow p_{ab} \geqq p_{ba}$$
$$\Rightarrow aWb.$$

To demonstrate (6.50), note that

$$xI_\lambda y \Leftrightarrow \left[p_{xy} \leqq \lambda \text{ and } p_{yx} \leqq \lambda \right].$$

Suppose $aWbWc$ and $aI_\lambda c$. Then $p_{ab} \geqq p_{ba}, p_{bc} \geqq p_{cb}$, and $p_{ac} \leqq \lambda$. By SST, $p_{ac} \geqq p_{ab}$ and $p_{ac} \geqq p_{bc}$. Thus, $p_{ab} \leqq \lambda$ and $p_{bc} \leqq \lambda$. Since $p_{ba} \leqq p_{ab}$ and $p_{cb} \leqq p_{bc}$, $aI_\lambda b$ and $bI_\lambda c$ follow. This completes the proof of Luce's result.

The family of semiorders

$$\mathscr{F} = \left\{ (A, R_\lambda): \lambda \in \left[\tfrac{1}{2}, 1 \right) \right\}$$

is special because the same weak order is compatible with each (A, R_λ). Such a family of semiorders is called *homogeneous*. Thus, SST implies that $\mathscr{F}$ is a homogeneous family of semiorders. The converse is also true, under a special assumption: for all $a \neq b$ in A, $p_{ab} \neq \frac{1}{2}$. A forced choice pair comparison system satisfying this assumption will be called *discriminated*. Discrimination is not a very special assumption, since if a large number of choices are presented between a and b, it is unlikely that exactly half the time a will be chosen over b. This assumption can be replaced by the even weaker assumption,

$$p_{ab} = \tfrac{1}{2} \Rightarrow (\forall c)(p_{ac} = p_{bc}).$$

THEOREM 6.15 (Roberts [1971a]). *Suppose (A, p) is a discriminated forced choice pair comparison system. Then (A, p) satisfies SST if and only if*

$$\mathscr{F} = \left\{ (A, R_\lambda): \lambda \in \left[\tfrac{1}{2}, 1 \right) \right\}$$

is a homogeneous family of semiorders.

Proof. We have proved one direction. The proof of the other direction will be omitted.

The condition that $\mathscr{F}$ is a homogeneous family of semiorders is a testable condition if A is finite. For in this case, $\mathscr{F}$ has only finitely many *different* relations (A, R_λ).

The condition that (A, R_λ) is a semiorder, or that $\mathscr{F}$ is a family of semiorders, is again a condition of probabilistic consistency, stated in terms of relations. For other related results about probabilistic consistency, see Block and Marschak [1959], Luce and Suppes [1965], Tversky and Russo [1969], Roberts [1971a], and Fishburn [1973].

6.2.5 Beyond Binary Choices

The probabilistic consistency models presented here, and the measurement models presented throughout this book, are special in the following sense: they restrict the basic data to "binary" choices. However, more can be learned about an individual's judgments by presenting him with a set B of more than two alternatives, and asking him to select that element a from B that is most preferred, loudest, etc. Suppose $p(a, B)$ represents the frequency with which a is chosen when B is presented. Then $p_{ab} = p(a, \{a, b\})$. A number of probabilistic consistency models starting with the basic data $p(a, B)$, rather than the basic data p_{ab}, have been developed. For a summary, the reader is referred to Luce and Suppes [1965] or Krantz et al. [to appear].

Exercises

1. (a) Show that the data of Table 6.5 violates SST.
   (b) What about MST?
   (c) What about WST?

2. Does the data of Table 6.6 satisfy the weak utility model? If so, find a function f satisfying Eq. (6.34).

3. Suppose an individual is considering the alternative (a) of taking a new job at a salary of $30,000, as opposed to the alternative (b) of keeping his present job at his present salary, and suppose $p_{ab} \approx \frac{1}{2}$ (he is indifferent). Consider the new alternative a' of taking the new job at a salary of $30,001. Argue that SST is violated.

4. If the scale values $f(a)$ satisfying the strict utility model represent response strength, show that the following statements made in terms of these values are meaningful provided we allow only positive scale values, but not if we allow negative ones:
   (a) a responds more strongly than b.
   (b) a responds twice as strongly as b.
   (c) a and b both respond more strongly than c.
   (d) The response strength of a is greater than the sum of the response strengths of b and c.

5. (a) Suppose $A = \{x, y, z, w\}$ and let p_{ab} be defined by

$$
(p_{ab}) = \begin{array}{c} \\ x \\ y \\ z \\ w \end{array}
\begin{array}{cccc} x & y & z & w \\
\left[\begin{array}{cccc}
\frac{1}{2} & \frac{7}{8} & \frac{5}{8} & \frac{3}{4} \\
\frac{1}{8} & \frac{1}{2} & \frac{3}{8} & \frac{5}{8} \\
\frac{3}{8} & \frac{5}{8} & \frac{1}{2} & \frac{3}{4} \\
\frac{1}{4} & \frac{3}{8} & \frac{1}{4} & \frac{1}{2}
\end{array}\right]
\end{array}.
$$

Show that (A, p) satisfies the weak utility model, but not the strong utility model or SST.

(b) Suppose $A = \{x, y, u, v, w\}$ is a set of sounds and p_{ab} as defined in Exer. 6, Section 3.1, gives the proportion of times a subject says a is louder than b. Which of the following are satisfied?

     (i) WST.
    (ii) MST.
   (iii) SST.
   (iv) Weak Utility Model.
    (v) Strong Utility Model.
   (vi) Fechnerian Utility Model.
  (vii) Strict Utility Model.

6. (Luce and Suppes [1965]) Suppose $A = Re \times Re$, let $\alpha \geq \beta \geq \frac{1}{2}$, and define

$$p_{ab} = \begin{cases} \alpha & \text{if } a_1 > b_1, \\ \beta & \text{if } a_1 = b_1 \text{ and } a_2 > b_2, \\ \frac{1}{2} & \text{if } a_1 = b_1 \text{ and } a_2 = b_2, \\ 1 - p_{ba} & \text{otherwise.} \end{cases}$$

(a) Show that (A, p) satisfies SST (and hence MST and WST).
(b) Show that (A, p) does not satisfy the weak utility model.

7. (Block and Marshak [1959]) Suppose $A = \{1, 2, 3, 4, 5\}$ and

$$\tfrac{1}{2} < p_{21} < p_{54} < p_{32} < p_{43} < p_{53} < p_{31} < p_{42} < p_{41} < p_{52} < p_{51}.$$

Show that the strong utility model fails.

8. Give examples to show that
(a) The strict utility model does not imply the Fechnerian utility model.
(b) WST does not imply MST.
(c) MST does not imply SST.

9. One of the difficulties with assessing the various conditions of probabilistic consistency is how to decide if they are satisfied or violated. For example, suppose $p_{ab} = .7$, $p_{bc} = .6$, and $p_{ac} = .68$. Is this really a violation of SST, or is it a statistically insignificant aberration, which arises only because we are using observed proportions to estimate probabilities? A possible statistical test of the strict utility model is based on the observation that if the strict utility model holds, then for all a and b in A,

$$f(a)/f(b) = p_{ab}/p_{ba}.$$

Thus, suppose the elements of A are ordered as $a_1, a_2, \ldots, a_n$, and suppose $f(a_n)$ is set at 1. Then $f(a_{n-1})$ can be estimated as $p_{a_{n-1}a_n}/p_{a_na_{n-1}}$.

Table 6.7. Entry a, b Is the Proportion of Subjects Who Preferred Spending an Hour with a to Spending an Hour with b*

	LJ	HW	CD	JU	CY	AF	BB	ET	SL
Lyndon Johnson		.68	.70	.75	.78	.76	.74	.68	.61
Harold Wilson			.59	.70	.74	.68	.67	.52	.52
Charles DeGaulle				.62	.67	.59	.60	.52	.51
Johnny Unitas					.75	.49	.53	.37	.26
Carl Yastrzemski						.33	.41	.31	.26
A. J. Foyt							.57	.39	.30
Brigitte Bardot								.29	.21
Elizabeth Taylor									.37
Sophia Loren									

*Data from Rumelhart and Greeno [1968].

Similarly, $f(a_{n-2})$ can be estimated as $f(a_{n-1}) \times (p_{a_{n-2}a_{n-1}}/p_{a_{n-1}a_{n-2}})$. And so on. Using the numbers $f(a)$ calculated in this way, one can calculate the numbers p_{ab} for all $(a, b) \neq (a_k, a_{k-1})$ and compare the observed p_{ab} with the calculated p_{ab}, using a statistical test of goodness of fit. Unfortunately, estimating the values $f(a)$ by this procedure can bias the results, for the estimates may be highly dependent on the particular order $a_1, a_2, \ldots, a_n$ chosen. More sophisticated statistical techniques for estimating the values $f(a)$ are discussed in the literature. See, for example, Bradley and Terry [1952], Davidson [1970], Beaver and Gokhale [1975], or Beaver [1977].

(a) Table 6.7 shows the results of an experiment due to Rumelhart and Greeno [1968] (Restle and Greeno [1970]). Subjects were asked to choose which of two individuals they would rather spend an hour with. The reported p_{ab} is the proportion of subjects who expressed a preference for a over b. If the order of alternatives is chosen as the order in which they are listed in Table 6.7, and f(Sophia Loren) is set equal to 1, estimate the values $f(a)$ using the procedure described above and check that these $f(a)$ lead to the predicted numbers p_{ab} shown in Table 6.8. Identify those entries of Table 6.8 that differ significantly from entries in Table 6.7.

(b) Table 6.9 shows data collected by Estes (see Atkinson, Bower, and Crothers [1965, pp. 146–150]). The entry p_{ab} is the proportion of subjects who expressed a preference for meeting and talking with a over meeting and talking with b.* Calculate several tables similar to Table 6.8 by choosing several orders of points in the set A of alternatives, and then compare the results.

(c) Table 6.10 shows data obtained by Davidson and Bradley [1969] in comparisons of the taste or flavor of different vanilla puddings. The entry p_{ab} is the proportion of subjects who expressed a preference for the taste of a over the taste of b.† Beaver [1977] has estimated the values of the parameters $f(a)$ using a weighted least-squares analysis. The values ob-

*This is the data that was used to generate Table 6.3.
†This is the data that was used to generate Table 6.4.

Table 6.8. Predicted Proportion of Subjects Who Prefer Spending an Hour with a to Spending an Hour with b

	LJ	HW	CD	JU	CY	AF	BB	ET	SL
LJ		.68	.75	.83	.94	.88	.91	.80	.71
HW			.59	.70	.88	.78	.83	.65	.53
CD				.62	.83	.71	.76	.57	.44
JU					.75	.60	.67	.45	.32
CY						.33	.40	.21	.14
AF							.57	.35	.24
BB								.29	.19
ET									.37
SL									

tained are as follows:

$$f(1) = 0.22, \quad f(2) = 0.21, \quad f(3) = 0.17, \quad f(4) = 0.14, \quad f(5) = 0.25.$$

Compute the numbers p_{ab} that correspond to these $f(a)$, compare with the values in Table 6.10, and compare with the numbers obtained if the $f(a)$ are obtained by a method such as that described above.

10. Prove that if an imperfect forced choice pair comparison system satisfies the strict utility model, then it satisfies the product rule.

11. (Luce and Suppes [1965]) A forced choice pair comparison system (A, p) satisfies the *quadruple condition* if for all a, b, c, d in A,

$$p_{ab} \geqq p_{cd} \Rightarrow p_{ac} \geqq p_{bd}.$$

This condition was introduced by Davidson and Marschak [1959]. Show the following:

(a) The quadruple condition follows from the strong utility model.

(b) For imperfect forced choice pair comparison systems, the product rule implies the quadruple condition.

(c) The quadruple condition implies SST.
(*Note*: The converses of all these implications are false. Statistical tests of the quadruple condition have been developed by Falmagne [1976].)

Table 6.9. Entry a, b Is the Proportion of Subjects Who Preferred Meeting and Talking with a to Meeting and Talking with b*

	Dwight Eisenhower	Winston Churchill	Dag Hamerskjold	William Faulkner
Dwight Eisenhower		.57	.80	.82
Winston Churchill			.76	.80
Dag Hamerskjold				.60
William Faulkner				

*Data from Estes (Atkinson, Bower, and Crothers [1965, pp. 146–150], Restle and Greeno [1970, p. 241]).

Table 6.10. Preferences for Vanilla Puddings*
(Entry a, b is the proportion of subjects who preferred the taste of pudding a to pudding b.)

	1	2	3	4	5
1		.20	.64	.79	.41
2			.25	.54	.47
3				.36	.50
4					.30
5					

*Data from an experiment of Davidson and Bradley [1969].

12. (Luce and Suppes [1965]) Show that if an imperfect forced choice pair comparison system (A, p) satisfies the product rule, then for any set of $n \geq 3$ distinct elements $a_1, a_2, \ldots, a_n$ from A,

$$P_{a_1 a_2} P_{a_2 a_3} \cdots P_{a_n a_1} = P_{a_1 a_n} P_{a_n a_{n-1}} \cdots P_{a_2 a_1}.$$

13. (Restle and Greeno [1970]) According to Restle [1961], one may think of a complex alternative as a set of objects or outcomes, and the choice between two alternatives as depending on the elements they do not have in common. Then, if A and B are distinct sets (of objects) and $n(A)$ is the number of elements of A, the probability of choosing A over B is given by

$$P_{AB} = \frac{n(A - B)}{n(A - B) + n(B - A)}. \tag{6.52}$$

(a) Show from the Restle model [Eq. (6.52)] that

$$P_{AB} - P_{BA} = \frac{n(A) - n(B)}{n(A - B) + n(B - A)}.$$

(b) In the strict utility model, if $n(A)$ is the measure of the strength of the response A, then

$$P_{AB} = \frac{n(A)}{n(A) + n(B)}.$$

Show from this that

$$P_{AB} - P_{BA} = \frac{n(A) - n(B)}{n(A) + n(B)}.$$

(c) Show from the Restle model that if A and B are two amounts of money, and A is a larger amount (that is, $A \supsetneq B$), then $p_{AB} = 1$.

(d) Let $S = \{A, B, C\}$, $n(A) = 5$, $n(B) = 4$, $n(C) = 3$, $n(A \cap B) = 2$, $n(A \cap C) = 0$, and $n(B \cap C) = 1$. Show that if p_{AB} is calculated from Eq. (6.52), (S, p) violates the product rule (and hence the strict utility model), but satisfies SST.

(e) On the other hand, if $n(A) = 10$, $n(B) = 9$, $n(C) = 5$, $n(A \cap B) = 6$, $n(A \cap C) = 0$, and $n(B \cap C) = 3$, and p_{AB} is calculated from Eq. (6.52), show that both the product rule and SST are violated by (S, p).

(f) Show that if p is defined using Eq. (6.52) on all sets in a family S, then (S, p) satisfies the weak utility model (and hence WST).

(g) Does (S, p) satisfy MST?

(h) Show that if $n(A \cap B) = n(C \cap D)$ whenever $A \neq B$ and $C \neq D$, and if p is defined using Eq. (6.52), then (S, p) satisfies the strict utility model.

14. Show that if a forced choice pair comparison system satisfies the strong utility model, and W is defined on A by Eq. (6.35), then

$$aWb \Rightarrow (\forall c)(p_{ac} \geq p_{bc}).$$

15. Luce [1958, 1959] defines the *trace* of a forced choice pair comparison system (A, p) as the binary relation T on A defined by

$$aTb \Leftrightarrow (\forall c)(p_{ac} \geq p_{bc}). \qquad (6.53)$$

Show that (A, T) is reflexive and transitive and, moreover,

$$aTb \ \& \ bTa \Rightarrow (\forall c)(p_{ac} = p_{bc}).$$

16. (Roberts [1971a]). Show that if a forced choice pair comparison system satisfies the strong utility model, then the trace of Eq. (6.53) is a weak order, and it is the same weak order as the relation W defined by Eq. (6.35).

17. (Roberts [1971a]). Suppose (A, p) is a discriminated forced choice pair comparison system. Show that SST holds if and only if the trace T of Eq. (6.53) is a weak order.

18. If (A, p) is the example in Exer. 5, show that

$$\mathcal{F} = \left\{ (A, R_\lambda) \colon \lambda \in \left[\tfrac{1}{2}, 1 \right) \right\}$$

is a family of semiorders, but not a homogeneous family.

19. Show that Theorem 6.15 is false without the assumption that (A, p) is discriminated.

20. (Roberts [1971a]). A. A. J. Marley (personal communication) has suggested several conditions on the family

$$\mathcal{F} = \left\{ (A, R_\lambda) \colon \lambda \in \left[\tfrac{1}{2}, 1 \right) \right\},$$

motivated by the semiorder axioms. Assume that (A, p) is a discriminated forced choice pair comparison system. $\mathscr{F}$ satisfies the *First Semiorder Condition* if, for all a, b, c, d in A and all γ, δ in $[\frac{1}{2}, 1)$,

$$(aR_\gamma b \ \& \ cR_\delta d) \Rightarrow (aR_\gamma d \ \text{or} \ cR_\delta b).$$

$\mathscr{F}$ satisfies the *Second Semiorder Condition* if, for all a, b, c, d in A and all γ, δ in $[\frac{1}{2}, 1)$,

$$(aR_\gamma b \ \& \ bR_\delta c) \Rightarrow (aR_\gamma d \ \text{or} \ dR_\delta c).$$

$\mathscr{F}$ satisfies the *Weak Second Semiorder Condition* if, for all a, b, c in A and all γ, δ in $[\frac{1}{2}, 1)$,

$$(aR_\gamma b \ \& \ bR_\delta c) \Rightarrow (aR_\gamma c \ \& \ aR_\delta c).$$

Show the following:

(a) The Weak Second Semiorder Condition follows from the Second Semiorder Condition, taking $d = a$ and $d = c$ and using irreflexivity of R_γ and R_δ.

(b) The Second Semiorder Condition is equivalent to the statement that $\mathscr{F}$ is a homogeneous family of semiorders.

(c) Using (b), the Second Semiorder Condition implies the First.

(d) However, the First Semiorder Condition does not imply the Second.

21. (Tversky and Russo [1969], Roberts [1971a]) We say that a discriminated forced choice pair comparison system (A, p) satisfies *partial substitutability* if

$$(\forall a, b, c)\left[p_{ac} = p_{bc} \Rightarrow p_{ab} = \tfrac{1}{2} \right].$$

We say that (A, p) satisfies the *strong version of SST* if

$$(\forall a, b, c)\left[p_{ab} \geqq 1/2 \ \& \ p_{bc} \geqq 1/2 \Rightarrow p_{ac} \geqq \max\{ p_{ab}, p_{bc}\} \right],$$

where strict inequality in both hypotheses implies strict inequality in the conclusion.

(a) Show that for discriminated forced choice pair comparison systems, the strong version of SST implies partial substitutability.

(b) Show that for discriminated forced choice pair comparison systems, SST does not imply partial substitutability.

22. (Tversky and Russo [1969], Roberts [1971a]). The following conditions of probabilistic consistency on a discriminated forced choice pair comparison system are studied by Tversky and Russo [1969]:

Substitutability:

$$(\forall a, b, c)\left[p_{ab} \geqq \tfrac{1}{2} \Leftrightarrow p_{ac} \geqq p_{bc} \right].$$

(In particular, substitutability says that if $p_{ac} = p_{bc}$ for any c, then $p_{ab} = \frac{1}{2}$.)
   Independence:

$$(\forall a, b, c, d)[\, p_{ac} \geqq p_{bc} \Leftrightarrow p_{ad} \geqq p_{bd}\,].$$

Strong Version of SST (see Exer. 21).
   Show the following:
   (a) For discriminated forced choice pair comparison systems, these three conditions are equivalent.
   (b) Each implies SST, but SST does not imply any of them.
   23. A forced choice pair comparison system (A, p) satisfies *simple scalability* if there is a real-valued function f on A and a function $\phi: Re \times Re \to Re$, with ϕ strictly increasing in the first argument and strictly decreasing in the second, so that

$$p_{ab} = \phi[\, f(a), f(b)\,].$$

That is, the choice probabilities are a function of scale values $f(a)$ and $f(b)$. Show that the Fechnerian utility model implies simple scalability. (Tversky [1972] proved that if A is finite, then (A, p) satisfies simple scalability if and only if it satisfies the independence condition of Exer. 22. More recent results along these lines appear in Smith [1976].)
   24. Exercises 12 of Section 2.6 and 27 of Section 5.4 discuss an experiment aimed at rating several stereo speakers. The top four speakers from the (first) ranking in Table 2.8b were used in a pair comparison experiment modified to allow ties, and with judgments gathered from each of four experts or assessors. The results are shown in Table 6.11a. Entry A means model A was preferred, B means model B was preferred, and = means indifference. Speakers were ranked on the basis of the number of preferences for them. Table 6.11b shows the resulting ranking.
   (a) Note that one of the assessors was nontransitive in his preferences.
   (b) If each assessor's preferences are transitive, and define a strict weak order, a well-known group decisionmaking procedure is the following. Let $f_i(x)$ be the number of elements x is strictly preferred to (ranked above) by the ith assessor. Rank x over y if and only if

$$\sum_i f_i(x) > \sum_i f_i(y).$$

In case of equality, declare a tie. The number $B(x) = \sum_i f_i(x)$ is called the *Borda count* of x, after Jean–Charles de Borda, an eighteenth-century soldier and sailor, and one of the discoverors of the so-called voter's paradox.* A problem with the Borda count is that it is possible to have

---

*The paradox says that even if each assessor's preferences define a strict weak order, there may be no strict weak order so that x is ranked above y in the order if and only if a majority of assessors rank x over y. (See Riker and Ordeshook [1973] for a recent discussion.)

Table 6.11. Pair Comparison of Stereo Speakers*

(a)

Test No.	Model A	Model B	Assessors			
			1	2	3	4
1	Omal	Goodmans	A	A	A	B
2	Omal	Marsden-Hall	A	A	=	A
3	Goodmans	Marsden-Hall	A	A	A	A
4	Goodmans	Quasar	B	B	B	A
5	Omal	Quasar	A	B	A	A
6	Marsden-Hall	Quasar	B	B	=	B

(b)

Rank Order of Models on Basis of Number of Preferences

Model	Number of Preferences
1. Omal	9
2. Quasar	7
3. Goodmans	6
4. Marsden-Hall	0

*From *Hi-Fi & Record News*, July 1975, p. 107.

$B(x) > B(y)$ while $p_{xy} < p_{yx}$, where p_{xy} = the number of assessors ranking x over y divided by the total number of assessors. Give an example to illustrate this. (For further reference on group decisionmaking, see Luce and Raiffa [1957], Sen [1970], Fishburn [1972], or Arrow [1951].)

(c) Restricted to A = {Goodmans, Omal, Quasar}, all the assessors are transitive. Show that on the set A, if p_{xy} is defined as in (b), then the weak utility model is satisfied, SST is satisfied, but the strong utility model fails. Show that the number of preferences for a speaker defines a function that satisfies the weak utility model for (A, p).

(d) Note that the ranking of speakers obtained in Table 6.11b is quite different from that obtained in Table 2.8b. This raises questions about the procedures used to obtain either of these tables. One possible approach to the discrepancy is to eliminate the judgments of the nontransitive assessor. Investigate how this changes the ranking of speakers obtained under the method of Section 2.6 and under the Borda count.

25. (a) Show that if A is finite, (A, R_λ) is a semiorder, and I_λ is defined on A by Eq. (6.51), then (A, I_λ) is an indifference graph (Exer. 23, Section 6.1).

(b) Suppose A is a set of sounds and q_{ab} represents the frequency of times that a and b are judged equally loud. Suppose J_λ is defined as follows:

$$aJ_\lambda b \Leftrightarrow q_{ab} \geqq \lambda. \tag{6.54}$$

and I_λ is defined on A from R_λ by Eq. (6.51). Suppose

(c) Suppose (A, p) is a forced choice pair comparison system,

$$aR_\lambda b \Leftrightarrow p_{ab} > \lambda.$$

and I_λ is defined on A from R_λ by Eq. (6.51). Suppose

$$q_{ab} = 1 - [p_{ab} + p_{ba}],$$

and J_λ is defined by Eq. (6.54). Is it possible for (A, I_λ) to be an indifference graph while (A, J_λ) is not?

(d) Table 6.12 shows data q_{ab} obtained from an experiment performed by Rothkopf [1957]. Find (A, J_λ) as defined by Eq. (6.54) for

    (i) $\lambda = .1$
    (ii) $\lambda = .2$
    (iii) $\lambda = .3$.

(e) For each value of λ in part (d), determine if (A, J_λ) is an indifference graph.

Table 6.12. Judgments of Equal Loudness[*] (Entry a, b is the proportion of subjects who judged sound a and sound b to be equally loud.)

	A	B	C	D	E
A	.92	.04	.05	.11	.05
B	.04	.84	.37	.46	.09
C	.05	.37	.87	.17	.09
D	.11	.46	.17	.88	.07
E	.05	.09	.09	.07	.96

[*]Data from Rothkopf [1957].

References

Adams, E. W., "Elements of a Theory of Inexact Measurement," *Phil. Sci.*, **32** (1965), 205–228.

Armstrong, W. E., "The Determinateness of the Utility Function," *Econ. J.*, **49** (1939), 453–467.

Armstrong, W. E., "Uncertainty and the Utility Function," *Econ. J.*, **58** (1948) 1–10.

Armstrong, W. E., "A Note on the Theory of Consumer's Behavior," *Oxford Economic Papers*, **2** (1950), 119–122.

Armstrong, W. E., "Utility and the Theory of Welfare," *Oxford Economic Papers*, **3** (1951), 259–271.

Arrow, K., *Social Choice and Individual Values*, Cowles Commission Monograph 12, Wiley, New York, 1951, 2nd ed., 1963.

Atkinson, R. C., Bower, G. H., and Crothers, E. J., *An Introduction to Mathematical Learning Theory*, Wiley, New York, 1965.

Baker, K. A., Fishburn, P. C., and Roberts, F. S., "Partial Orders of Dimension 2, Interval Orders, and Interval Graphs," Paper P-4376, The RAND Corporation, Santa Monica, California, June 1970.

Baker, K. A., Fishburn, P. C., and Roberts, F. S., "Partial Orders of Dimension 2," *Networks*, **2** (1971), 11–28.

Beaver, R. J., "Weighted Least Squares Analysis of Several Uivariate Bradley-Terry Models, *J. Amer. Stat. Assoc.*, **72** (1977), 629–634.

Beaver, R. J., and Gokhale, D. V., "A Model To Incorporate Within-Pair Order Effects in Paired Comparison Experiments," *Communications in Statistics*, **4** (1975), 923–939.

Block, H. D., and Marschak, J., "Random Orderings and Stochastic Theories of Responses," in I. Olkin, S. Ghurye, W. Hoefding, W. Madow, and H. Mann (eds.), *Contributions to Probability and Statistics*, Stanford University Press, Stanford, California, 1959.

Bogart, K., Rabinovitch, I., and Trotter, W. T., "A Bound on the Dimension of Interval Orders," *J. Comb. Theory*, **21** (1976), 319–328.

Bradley, R. A., "Incomplete Block Rank Analysis: On the Appropriateness of the Model for a Method of Paired Comparisons," *Biometrics*, **10** (1954a), 375–390.

Bradley, R. A., "Rank Analysis of Incomplete Block Designs. II. Additional Tables for the Method of Paired Comparisons," *Biometrika*, **41** (1954b), 502–537.

Bradley, R. A., "Rank Analysis of Incomplete Block Designs. III. Some Large Sample Results on Estimation and Power for a Method of Paired Comparisons," *Biometrika*, **42** (1955), 450–470.

Bradley, R. A., and Terry, M. E., "Rank Analysis of Incomplete Block Designs. I. The Method of Paired Comparisons," *Biometrika*, **39** (1952), 324–345.

Cohen, J. E., *Food Webs and Niche Space*, Princeton University Press, Princeton, New Jersey, 1978.

Coombs, C. H., "Inconsistency of Preferences as a Measure of Psychological Distance," in R. W. Churchman and P. Ratoosh (eds.), *Measurement: Definitions and Theories*, Wiley, New York, 1959, pp. 221–232.

Coombs, C. H., Raiffa, H., and Thrall, R. M., "Some Views on Mathematical Models and Measurement Theory," in R. M. Thrall, C. H. Coombs, and R. L. Davis (eds.), *Decision Processes*, Wiley, New York, 1954, pp. 19–37.

Coombs, C. H., and Smith, J. E. K., "On the Detection of Structures in Attitudes and Developmental Processes," *Psych. Rev.*, **80** (1973), 337–351.

Davidson, D., and Marschak, J., "Experimental Tests of a Stochastic Decision Theory," in C. W. Churchman and P. Ratoosh (eds.), *Measurement: Definitions and Theories*, Wiley, New York, 1959, pp. 233–269.

Davidson, R. R., "On Extending the Bradley–Terry Model to Accommodate Ties in Pair Comparison Experiments," *J. Amer. Stat. Assoc.*, **65** (1970), 317–328.

Davidson, R. R., and Bradley, R. A., "Multivariate Paired Comparisons: The Extension of a Univariate Model and Associated Estimation and Test Procedures," *Biometrika*, **56** (1969), 81–95.

Domotor, Z., *Probabilistic Relational Structures and Their Applications*, Tech. Rept. 144, Institute for Mathematical Studies in the Social Sciences, Stanford University, Stanford, California, 1969.

Domotor, Z., and Stelzer, J. H., "Representation of Finitely Additive Semiordered Qualitative Probability Structures," *J. Math. Psychol.*, **8** (1971), 145–158.

Ducamp, A., "A Note on an Alternative Proof of the Representation Theorem for Bi-Semiorder," *J. Math. Psychol.*, **18** (1978), 100–104.

Ducamp, A., and Falmagne, J. C., "Composite Measurement," *J. Math. Psychol.*, **6** (1969), 359–390.

Dushnik, B., and Miller, E. W., "Partially Ordered Sets," *Amer. J. Math.*, **63** (1941), 600–610.

Falmagne, J. C., "Foundations of Fechnerian Psychophysics," in D. H. Krantz, R. C. Atkinson, R. D. Luce, and P. Suppes (eds.), *Contemporary Developments in Mathematical Psychology*, Freeman, San Francisco, California, 1974.

Falmagne, J. C., "Statistical Tests in Measurement Theory: Two Methods," mimeographed, New York University, Department of Psychology, paper presented at the Spring Meeting of the Psychometric Society, April 1976.

Fishburn, P. C., "Weak Qualitative Probability on Finite Sets," *Ann. Math. Stat.*, **40** (1969), 2118–2126.

Fishburn, P. C., "Intransitive Indifference in Preference Theory: A Survey," *Operations Research*, **18** (1970a), 207–228.

Fishburn, P. C., "Intransitive Indifference with Unequal Indifference Intervals," *J. Math. Psychol.*, **7** (1970b), 144–149.

Fishburn, P. C., *Utility Theory for Decisionmaking*, Wiley, New York, 1970c.

Fishburn, P. C., *The Theory of Social Choice*, Princeton University Press, Princeton, New Jersey, 1972.

Fishburn, P. C., "Binary Choice Probabilities: On the Varieties of Stochastic Transitivity," *J. Math. Psychol.*, **10** (1973), 329–352.

Folgmann, E. E. E., "An Experimental Study of Composer-Preferences of Four Outstanding Symphony Orchestras," *J. Exp. Psychol.*, **16** (1933), 709–724.

Fulkerson, D. R., and Gross, O. A., "Incidence Matrices and Interval Graphs," *Pacific J. Math.*, **15** (1965), 835–855.

Gabai, H., "Bounds for the Boxicity of a Graph," mimeographed, York College, City University of New York, 1976.

Ghouila–Houri, A., "Caractérisation des graphes Nonorientés dont on peut Orienter les Arêtes de Manière à Obtenir le Graphe d'une Relation d'Ordre," *C. R. Acad. Sci.*, **254** (1962), 1370–1371.

Gilmore, P. C., and Hoffman, A. J., "A Characterization of Comparability Graphs and of Interval Graphs," *Canadian J. Math.*, **16** (1964), 539–548.

Goodman, N., *Structure of Appearance*, Harvard University Press, Cambridge, Massachusetts, 1951.

Greenough, T. L., and Bogart, K. P., "The Representation and Enumeration of Interval Orders," *Discrete Math.*, 1979, to appear.

Griswold, B. J., and Luce, R. D., "Choices Among Uncertain Outcomes: A Test of a Decomposition and Two Assumptions of Transitivity," *Am. J. Psychol.*, **75** (1962), 35–44.

Guilford, J. P., *Psychometric Methods*, 2nd ed., Macmillan, New York, 1954.

Hamming, R. W., "Numerical Analysis: Pure or Applied Mathematics?" *Science*, **148** (1965), 473–475.

Hubert, L., "Some Applications of Graph Theory and Related Non-metric Techniques to Problems of Approximate Seriation: The Case of Symmetric Proximity Measures," *Brit. J. Math. Stat. Psych.*, **27** (1974), 133–153.

Kendall, D. G., "A Statistical Approach to Flinders Petrie's Sequence Dating," *Bull. Int. Stat. Inst.*, **40** (1963), 657–680.

Kendall, D. G., "Incidence Matrices, Interval Graphs, and Seriation in Archaeology," *Pacific J. Math.*, **28** (1969a), 565–570.

Kendall, D. G., "Some Problems and Methods in Statistical Archaeology," *World Archaeology*, **1** (1969b), 61–76.

Kendall, D. G., "A Mathematical Approach to Seriation," *Phil. Trans. Roy. Soc. Ser. A*, **269** (1971a), 125–135.

Kendall, D. G., "Abundance Matrices and Seriation in Archaeology," *Z. Wahrscheinlichkeitstheorie Verw. Gebiete*, **17** (1971b), 104–112.

Kendall, D. G., "Seriation from Abundance Matrices," in F. R. Hodson *et al.* (eds.), *Mathematics in the Archaeological and Historical Sciences*, Edinburgh University Press, Edinburgh, 1971c.

Kramer, G. H., "An Impossibility Result Concerning the Theory of Decision-Making," in J. L. Bernd (ed.), *Mathematical Applications in Political Science*, Vol. III, University of Virginia Press, Charlottesville, Virginia, 1968, pp. 39–51.

Krantz, D. H., "Extensive Measurement in Semiorders," *Phil. Sci.*, **34** (1967), 348–362.

Krantz, D. H., "A Survey of Measurement Theory," in G. B. Dantzig and A. F. Veinott, Jr. (eds.), *Mathematics of the Decision Sciences*, Part 2, Vol. 12, Lectures in Applied Mathematics, American Mathematical Society, Providence, Rhode Island, 1968, pp. 314–350.

Krantz, D. H., Luce, R. D., Suppes, P., and Tversky, A., *Foundations of Measurement*, Vol. II,

Academic Press, New York, to appear.

Lekkerkerker, C. B., and Boland, J. Ch., "Representation of a Finite Graph by a Set of Intervals on the Real Line," *Fund. Math.*, 51 (1962), 45–64.

Luce, R. D., "Semiorders and a Theory of Utility Discrimination," *Econometrica*, 24 (1956), 178–191.

Luce, R. D., "A Probabilistic Theory of Utility," *Econometrica*, 26 (1958), 193–224.

Luce, R. D., *Individual Choice Behavior*, Wiley, New York, 1959.

Luce, R. D., "Three Axiom Systems for Additive Semiordered Structures," *SIAM J. Appl. Math.*, 25 (1973), 41–53.

Luce, R. D., and Galanter, E., "Discrimination," in R. D. Luce, R. R. Bush, and E. Galanter (eds.), *Handbook of Mathematical Psychology*, Vol. 1, Wiley, New York, 1963, pp. 191–243.

Luce, R. D., and Raiffa, H., *Games and Decisions*, Wiley, New York, 1957.

Luce, R. D., and Suppes, P., "Preference, Utility, and Subjective Probability," in R. D. Luce, R. R. Bush, and E. Galanter (eds.), *Handbook of Mathematical Psychology*, Vol. 3, Wiley, New York, 1965, pp. 249–410.

Manders, K., "A Model-Theoretic Treatment of Semiorder Representation," manuscript, Department of Philosophy, University of California, Berkeley, California, 1977.

McLaughlin, D. H., and Luce, R. D., "Stochastic Transitivity and Cancellation of Preferences between Bitter-sweet Solutions," *Psychon. Sci.*, 2 (1965), 89–90.

Menger, K., "Probabilistic Theories of Relations," *Proc. Nat. Acad. Sci.*, 37 (1951), 178–180.

Pfanzagl, J., *Theory of Measurement*, Wiley, New York, 1968.

Poston, T., "Fuzzy Geometry," Tech. Rept., Mathematics Institute, University of Warwick, England, June 1971.

Rabinovitch, I., "The Scott–Suppes Theorem on Semiorders," *J. Math. Psychol.*, 15 (1977), 209–212.

Rabinovitch, I., "The Dimension of Semiorders," *J. Comb. Theory*, A25 (1978), 50–61.

Raiffa, H., *Decision Analysis*, Addison-Wesley, Reading, Massachusetts, 1968.

Restle, F., *The Psychology of Judgment and Choice*, Wiley, New York, 1961.

Restle, F., and Greeno, J. G., *Introduction to Mathematical Psychology*, Addison-Wesley, Reading, Massachusetts, 1970.

Riker, W. H., and Ordeshook, P. C., *Positive Political Theory*, Prentice-Hall, Englewood Cliffs, New Jersey, 1973.

Roberts, F. S., "Indifference Graphs," in F. Harary (ed.), *Proof Techniques in Graph Theory*, Academic Press, New York, 1969a, pp. 139–146.

Roberts, F. S., "On the Boxicity and Cubicity of a Graph," in W. T. Tutte (ed.), *Recent Progress in Combinatorics*, Academic Press, New York, 1969b, 301–310.

Roberts, F. S., "Notes on Perceptual Geometries," Paper P-4423, The RAND Corporation, Santa Monica, California, July 1970.

Roberts, F. S., "Homogeneous Families of Semiorders and the Theory of Probabilistic Consistency," *J. Math. Psychol.*, 8 (1971a), 248–263.

Roberts, F. S., "On the Compatibility between a Graph and a Simple Order," *J. Comb. Theory*, 11 (1971b), 28–38.

Roberts, F. S., "Tolerance Geometry," *Notre Dame J. Formal Logic*, 14 (1973), 68–76.

Roberts, F. S., *Discrete Mathematical Models, with Applications to Social, Biological, and Environmental Problems*, Prentice-Hall, Englewood Cliffs, New Jersey, 1976.

Roberts, F. S., "Food Webs, Competition Graphs, and the Boxicity of Ecological Phase Space," in Y. Alavi, and D. Lick (eds.), *Theory and Applications of Graphs*, Springer-Verlag, New York, 1978a, pp. 477–490.

Roberts, F. S., *Graph Theory and Its Applications to Problems of Society*, CBMS-NSF Monograph Number 29, Society for Industrial and Applied Mathematics, Philadelphia, Pennsylvania, 1978b.

Roberts, F. S., "Indifference and Seriation," in F. Harary (ed.), *Advances in Graph Theory*, New York Academy of Sciences, New York, 1979, pp. 171–180.

Rothkopf, E. Z., "A Measure of Stimulus Similarity and Errors in Some Paired Associate Learning Tasks," *J. Exp. Psychol.*, **53** (1957), 94–101.

Rothstein, J., "Numerical Analysis: Pure or Applied Mathematics?" *Science*, **149** (1965), 1049–1050.

Rumelhart, D. L., and Greeno, J. G., "Choices between Similar and Dissimilar Objects: An Experimental Test of the Luce and Restle Choice Models," paper read at Midwestern Psychological Association, Chicago, Illinois, 1968.

Scott, D., and Suppes, P., "Foundational Aspects of Theories of Measurement," *J. Symbolic Logic*, **23** (1958), 113–128.

Sen, A. K., *Collective Choice and Social Welfare*, Holden-Day, San Francisco, 1970.

Smith, T. E., "Scalable Choice Models," *J. Math. Psychol.*, **14** (1976), 239–243.

Stelzer, J. H., "Some Results Concerning Subjective Probability Structures with Semiorders," Tech. Rept. 115, Institute for Mathematical Studies in the Social Sciences, Stanford University, Stanford, California, 1967.

Suppes, P. and Zinnes, J., "Basic Measurement Theory," in R. D. Luce, R. R. Bush, and E. Galanter (eds.), *Handbook of Mathematical Psychology*, Vol. I, Wiley, New York, 1963, pp. 1–76.

Thurstone, L. L., "A Law of Comparative Judgment," *Psychol. Rev.*, **34** (1927a), 273–286.

Thurstone, L. L., "Psychophysical Analysis," *Amer. J. Psychol.*, **38** (1927b), 368–389.

Trotter, W. T., "A Characterization of Roberts' Inequality for Boxicity," mimeographed, Department of Mathematics, University of South Carolina, Columbia, South Carolina, 1977.

Tversky, A., "Intransitivity of Preferences," *Psychol. Rev.*, **76** (1969), 31–48.

Tversky, A., "Choice by Elimination," *J. Math. Psychol.*, **9** (1972), 341–367.

Tversky, A., and Russo, J. E., "Similarity and Substitutability in Binary Choices," *J. Math. Psychol.*, **6** (1969), 1–12.

Zeeman, E. C., "The Topology of the Brain and Visual Perception," in M. K. Fort (ed.), *The Topology of 3-Manifolds*, Prentice-Hall, Englewood Cliffs, New Jersey, 1962, pp. 240–256.

Zermelo, E., "Die Berechnung der Turnierergebnisse als ein Maximumproblem der Wahrscheinlichkeitsrechnung," *Math. Z.*, **29** (1929), 436–460.

CHAPTER 7

Decisionmaking under Risk or Uncertainty

7.1 The Expected Utility Rule and the Expected Utility Hypothesis

In this chapter, we consider for the first time a situation of decisionmaking under uncertainty. We allow for the possibility that one of a set of uncertain events may occur, each with a certain probability, and each with a known consequence. We face such decisionmaking problems often in our lives. For example, when we consider whether or not to buy insurance before we are 30 years old, we consider the possible but uncertain event of death. A doctor often faces a choice among alternative treatments, with different uncertain side effects. A government must spend money on one of several technologies designed to solve a problem, each of which has only a certain probability of providing the desired results. (For example: What design should we use for a rapid transit system? Should we invest large amounts of money on breeder reactor research, rather than on solar power? And so on.)

To motivate our discussion of such decisionmaking problems, let us consider a simple gambling situation.* Suppose you are attending a business meeting and have a choice between parking your car at a meter or putting it in a parking lot. The meter can take coins for up to 2 hours—the maximum would require $1. The meter is monitored almost constantly, and the fine for overtime parking is $25. The lot would be $5, a flat fee. You think that the chance the meeting will be over within 2 hours is 80%. Should you put the car in the lot? The decision you face can be

---

*This example is based on Lighthill [1978] and Welford [1970].

ENCYCLOPEDIA OF MATHEMATICS and Its Applications, Gian-Carlo Rota (ed.). Vol. 7: Fred S. Roberts, Measurement Theory

Table 7.1.

Event	Probability	Payoff or Consequence to You
Act I: Put Car on Meter		
Meeting over within 2 hours	.8	$-\$1$
Meeting lasts more than 2 hours	.2	$-(\$1 + \$25) = -\$26$
Act II: Put Car in Lot		
Meeting over within 2 hours	.8	$-\$5$
Meeting lasts more than 2 hours	.2	$-\$5$

summarized in Table 7.1, which gives the amount of money you have lost under each of the two possible actions, depending on the outside event of how long the meeting lasts. One way to make the decision is to compute the expected value of the payoff under each possible action, and choose the action with the larger expected value. In this case, the expected value of act I is

$$.8(-\$1) + .2(-\$26) = -\$6,$$

and the expected value of act II is

$$.8(-\$5) + .2(-\$5) = -\$5.$$

Since $-\$5$ is larger than $-\$6$, you would choose the second act, and put your car in the lot.

Let us change the numbers in this illustration. Suppose the lot would cost you $7 instead of $5. Then putting the car in the lot would no longer have the highest expected value—its expected value is now $-\$7$. However, you might still be tempted to put the car in the lot, since by putting the car at a meter you risk having a large loss. This suggests, as we have pointed out in Sections 4.3.4 and 5.4.2, that the "value" of a sum of money is not necessarily proportional to its dollar amount. Rather, we should assess your utility of n dollars and compute for each act your *expected utility* (expected value of your utility function) given that the act is chosen. Here, we see that act I has as its expected utility

$$.8u(-\$1) + .2u(-\$26),$$

while act II has as its expected utility $u(-\$7)$, where u is utility. If, for example, $u(-\$1) = -1$ and $u(-\$7) = -10$ and $u(-\$26) = -100$, then act I has expected utility -20.8 and act II has expected utility -10. If we were to choose on the basis of expected utility, we would still choose act II.

Speaking very generally, suppose we think of an *act* or *choice* or *gamble* as leading to the occurrence of one of several events, $A_1, A_2, \ldots, A_n$. Suppose that the events are mutually exclusive and exhaustive, that is, that

$$p(A_i \cap A_j) = 0 \quad \text{if} \quad i \neq j$$

and

$$\Sigma p(A_i) = 1,$$

where $p(B)$ is the probability that event B will occur. Associated with each event A_i is a reward or consequence c_i, not necessarily money, to which you associate some value or utility $u(c_i)$. The reward c_i could be a kewpie doll, a pass to gamble again, and so on. The *expected utility* of the act or choice is given by

$$E = \sum_{i=1}^{n} p(A_i)u(c_i). \tag{7.1}$$

We anticipate choosing an act with the largest possible expected utility.

The notion of expected utility of an act or choice applies to a very broad range of situations*. Let us apply it to a hypothetical medical decisionmaking problem. A physician is considering two possible treatments (acts or choices), x and y. If treatment x is used, two possible events can occur. They are:

A_1: the treatment works,
A_2: the treatment doesn't work.

Associated with these events are the consequences:

c_1: the patient is completely cured and achieves good health,
c_2: the patient dies.

Let us suppose that past records indicate that treatment x works one-third of the time. Thus, we can assign probabilities

$$p(A_1) = \tfrac{1}{3},$$

$$p(A_2) = \tfrac{2}{3}.$$

*This set-up still does not account for all the possible uncertainties in decisionmaking. The probabilities $p(A_i)$ may not be known. There may be several possible consequences c_i associated with an event A_i, and perhaps covered by some (unknown?) probability distribution. And so on. The set-up also does not take into account changing information, and the impact of this new information on decisions.

Suppose that a patient assesses his utility of death as 0 and his utility of a healthy life as 100. (We shall discuss the utility of a life below.) Then

$$u(c_1) = 100,$$
$$u(c_2) = 0,$$

and the expected utility of treatment x for this patient is

$$E(x) = p(A_1)u(c_1) + p(A_2)u(c_2)$$
$$= \left(\tfrac{1}{3}\right)(100) + \tfrac{2}{3}(0)$$
$$= 33\tfrac{1}{3}.$$

Let us now compare treatment y. Here, there are three possible events:

A_1': the treatment works completely,
A_2': the treatment doesn't work,
A_3': the treatment works partially.

Associated with these events are the consequences:

c_1': the patient is completely cured and achieves good health,
c_2': the patient dies,
c_3': the patient becomes a cripple for life.

According to past records, the various events following treatment y have the following probabilities:

$$p(A_1') = .0001,$$
$$p(A_2') = .4999,$$
$$p(A_3') = .5000.$$

Suppose the patient's utilities are as follows:

$$u(c_1') = 100,$$
$$u(c_2') = 0,$$
$$u(c_3') = -200.$$

(This patient would rather die than be a cripple for life. Note that $u(c_1') = u(c_1)$, $u(c_2') = u(c_2)$, for consistency.) The expected utility of treatment for this patient is

$$E(y) = p(A_1')u(c_1') + p(A_2')u(c_2') + p(A_3')u(c_3')$$
$$= (.0001)(100) + (.4999)(0) + (.5000)(-200)$$
$$= -99.99.$$

It is not unreasonable to choose between the two treatments on the basis of the expected utilities. That is, one chooses treatment x over treatment y if $E(x) > E(y)$, and treatment y over treatment x if $E(y) > E(x)$. In our hypothetical example, one chooses treatment x. This is the case even though treatment x leads to death with a higher probability. The reason for this result is that the patient would rather die than be a cripple for life. For a patient with a different utility function, the choice might be different.*

The use of expected utilities to make choices can be viewed as a prescriptive or normative rule. We shall call it the *Expected Utility Rule* (*EU Rule*). This rule goes back to Daniel Bernoulli [1738]. It can be formalized as follows: If x and y are gambles, then

$$x \text{ is preferred to } y \quad \text{iff} \quad \Sigma p(A_i)u(c_i) > \Sigma p(A_i')u(c_i'), \qquad (7.2)$$

where the first sum is over events and consequences of gamble x and the second is over events and consequences of gamble y. The use of expected utilities to make choices can also be viewed as a descriptive rule. Then it is asserted that individuals make choices *as if* they were maximizing expected utility. It is not necessarily claimed that the calculation of expected utility is conscious. Rather, given preferences among acts or choices under risk or uncertainty, it is claimed that we can account for the preferences by finding utilities and probabilities so that (7.2) holds.

The assertion that individuals choose *as if* they maximize expected utility is known as the *Expected Utility Hypothesis* (*EU Hypothesis*). If the probabilities are based on subjective-assessments, then the hypothesis is sometimes called the *Subjective Expected Utility Hypothesis* (*SEU Hypothesis*). We shall have more to say about subjective assessments of probability in Chapter 8. Both the EU and SEU Hypotheses will be regarded, as have other descriptive measurement models, as hypotheses that are subject to test. In any case, the relation between the EU Hypothesis and the EU Rule

---

*Whether or not the choice of treatment should depend on the *patient's* utilities is a moral issue for medical decisionmakers. (Slack [1972] argues that, since it is the patient who has the major stake in the outcome of the decision, it is the patient whose utilities should be used. For a discussion of the use of decision theory in medical decisionmaking, see Aitchison [1970], Forst [1972], Ginsberg [1971], Ledley and Lusted [1959], Lusted [1968] and Schwartz et al. [1973]. A survey of the literature is contained in Fryback [1974].)

is the following: The EU Hypothesis holds if we make choices as if we are following the EU Rule.

In this chapter, we shall take both a prescriptive and a descriptive approach. We shall describe in Section 7.2 various ways to make use of the EU Rule or Hypothesis in making decisions. In Section 7.3, we shall discuss how the EU Rule or Hypothesis can be used to calculate utility functions over multidimensional sets of consequences. Then, in Section 7.4, we shall discuss a representation theorem which starts with preference among acts and gives conditions sufficient for these preferences to arise from comparisons of expected utilities. This theorem gives conditions on preferences among acts for there to be utility (and probability) functions satisfying Eq. (7.2).

Before closing this section, let us make one important comment. Suppose probabilities $p(A_i)$ are fixed. If utilities of consequences are just measured on an ordinal scale, then decisions using the EU Rule are not meaningful (and should not be made). For example, suppose choice x has outcomes a or b, each with probability $\frac{1}{2}$, while choice y has outcomes c or d, each with probability $\frac{1}{2}$. Suppose $u(a) = 100$, $u(b) = 200$, $u(c) = 50$, and $u(d) = 300$. Then $E(x) = 150$ and $E(y) = 175$, so y is chosen over x. However, if $u'(a) = 200$, $u'(b) = 400$, $u'(c) = 10$, and $u'(d) = 500$, then using u' rather than u, $E'(x) = 300$ and $E'(y) = 255$, so x is chosen over y. If u is just an ordinal utility function, then u' is obtained from u by an admissible transformation. Comparisons of expected utilities using utility functions on an interval scale are meaningful. For

$$\Sigma p(A_i)[\alpha u(c_i) + \beta] > \Sigma p(A_i')[\alpha u(c_i') + \beta]$$

$$\Leftrightarrow$$

$$\alpha \Sigma p(A_i)u(c_i) + \beta \Sigma p(A_i) > \alpha \Sigma p(A_i')u(c_i') + \beta \Sigma p(A_i')$$

$$\Leftrightarrow$$

$$\Sigma p(A_i)u(c_i) > \Sigma p(A_i')u(c_i'),$$

using $\alpha > 0$ and $\Sigma p(A_i) = \Sigma p(A_i') = 1$.

Exercises

1. In comparing two television sets of equal price, we consider three possible events: The set will last a long time with satisfactory performance, the set will last a moderate amount of time with satisfactory performance, or the set will never work satisfactorily. For Brand x, we assign these events the probabilities $\frac{1}{2}$, $\frac{1}{3}$, and $\frac{1}{6}$, respectively. For Brand y, we assign them probabilities $\frac{5}{8}$, $\frac{1}{8}$, and $\frac{1}{4}$, respectively. If the consequences corresponding to these events have utilities 100, 10, and -200, respectively, then according to the EU Rule, which brand should we buy?

2. As we have observed, most economists believe that the utility of money is not a linear function of the dollar amount. Adding the same amount of money becomes less and less important as you have more and more, it is frequently argued. As we pointed out in Chapters 4 and 5, Bernoulli [1738] hypothesized that the utility of money is a logarithmic function of the dollar amount, and Gabriel Cramer in 1728 argued that the utility of money is a power function of the dollar amount. Let us consider as an example to compare these possibilities the question of whether to buy insurance. You have a $100,000 home. Fire insurance would cost $200. Suppose you think the probability of your house's burning is .001. Thus, with no insurance, you lose $100,000 with probability .001, and otherwise you break even. With insurance, you lose $200 with certainty. Under the EU Rule, which of the following utility functions implies purchase of the insurance?

    (a) $u(n) = n - 1$,

    (b) $u(n) = -\log_{10}|n - 1|$ if $n \leq 0$,

    (c) $u(n) = -2^{|n-1|}$ if $n \leq 0$.

(More on the utility of money in the next exercise and in Section 7.2.4.)

3. Suppose you are indifferent between the following two gambles:

    Gamble 1. Play a game in which you are sure to win $15.

    Gamble 2. Play a game in which you have a 50% chance of
                 winning $20 and a 50% chance of losing $12.

Suppose u is a utility function over the set of consequences. If the EU Hypothesis holds with u, and $u(\$1) \neq 0$, show that $u(\$n)$ could not always be $nu(\$1)$.

4. An individual faces a choice of careers. If he chooses career A, he will either make it big or not. He has a 50–50 chance, as he sees it, of making it big, in which case he will have steady earnings of $30,000 annually over his entire career. If he doesn't make it big, he will have steady earnings of only $15,000 annually over his entire career. If the individual chooses career B, he will either make it big early and fade later, or struggle early and make it big later. He thinks there is a 50–50 chance of either alternative. In the former case, he will earn about $30,000 annually over the first half of his career, but only $15,000 for the second half. In the latter case, the situation is reversed. Thus, all other things being equal, the individual faces in each career a 50–50 chance between two income patterns. In career A, the choice is between ($30,000, $30,000) and ($15,000, $15,000), and in career B between ($30,000, $15,000) and ($15,000, $30,000). If the decision is purely on the basis of income pattern, Fishburn [1970] suggests many individuals would choose career B.

    (a) Show that if the EU Hypothesis holds with the utility function u, then u for such individuals would satisfy

$$u(\$30,000, \$15,000) + u(\$15,000, \$30,000) >$$
$$u(\$30,000, \$30,000) + u(\$15,000, \$15,000). \quad (7.3)$$

    (b) Show that for such individuals, there can be no function f so that $u(a, b) = f(a) + f(b)$.

(c) Indeed, show that there are not even functions f_1 and f_2 so that $u(a, b) = f_1(a) + f_2(b)$.

(For further discussion of multiperiod income streams, see, for example, Bell [1974], Fishburn [1973], Keeney and Raiffa [1976], and McGuire and Radner [1972]. For further discussion of additivity of utility functions, see Section 7.3.2.)

5. (a) In Exer. 4, suppose $A_1 = A_2 = \{\$15,000, \$30,000\}$, R is the binary relation of preference on $A_1 \times A_2$, and u is an ordinal utility function for R satisfying (7.3). What possible R's satisfy these conditions?

(b) Which of the above R's define an additive conjoint structure (Section 5.4)?

6. (a) Suppose you have invested in a stock of XYZ Company, and the stock has gone down in price. You could sell it now at a loss of $225. Let us call this Strategy I. You also have two alternative investment strategies (involving different choice of lower and upper selling prices for the stock). Strategy II, according to your assessment, would lead to an ultimate loss of $425 with probability .65, a loss of $125 with probability .25, and an ultimate profit of $275 with probability .10. Strategy III would lead to an ultimate loss of $325 with probability .75, an ultimate loss of $125 with probability .2, and an ultimate gain of $275 with probability .05. All probabilities are your subjective assessments and may not be completely accurate. The dollar amounts are small enough so that it might be reasonable to assume that the utility of n is $nu(\$1)$, or n units. Discuss the alternative strategies, both from your intuitive feeling and from the point of view of expected utility. Note that subjective probabilities are only estimates, and expected gains are close, and there is a measure of risk in both Strategies II and III. This might affect choice of strategy.

(b) It is interesting to analyze these three strategies from your *present* position. With Strategy I, you break even for sure. With Strategy II, you lose $200 with probability .65, gain $125 with probability .25, and gain $500 with probability .1. With Strategy III, you lose $100 with probability .75, gain $125 with probability .2, and gain $500 with probability .05. If the decision is considered from this new point of view, observe that one's preferences may be different from those expressed about the previously formulated problem. What does that say about utility if the EU Hypothesis holds?

7.2 Use of the EU Rule and Hypothesis in Decisionmaking*

7.2.1 *Lotteries*

In the previous section, we introduced the notion of expected utility and the Expected Utility Rule and Hypothesis. In this section, we present several tools or methods for help in making complex decisions, assuming

---

*The discussion in Sections 7.2 and 7.3 is based heavily on Raiffa [1968, 1969].

the EU Rule or Hypothesis holds. In particular, we shall introduce the notion of a lottery, we shall see how to use the EU Hypothesis and lotteries to calculate a utility function, we shall introduce the notion of a basic reference lottery ticket (brlt) and we shall apply lotteries and brlt's to the calculation of the utility of money and to certain public health questions.

In Section 7.1, we associated with each act or choice a collection of possible events (mutually exclusive and exhaustive) and a consequence corresponding to each event. We spoke of the probability of an event and the utility of a consequence. We can suppress all mention of the events and speak only of the consequences. Then we speak of the probability of a consequence as well as its utility. It is convenient to think of an act or choice as a "lottery" with consequences $c_1, c_2, \ldots, c_n$, consequence c_i attained with probability p_i. We might represent such a lottery using a tree diagram with the end points of the branches labeled with the consequences and the branches themselves labeled with the probabilities. Such a diagram is shown below:

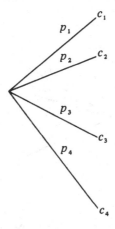

By our assumptions about the events associated with a given act or choice, $p_1 + p_2 + \cdots + p_n = 1$, an assumption we shall always make for our lotteries.

Suppose K is a set of consequences and L is a collection of lotteries with consequences in K. We let R be a binary relation on L, the relation of (strict) preference. (We assume R is the preference relation of some decisionmaker or decisionmaking body.) We define weak preference S and indifference E as usual, namely, by

$$\ell S \ell' \Leftrightarrow \sim \ell' R \ell, \tag{7.4}$$

$$\ell E \ell' \Leftrightarrow \sim \ell R \ell' \ \& \sim \ell' R \ell. \tag{7.5}$$

Any function $u:K \to Re$ will be called a *value function* on K, and it will be convenient to distinguish value functions on K from utility functions, which are value functions on K with certain special properties, for example, the property of preserving certain observed relations on K.

We say that the triple (K, L, R) satisfies the *Expected Value (EV) Hypothesis* if there is a value function u on K with the following property: Whenever ℓ and ℓ' are in L and ℓ is the lottery

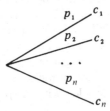

and ℓ' is the lottery

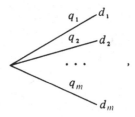

then

$$\ell R \ell' \Leftrightarrow \sum_{i=1}^{n} p_i u(c_i) > \sum_{i=1}^{m} q_i u(d_i). \qquad (7.6)$$

We shall say u is a *value function satisfying the Expected Value (EV) Hypothesis*. The first sum in Eq. (7.6) is denoted $E(\ell)$ and is called the *expected value of the lottery* ℓ. If there is a utility function u on K satisfying the EV Rule, we say that the *Expected Utility (EU) Hypothesis* holds, that u *satisfies the Expected Utility (EU) Rule*, and that $E(\ell)$ is the *expected utility of the lottery* ℓ.

Recall that the statement $E(\ell) > E(\ell')$ is meaningless if utility u is only an ordinal scale. That does not mean that the EU Hypothesis is meaningless or false if utility is only an ordinal scale. For the EU Hypothesis simply says that individuals act by making such comparisons (for a particular utility function); the EU Hypothesis is not concerned with the meaningfulness of the comparisons, nor is it affected by the issue of meaningfulness. In other words, even though the statement $E(\ell) > E(\ell')$ might be meaningless, it is still interesting to explore this statement and the EU Hypothesis if u just defines an ordinal scale.

Sometimes it is convenient to speak of complex lotteries. For example, suppose lottery ℓ is as follows:

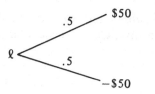

Lottery ℓ gives a 50% chance of winning $50 and a 50% chance of losing $50. Consider next the lottery*

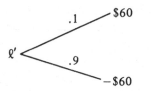

Suppose you are offered a gamble in which, with probability .5, you win $50, and with probability .5, you are entered in the lottery ℓ'. We can represent this complex lottery as a lottery ℓ'' in two ways:

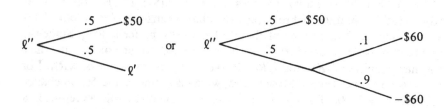

By using standard properties of tree diagrams, which we shall at this point assume hold for our lotteries, we may translate ℓ'' into a simple lottery with the same expected utility.

*The reader should note that ℓ used in the text and ℓ used in the art are the same symbol, even though they differ slightly in appearance.

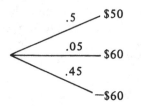

Before closing this subsection, we observe that preference among lotteries may not be transitive. Consider the following lotteries:

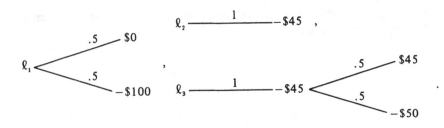

In ℓ_1, there is a 50% chance of breaking even and a 50% chance of losing $100. In ℓ_2 you lose $45 for certain. In ℓ_3, you lose $45 for certain and then enter a lottery that gives you $45 with probability .5 and takes $50 with probability .5. According to Raiffa [1968, p. 75], many subjects prefer ℓ_1 to ℓ_2, even though, if utility satisfies $u(\$n) = nu(\$1)$, $E(\ell_1) < E(\ell_2)$. They would rather gamble on breaking even than losing $45 for certain. This violates the EU Rule. Many of these subjects also prefer ℓ_2 to ℓ_3. Since the gamble for $45 versus $-$50 is unfair, they would rather not take it. Thus, for many subjects, $\ell_1 R \ell_2$ and $\ell_2 R \ell_3$. However, for many subjects, $\ell_3 R \ell_1$. For ℓ_3 gives a 50–50 chance at $0 or $-$95, which is better than a 50–50 chance for $0 or $-$100. This is a violation of transitivity of preference, a condition that follows from the EU Rule. When non-transitivity is pointed out to such subjects, some become uncomfortable with their choices, whereas others will not change their minds, saying that these original choices represent their true preferences, even though this seems illogical.

We mention another example of violation of the EU Rule in Section 7.2.4, and other tests of the EU Rule and Hypothesis in Sections 7.2.2 and

7.3.2. For summaries of tests of the EU Rule and Hypothesis, see Becker and McClintock [1967], Coombs, Dawes, and Tversky [1970, Chapter 5], Edwards [1954, 1961], and Luce and Suppes [1965].

7.2.2 Utility from Expected Utility

In Section 3.1, we studied *ordinal utility functions*, real-valued functions u on relational systems (A, R) so that for all $a, b \in A$,

$$aRb \Leftrightarrow u(a) > u(b). \tag{7.7}$$

For this chapter, we shall call a function u satisfying (7.7) an *order-preserving utility function*, so as not to confuse such functions with ordinal scales; the order-preserving utility functions we shall study might have properties in addition to satisfying (7.7), and so might define scale types stronger than ordinal. We shall also speak of *full utility functions*, utility functions satisfying all the properties we want to demand of them, and we note that order-preserving utility functions might not be full. In Section 3.1, we observed that an order-preserving utility function u could be calculated by using a pair comparison experiment and then setting

$$u(x) = \text{number of } y \text{ so that } x \text{ is preferred to } y.$$

If there are many alternatives, performing a pair comparison experiment becomes unworkable: With n alternatives, $n(n - 1)$ ordered pairs must be tested, or $n(n - 1)/2$ unordered pairs[*]. In this section, we shall see that if the EU Hypothesis is true, then we can use it to find an order-preserving utility function among consequences more rapidly than by performing a pair comparison experiment.

Suppose $\succ$ is a strict preference relation on the set K. We would like to find an order-preserving utility function on $(K, \succ)$, that is, a function $u: K \rightarrow Re$ satisfying

$$c \succ d \Leftrightarrow u(c) > u(d). \tag{7.8}$$

---

[*]Because of this, experimenters take various shortcuts in performing a pair comparison experiment. For instance, they assume such properties of preference as transitivity, and use transitivity to extrapolate some preferences from those previously expressed. See Warfield [1973, 1974a,b, 1976] and Baldwin [1975] for examples of this approach. There has been a fair amount of theoretical work devoted to this question of how to design a pair comparison experiment so as to minimize the total number of questions which must be asked if, for example, it is assumed that preferences form a strict simple order. For some discussion of this problem, see for example Knuth [1973], Wells [1971 pp.206ff.], Busacker and Saaty [1965, pp. 228ff], Ford and Johnson [1959], and Steinhaus [1950, pp. 37–40].

We shall define weak preference $\succeq$ and indifference $\sim$ on K by

$$c \succeq d \Leftrightarrow \sim d \succ c$$

and

$$c \sim d \Leftrightarrow \sim c \succ d \,\&\, \sim d \succ c.$$

In a sense, each element in the set K is a lottery, for we can identify a consequence c in K with the lottery that gives c with probability 1. This lottery will be denoted $\ell(c)$. *Throughout this chapter, we shall assume that each such lottery is in L.* If R is a strict preference relation on L, we shall also assume that for all $c, d \in K$,

$$c \succ d \Leftrightarrow \ell(c) R \ell(d). \tag{7.9}$$

If these two assumptions hold, we say (L, R) *extends* $(K, \succ)$. If (L, R) extends $(K, \succ)$, then any value function u on K satisfying the EV Rule is an order-preserving utility function for $(K, \succ)$. For

$$c \succ d \Leftrightarrow \ell(c) R \ell(d)$$
$$\Leftrightarrow E[\ell(c)] > E[\ell(d)]$$
$$\Leftrightarrow u(c) > u(d).$$

We shall assume the EV Hypothesis, that is, that there is a value function u on K satisfying the EV Rule. We shall not assume that u is known, but only that it exists.

Assume that there are two elements of K, c_*, and c^*, so that

$$c^* \succ c_* \tag{7.10}$$

and so that for all c in K,

$$\sim c \succ c^* \quad \text{and} \quad \sim c_* \succ c. \tag{7.11}$$

That is, c^* is strictly preferred to c_*, nothing is strictly preferred to c^*, and c_* is not strictly preferred to anything. The consequences c^* and c_* can be thought of as the "best" and "worst" consequences in K. Since $c^* \succ c_*$ and u is an order-preserving utility function, we have $u(c^*) > u(c_*)$, and hence $u(c^*) - u(c_*) > 0$.

Given c in K, let $\ell(c)$ be the lottery with only one consequence, c. We are assuming that $\ell(c)$ always belongs to L. Consider the lottery

$$\tag{7.12}$$

Since (L, R) extends $(K, \succ)$, if π is 1, you either prefer ℓ_π to $\ell(c)$ or are indifferent between these lotteries. If π is 0, you either prefer $\ell(c)$ to ℓ_π or are indifferent between these lotteries. As we let π gradually increase from 0 to 1, it is reasonable to assume that we can find some number $\pi = \pi(c)$ so that you are indifferent between $\ell_{\pi(c)}$ and $\ell(c)$.* Then $\pi(c)$ defines an order-preserving utility function over K. For

$$E(\ell_{\pi(c)}) = \pi(c)u(c^*) + [1 - \pi(c)]u(c_*)$$
$$= \pi(c)[u(c^*) - u(c_*)] + u(c_*).$$

Since you are indifferent between $\ell_{\pi(c)}$ and $\ell(c)$, the EV Hypothesis implies that

$$E[\ell(c)] = E[\ell_{\pi(c)}],$$

or

$$u(c) = \pi(c)[u(c^*) - u(c_*)] + u(c_*).$$

Since $u(c^*) - u(c_*) \neq 0$, we may divide by $u(c^*) - u(c_*)$, and we obtain

$$\pi(c) = \frac{u(c) - u(c_*)}{u(c^*) - u(c_*)}$$
$$= \alpha u(c) + \beta,$$

where

$$\alpha = \frac{1}{u(c^*) - u(c_*)}$$

and

$$\beta = \frac{-u(c_*)}{u(c^*) - u(c_*)}.$$

Since $\alpha > 0$, it follows that for all c and d in A,

$$u(c) > u(d) \Leftrightarrow \pi(c) > \pi(d).$$

Since u is an order-preserving utility function over K, we have

$$c \succ d \Leftrightarrow \pi(c) > \pi(d).$$

---

*The existence of such a number π requires an assumption about preferences among lotteries, which we shall formalize below.

Thus, π is an order-preserving utility function over K. Notice that computation of the utility function π only assumes that u exists, and does not require knowledge of u. Also, there is no need to assume that the individual whose utility function is being calculated actually computes expected values or utilities to make decisions, but only to assume that he acts *as if* he makes decisions on the basis of expected utilities. Finally, calculation of the number $\pi(c)$ is required only n times if there are n consequences being compared.

In principle, the procedure is simple. To illustrate it, suppose the "best" and "worst" consequences of a given surgical procedure are "cure" (c^*) and "death" (c_*). To calibrate an individual's utility function for various complications—for example, loss of a leg—we would consider gambles with a probability π of cure and $1 - \pi$ of death. We would gradually increase π until the individual told us he was indifferent between the gamble on the one hand and the complication—loss of leg, for example—with certainty. (This involves some hard choices!) The value of π for which he is indifferent can be used as his utility for the complication, and used in later decisionmaking.

Perhaps a more down-to-earth example is the following: Suppose an individual is considering how much money to invest on a risky venture if his payoff is \$1000 if things go well and \$0 otherwise. He can calculate his utility of, say, \$400 by finding that value of $\pi = \pi(\$400)$ for which he is indifferent between having \$400 for certain and having a π-probability of obtaining \$1000, a $(1 - \pi)$-probability of \$0. Note that $\pi(\$n)$ is almost certainly not $n\pi(\$1)$. For example, $\pi(\$500)$ is almost certainly greater than $1/2$, so $\pi(\$1000) < 2\pi(\$500)$. We shall ask the reader to think about the meaningfulness of this conclusion. See Section 7.2.4 for more on the utility of money.

To the best of the author's knowledge, the procedure we have described is due to Raiffa [1969]. Raiffa admits that the procedure is not really practical in complex decisionmaking problems. However, it contains the basic idea which he does find useful in decisionmaking, and which we turn to in the next subsection. Keeney and Raiffa [1976] argue that a more practical procedure to assess a utility function is to use this procedure to fix a few utility values and then to fit a curve to these data points. The general shape of the curve is estimated by studying an individual's attitudes toward risk. Some examples of computation of utility functions using variants of this idea are the following: Grayson [1960] computes utility functions of oil wildcatters involved in the search for gas and oil; Swalm [1966] and Spetzler [1968] compute utility functions of business executives; and Keeney [1972a] computes the utility function of operators of a hospital blood bank. For a discussion of these and other examples, see Keeney and Raiffa [1976].

Let us now formalize the result we have obtained. We say that $(K, \succ, L, R)$ satisfies *continuity* if, for all c, d, d' belonging to K, whenever $d \succsim c \succsim d'$, then for some $p \in [0, 1]$, the following lottery ℓ

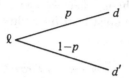

is in L and $\ell(c)$ is indifferent to ℓ. (We do not use the full force of continuity, but use it only with $d = c^*$ and $d' = c_*$. The reader should compare the notion of continuity used in Theorem 5.1.)

THEOREM 7.1 (Raiffa). *Suppose K is a set, $\succ$ is a binary relation on K, L is a collection of lotteries with consequences in K, and R is a binary relation on L. Suppose in addition that*

(a) (K, L, R) satisfies the EV Hypothesis;

(b) (L, R) extends $(K, \succ)$;

(c) $(K, \succ)$ has "best" and "worst" consequences c^* and c_* satisfying Eqs. (7.10) and (7.11); and

(d) $(K, \succ, L, R)$ satisfies continuity.

Then an order-preserving utility function π for $(K, \succ)$ is given by taking $\pi(c)$ equal to that number π so that $\ell(c)E\ell_\pi$, where E (indifference) is defined in Eq. (7.5) and ℓ_π is the lottery

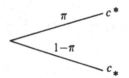

Recall again that comparisons of expected utilities are meaningless if the utility function over consequences is only an ordinal scale, but are meaningful if this utility function is an interval scale. We shall show that under reasonable assumptions, an order-preserving utility function actually defines an interval scale.

COROLLARY 1. *Suppose hypotheses (b) through (d) of Theorem 7.1 hold, and in addition the EU Hypothesis holds. Suppose every full utility function* on K is regular and defines an interval scale. Then the order-preserving utility*

---

*Recall that we use the adjective "full" to distinguish order-preserving utility functions from utility functions. The latter may be required to satisfy more conditions than just order-preservation.

function π *defined in Theorem 7.1 is a full utility function on* K, *and it defines a (regular) interval scale.*

Proof. By the proof of Theorem 7.1, if u is a full utility function on K satisfying the EU Hypothesis, then $\pi(x) = \alpha u(x) + \beta$, $\alpha > 0$. Since u defines an interval scale, π must be a full utility function for K. Hence, it defines an interval scale.   ∎

COROLLARY 2. *Suppose hypotheses* (b) *through* (d) *of Theorem 7.1 hold. Suppose in addition that every full utility function* u *on* K *satisfies the EU Hypothesis. Finally, suppose that every order-preserving value function on* K *that satisfies the EV Hypothesis is a full utility function. Then every full utility function* u *on* K *defines a (regular) interval scale.*

Proof. Given u, $v = \alpha u + \beta$ is again a full utility function, if $\alpha > 0$. For v clearly is order-preserving and satisfies the EV Hypothesis. Next, given u, suppose v is any other full utility function on K. By the proof of Theorem 7.1, there are constants α, β, γ, δ, with α, $\gamma > 0$, so that

$$\pi(x) = \alpha u(x) + \beta$$

and

$$\pi(x) = \gamma v(x) + \delta.$$

Thus, for all x,

$$v(x) = \frac{\alpha}{\gamma} u(x) + \frac{\beta - \delta}{\gamma}.$$

∎

In closing this subsection, let us note that there are various alternative ways the EU Hypothesis can be used to calculate utilities over consequences. Some of these methods have been developed experimentally and, incidentally, provide experimental tests of the EU Hypothesis. See, for example, Mosteller and Nogee [1951], Davidson, Suppes, and Siegel [1957], or Coombs and Komorita [1958].

7.2.3 *A Basic Reference Lottery Ticket*

Let us fix the same two consequences c_* and c^* described in Eqs. (7.10) and (7.11). A ticket to enter the lottery (7.12) will be called a π-*basic reference lottery ticket* or a π-*brlt* for short. To make a choice between two complicated lotteries ℓ and ℓ', Raiffa [1968, 1969] suggests that we reduce them to π-brlt's. Namely, if c is a consequence and $\ell(c)E\ell_{\pi(c)}$, then replace c in lotteries ℓ and ℓ' by the lottery $\ell_{\pi(c)}$. For example, the lottery

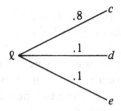

becomes the lottery

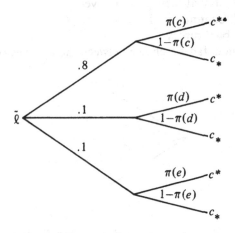

Using standard properties of tree diagrams (which we will again assume hold), we can replace $\tilde{\ell}$ by the simple λ-brlt

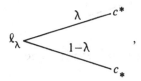

where $\lambda = (.8)\pi(c) + (.1)\pi(d) + (.1)\pi(e)$. We have $\ell E\ell_\lambda$. Similarly, we reduce an alternative lottery ℓ' to a δ-brlt ℓ_δ, with $\ell'E\ell_\delta$. Then, if preference R is strict weak, we should have

$$\ell R\ell' \Leftrightarrow \ell_\lambda R\ell_\delta.$$

If the EV Hypothesis holds, we should have

$$\ell_\lambda R \ell_\delta \Leftrightarrow \lambda > \delta,$$

since $c^* \succ c_*$. The point of this procedure is that you can make choices between lotteries or acts or choices without knowing a utility function over consequences.

To illustrate how to make decisions using the idea of a π-brlt, let us give some examples. Suppose a small businessman faces decreasing sales in his present location and considers moving to a new location. Suppose for want of further information, he thinks it is equally likely that one of the following will happen if he moves:

a = he will increase his sales,
b = his sales will stay at their present level,
c = his sales will continue to decrease,
d = he will face bankruptcy.

Thus, the businessman faces a choice between the lotteries $\ell(c)$ and

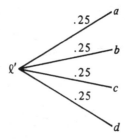

Suppose he chooses $c^* = a$ and $c_* = d$, and, on reflection, he is indifferent between $\ell(b)$ and

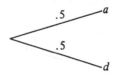

a .5-brlt; and he is indifferent between $\ell(c)$ and

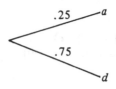

a .25-brlt. Then he is indifferent between ℓ' and

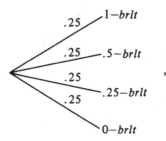

which is the same as a .4375-brlt, since

$$.25(1) + .25(.5) + (.25)(.25) + (.25)(0) = .4375.$$

Since .25 is smaller than .4375, the small businessman should choose ℓ' over $\ell(c)$—that is, he should choose to move.*

7.2.4 *The Utility of Money*

As we remarked at the beginning of this chapter, the utility of money is not necessarily a linear function of the dollar amount. If it were, most people would not buy insurance. For the expected dollar loss from buying insurance is usually greater than the expected dollar loss from not buying it. (See Exer. 2, Section 7.1.) As the same dollar amount is added, at higher and higher levels of holdings, the utility increases by less and less. (We say the marginal utility is decreasing.) This was first observed by Bernoulli [1738], who suggested that the utility of $\$n$ is a logarithmic function of n. (See Exer. 2 below.)

How might we estimate an individual's utility function for money? The most direct method is to use a procedure like that of Section 7.2.2. We fix the lowest and highest dollar amounts in reason, for example, $\$0$ and $\$1,000,000$. We then find that number $\pi(n)$ so that an individual would pay exactly $\$n$ to enter a lottery of the form

*The point of this approach is that this is how the businessman *should* behave (according to Raiffa), not how he *does* behave. The method of π-brlt's is very much a prescriptive one.

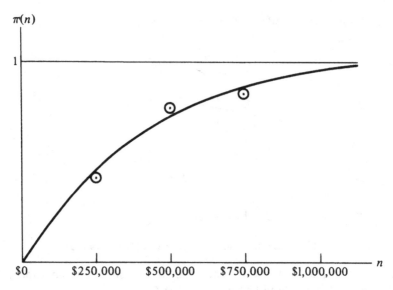

Figure 7.1. Estimation of a utility function for money

Then $\pi(n)$ is the utility of $\$n$. Often it is sufficient to estimate $\pi(n)$ for several intermediate values of n, and then fit a curve to these values, provided one also checks the general shape of this curve. For most individuals, the shape will be concave-down, for most individuals tend to be *risk-averse*. (See Fig. 7.1 for a sample estimated utility curve for money for such an individual.) Risk-averseness means that if c, d, and a are monetary amounts and an individual is indifferent between $\ell(a)$ and

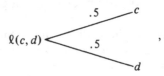

then $a \leq (c + d)/2$. If π is the function so that $\ell(x)$ is indifferent to $\ell_{\pi(x)}$, it follows that $\ell(c, d)$ is indifferent to a λ-brlt with $\lambda = .5\pi(c) + .5\pi(d)$. Thus,

$$\pi\left(\frac{c + d}{2}\right) \geq \pi(a) = \frac{\pi(c) + \pi(d)}{2}.$$

Hence, the function π, the *utility function for money*, is concave-down.

Stevens [1959, p. 55] argues that $\pi(a)$ is a power function αa^{β}, with β in the vicinity of .4 or .5. Such a function has the concave-down shape. Apparently the idea that the utility of money might be a power function goes back to Gabriel Cramer in the eighteenth century (see Bernoulli [1738] and Stevens [1959, p. 58].)

7.2.5 The Allais Paradox

Consider the following four lotteries:

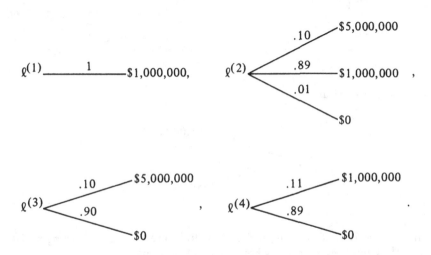

We shall pose two problems: Problem 1 is to choose between $\ell^{(1)}$ and $\ell^{(2)}$, and Problem 2 is to choose between $\ell^{(3)}$ and $\ell^{(4)}$. The French economist Allais [1953], and others, have reported that most subjects prefer $\ell^{(1)}$ to $\ell^{(2)}$ and $\ell^{(3)}$ to $\ell^{(4)}$. To quote Raiffa [1968, p. 80], most subjects reason as follows: "In Problem 1, I have a choice between $1,000,000 for certain and a gamble where I might end up with $0. Why gamble? In Problem 2, there is a good chance that I will end up with $0 no matter what I do. The chances of getting $5,000,000 are almost as good as getting $1,000,000, so I might as well go for the $5,000,000 and choose" $\ell^{(3)}$ over $\ell^{(4)}$.

Following Raiffa, let us try to analyze Problems 1 and 2 by using π-brlt's. We let c^* be $5,000,000 and c_* be $0. Then $\ell^{(1)}$ is (judged) indifferent to a π_1-brlt, for some π_1; that is, $\ell^{(1)}$ is indifferent to ℓ_{π_1}. Now

$\ell^{(2)}$ is indifferent to

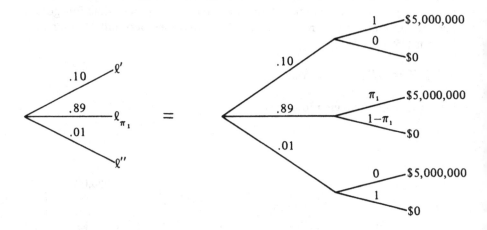

which is the same as

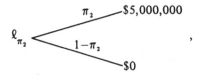

where $\pi_2 = .10 + .89\pi_1$. Similarly, $\ell^{(3)}$ is indifferent to ℓ_{π_3}, where $\pi_3 = .10$, and $\ell^{(4)}$ is indifferent to ℓ_{π_4}, where $\pi_4 = .11\pi_1$. Thus,

$$\ell^{(1)} R \ell^{(2)} \Leftrightarrow \pi_1 > \pi_2$$
$$\Leftrightarrow \pi_1 > .10 + .89\pi_1$$
$$\Leftrightarrow \pi_1 > \frac{10}{11}$$

and

$$\ell^{(3)} R \ell^{(4)} \Leftrightarrow \pi_3 > \pi_4$$
$$\Leftrightarrow .10 > .11\pi_1$$
$$\Leftrightarrow \pi_1 < \frac{10}{11}.$$

Thus, decisionmakers who make choices using the method of π-brlt's cannot make choices of *both* $\ell^{(1)}$ over $\ell^{(2)}$ and $\ell^{(3)}$ over $\ell^{(4)}$. Decisionmakers who would make these choices would violate the EU Hypothesis. For if u is a utility function on K satisfying the EU Hypothesis, then

$$\ell^{(1)} R \ell^{(2)} \Leftrightarrow u(\$1,000,000) > .10u(\$5,000,000) + .89u(\$1,000,000) + .01u(\$0)$$

$$\Leftrightarrow .11u(\$1,000,000) > .10u(\$5,000,000) + .01u(\$0)$$

and

$$\ell^{(3)} R \ell^{(4)} \Leftrightarrow .10u(\$5,000,000) + .90u(\$0) > .11u(\$1,000,000) + .89u(\$0)$$

$$\Leftrightarrow .10u(\$5,000,000) + .01u(\$0) > .11u(\$1,000,000).$$

Raiffa argues that it is exactly such examples that illustrate why people make mistakes in decisionmaking. Thus, they *should* make decisions by using a precise procedure such as the EU Hypothesis, or the method of π-brlt's, to avoid making "inconsistent" judgments. Others have tried to argue that the Allais Paradox is not really a paradox, for it does not really exhibit a violation of the EU Hypothesis. Morrison [1967] argues this way, arguing that $0 plays a different role in different parts of the two problems. Note that the Allais Paradox is a paradox regardless of what the utility function for money may be.

7.2.6 The Public Health Problem: The Value of a Life.*

In this subsection, we apply some of the ideas of earlier subsections. Suppose under present conditions a certain proportion p of the population will die in a given year of some disease—let us say, for concreteness, respiratory disease due to air pollution. Suppose a study has shown that by spending x dollars over the next year, we can reduce the proportion of people who die from p to p'. How does society decide whether to spend that x dollars? Let us call this question the *public health question*.

Naturally, the public health question is usually not considered in isolation—there are usually alternative possible uses for the same amount of money, and one would usually compare these alternative uses. However, one reasonable alternative is to keep the money for later spending, and one would at least like to be able to decide if it is worth not keeping it, that is, if it is worth spending x dollars to decrease p to p'. What is the tradeoff between health and (monetary) assets? This is the question we consider.

Suppose we let s be the sum total of society's assets at the present time. Assets will be measured, for the purposes of our discussion, in monetary terms. (They may, of course, include resources, capital, etc.) Let us

*The discussion in this subsection is motivated by that in Raiffa [1969, Section 8].

consider first the consequences for a particular individual of spending x dollars on air pollution reduction, if this will reduce p to p'. Suppose there are N people in all. A particular individual's share of the expense, in the most simple-minded model where wealth is shared equally, will be $(1/N)x$. His share of the initial assets is $(1/N)s$. The numbers $q = 1 - p$ and $q' = 1 - p'$ can be interpreted as the individual's "probability of life." Thus at present he faces the following lottery ℓ:

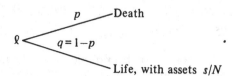

If x dollars are spent, he will face the lottery ℓ':

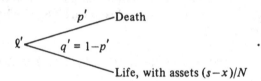

Assuming the EU Hypothesis, letting u be an order-preserving utility function which satisfies this hypothesis, we may assume that $u(\text{Death}) = 0$. (For given any constant λ, $u + \lambda$ is still an order-preserving utility function, and it satisfies the EU Hypothesis.) Using $u(\text{Death}) = 0$, we have

$$E(\ell) = p \cdot u(\text{Death}) + q \cdot u(\text{Life}, s/N) = q \cdot u(\text{Life}, s/N).$$

Similarly,

$$E(\ell') = q' \cdot u(\text{Life}, (s - x)/N).$$

Let us suppose that in fact the individual in question is indifferent between ℓ and ℓ'; that is, if he starts with assets s/N, then x/N is just at the borderline of the amount of money he would be willing to spend to raise the probability of life from q to q'. Let us suppose that $q''/q''' = q/q'$. It follows that under the same starting assets s/N, x/N is exactly the amount he would be willing to spend to raise q'' to q'''. For,

$$E(\ell) = E(\ell') \quad \text{iff} \quad E(\ell'') = E(\ell'''),$$

where ℓ'' and ℓ''' are the lotteries

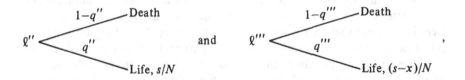

respectively. This conclusion is quite interesting. To illustrate it, let us take

$$q = \frac{999,900}{1,000,000},$$

$$q' = \frac{999,909.999}{1,000,000} \approx \frac{999,910}{1,000,000},$$

$$q'' = \frac{999,990}{1,000,000},$$

$$q''' = \frac{999,999.9999}{1,000,000} \approx 1.$$

Then $q/q' = q''/q''' = .99999$, and

$$p = \frac{100}{1,000,000},$$

$$p' \approx \frac{90}{1,000,000},$$

$$p'' = \frac{10}{1,000,000},$$

$$p''' \approx 0.$$

Thus, the conclusion says that, if it is worth x/N dollars to you to reduce deaths from air pollution from 100 people per million to approximately 90 people per million, then (at the same level of starting assets) it should be worth exactly the same amount to reduce the figure from 10 per million to approximately 0 per million. (This conclusion should be evaluated in the light of the argument that cutting the probability of death from a particular cause those last few steps down to a very small amount is sometimes proportionately more expensive. Of course, the conclusion is based on some very strong assumptions.)

(A similar analysis applies to an individual's preferences among alternative treatments for a disease. The question here is: If treatment a costs x dollars and has a probability q of success and treatment b costs x' dollars and has a probability q' of success, which treatment is preferred?)

So far our analysis has been from the point of view of an individual. Let us now make the analysis from the point of view of society. This situation is more difficult. We shall treat the set of consequences as multidimensional alternatives, as in Chapter 5. (More on multidimensional alternatives in the next section.) K will be $A_1 \times A_2$, where A_1 and A_2 are both the set of nonnegative real numbers. A pair (n, a) represents a state of society consisting of n individuals with total assets a. Let $\succ$ be society's relation of (strict) preference on K. To answer the public health question, it will suffice to calculate an order-preserving utility function u over $(K, \succ)$. For, suppose our current assets are s and our current population is N. If we do nothing, then after one year our population will be $qN + \alpha N - \beta N$, where α is the birth rate and β is the death rate due to all causes other than air pollution.* Our assets will be s. If we spend x dollars, our population will become $q'N + \alpha N - \beta N$ and our assets $s - x$. Thus, whether we spend x dollars or not depends on which we prefer,

$$(qN + \alpha N - \beta N, s) \quad \text{or} \quad (q'N + \alpha N - \beta N, s - x),$$

or equivalently on which is bigger,

$$u(qN + \alpha N - \beta N, s) \quad \text{or} \quad u(q'N + \alpha N - \beta N, s - x).$$

Of course, if calculation of the utility function u is based entirely on the relation $\succ$, as it was in Chapter 5, then we cannot know u unless we know $\succ$, and since $\succ$ is what we need to know to answer the public health question, we do not gain anything by going to utility functions. Fortunately, there are techniques, such as those using Theorem 7.1, for calculating an order-preserving utility function u over $(K, \succ)$ even without much knowledge about $\succ$.

The preference relation $\succ$ and the utility function will differ from society to society, as different societies have different points of view about human life, assets, etc. Thus we shall have to make some assumptions about a hypothetical society in order to go further in answering the public health question, and in particular in applying the result of Theorem 7.1. We shall consider two different societies.

Let us assume for the sake of discussion that, as before, societal wealth in our first society is spread equally among the population. Moreover, let us assume that the society has determined a certain level of assets C for an individual which is considered the comfort level—at level C he has all he can possibly need. Moreover, in our hypothetical society, let us suppose that too much wealth for an individual is also considered bad—it leads to decadence, decay, etc. Thus, for a given level a of the society's assets, there will be an optimal population n, namely that n so that $a/n = C$.

---

*We disregard the changes in α and β due to changing q.

For this hypothetical society, let us use the method of calculating a utility function described in Theorem 7.1. Let a_0 be an arbitrary number, and let a_1, n_1 be numbers such that $a_1/n_1 = C$. Pick $c_* = (0, a_0)$ and $c^* = (n_1, a_1)$. If c is any other consequence, then by our assumptions, $c^* \succsim c$. Moreover, it is reasonable to assume that $c \succsim c_*$: anything is better than having the entire population annihilated. Moreover, $c^* \succ c_*$. Assuming that continuity and the other hypotheses of Theorem 7.1 hold, we now measure utility of (n, a) as that number $\pi(n, a)$ so that we are indifferent between attaining (n, a) with certainty and the lottery

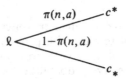

In the lottery ℓ, we have a $1 - \pi(n, a)$ chance of total annihilation (say by nuclear war) and a $\pi(n, a)$ chance of reaching the ideal spread of resources. Society should be as happy with this risk as with a guarantee of (n, a). Of course this method of calculating $\pi(n, a)$ is a bit unrealistic, but perhaps it will be helpful in thinking about the utility function $\pi(n, a)$. The properties of $\pi(n, a)$ for different societies are by no means well thought out. Once $\pi(n, a)$ has been calculated, it can be used to choose between

$$(qN + \alpha N - \beta N, s) \quad \text{and} \quad (q'N + \alpha N - \beta N, s - x),$$

and so to solve the public health problem.

Let us compare an alternative society, which in choosing between two alternatives (n, a) and (n, b), prefers that situation in which average assets per person are highest. Thus, for this society, if $n, m \neq 0$,

$$(n, a) \succ (m, b) \Leftrightarrow a/n > b/m.$$

We assume that $(0, a)$ is the worst possible situation, all a. Now there is no "best" consequence c^*, so Theorem 7.1 does not apply. However, it is easy to calculate a utility function u over $(K, \succ)$. Namely, we take

$$u(n, a) = \begin{cases} a/n & \text{if } n \neq 0 \\ 0 & \text{if } n = 0. \end{cases}$$

In this case, assuming $qN + \alpha N - \beta N \neq 0$ and $q'N + \alpha N - \beta N \neq 0$,

$$u(qN + \alpha N - \beta N, s) = \frac{s}{qN + \alpha N - \beta N} = \frac{1}{N}\frac{s}{q + \alpha - \beta},$$

$$u(q'N + \alpha N - \beta N, s - x) = \frac{1}{N}\frac{s - x}{q' + \alpha - \beta}.$$

Thus, it is "worth" spending x dollars to reduce from p to p' if and only if

$$\frac{s - x}{q' + \alpha - \beta} > \frac{s}{q + \alpha - \beta}.$$

Since $x > 0$ and $q' > q$, this is never true! This society never spends the money. Spending the money would only lead to more people and fewer assets. We return to the public health problem in the next section. For more on this problem, see, for example, Schelling [1968].

Exercises

1. Suppose two alternative medical treatments are being considered. Treatment x leads to complete cure with probability .1, paralysis of a leg with probability .4, and loss of a leg with probability .5. Treatment y leads to these outcomes with respective probabilities .2, .1, and .7. The patient is indifferent between losing a leg with certainty and a 50–50 chance between complete cure and death. He is also indifferent between having a paralyzed leg with certainty and an 80–20 chance of complete cure or death. According to the EU Hypothesis, which treatment would he prefer?

2. Bernoulli was led to consider a logarithmic function for $u(\$n)$ in part because of the so-called *St. Petersburg Paradox*. This paradox arises in the following game. A coin is tossed. If a head appears on the first trial, the gambler wins \$2. If not, the coin is tossed again. If a head appears on the next toss, the gambler wins \$4. In general, if a head appears for the first time on the nth toss, the gambler wins $\$2^n$. How much would you be willing to pay to play this game? Compare your answer to your expected winnings in dollars. This should suggest that the utility function for money is not linear in the dollar amount.

3. A candidate running for President is entered in a primary election with three other candidates. He figures that if he takes a strong stand on abortion, he will either end up first among the four candidates, or last, with the two events being equally likely. If he says nothing, he is assured of either second or third place, the former with probability 3/4. How might the candidate use the method of π-brlt's to determine whether or not to take the stand?

4. A student has applied to four colleges, and has been admitted to his third- and fourth-choice schools. He has to let his third-choice school know immediately whether he will come, but has not yet heard from his top two choice schools. How might the student decide whether to accept his third choice?

5. Is the method of π-brlt's relevant to the investment decision of Exer. 6 of Section 7.1?

6. Consider the meaningfulness of the statement $\pi(\$1000) < 2\pi(\$500)$ of Section 7.2.2.

7. (Raiffa [1968, p. 89]) Let ℓ be a lottery with dollar amounts as consequences. Let $b(\ell)$ be the price you would be willing to pay to enter

the lottery ℓ, $s(\ell)$ the price for which you would sell your ticket to enter ℓ.

(a) It is not necessarily the case that $b(\ell) = s(\ell)$. For you are indifferent between having $s(\ell)$ for certain and having the lottery ℓ. You are also indifferent between having nothing and entering a lottery ℓ' obtained from ℓ by subtracting $b(\ell)$ from each consequence in ℓ. Thus, $u(s) = E(\ell), u(0) = E(\ell')$. Can you think of lotteries ℓ and utility functions $u(\$n)$ where $b(\ell) \neq s(\ell)$?

(b) Show that if $u(\$n) = n$, all n, then $b(\ell)$ always equals $s(\ell)$.

(c) Show that if $u(\$n) = 1 - e^{-\lambda n}$, where $\lambda \neq 0$ is a measure of risk aversion, then $b(\ell)$ always equals $s(\ell)$.

8. Suppose $c \succ d \succ e$, (L, R) extends $(K, \succ)$, and the EU Hypothesis holds. Show that there is no $p \in [0, 1]$ so that you are indifferent between

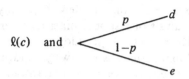

9. Suppose $K = \{1, 2, \ldots, n\}$, $\succ$ on K is $>$, L consists of *all* lotteries with consequences in K, and (L, R) extends $(K, \succ)$. Show that no matter what the function u on K is, if R is defined on L from u and the EU Rule, then $(K, \succ, L, R)$ satisfies continuity.

10. (a) If (K, L, R) satisfies the EU Hypothesis, does it follow that (L, R) is a strict weak order?

(b) What about $(K, \succ)$?

11. Suppose (L, R) extends $(K, \succ)$, u is a value function on K, and (K, L, R) satisfies the following modified version of the EV Hypothesis:

$$\ell R \ell' \Leftrightarrow \Sigma p_i u(c_i) > \Sigma q_i u(d_i) + \delta,$$

where δ is a fixed positive number and ℓ and ℓ' are

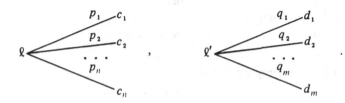

(a) Show that (L, R) is a semiorder.

(b) Is $(K, \succ)$?

7.3 Multidimensional Alternatives

7.3.1 *Reducing Number of Dimensions*

In this section, we consider choices among lotteries with multidimensional consequences, such as we encountered in the public health problem in Section 7.2.6. Thus, the set K of consequences will have a product structure $K = A_1 \times A_2 \times \ldots A_n$. See Farquhar [1977] and Keeney and Raiffa [1976] for recent surveys of the literature of this subject. Keeney and Raiffa also have an excellent summary of the growing number of applications of utility functions computed using lotteries with multidimensional consequences. Included in the applications are work on preference tradeoffs among instructional programs (Roche [1971]); decisionmaking concerning sulfur dioxide emissions in New York City (Ellis [1970], Ellis and Keeney [1972]); analysis of response times by emergency services such as fire departments (Keeney [1973b]); study of safety of landing aircraft (Yntema and Klem [1965]); analysis of sewage sludge disposal in Boston (Horgan [1972]); assessment of the risks in transport of hazardous substances (Kalelkar, Partridge, and Brooks [1974]); understanding options in control of the spruce budworm in Canadian forests (Bell [1975]); and developing alternatives for expansion of Mexico City's airport (de Neufville and Keeney [1972], Keeney [1973a]).

We shall begin by continuing the discussion of Section 5.2 on how to reduce the number of dimensions. We return to the medical decisionmaking problem discussed in Section 5.2. We considered the following dimensions for evaluating the result of a treatment:

a_1 = amount of money spent for treatment, drugs, etc.,

a_2 = number of days in bed with a high index of discomfort,

a_3 = number of days in bed with a medium index of discomfort,

a_4 = number of days in bed with a low index of discomfort,

$$a_5 = \begin{cases} 1 & \text{occurs,} \\ & \text{if complication } A \\ 0 & \text{does not occur,} \end{cases}$$

$$a_6 = \begin{cases} 1 & \text{occurs,} \\ & \text{if complication } B \\ 0 & \text{does not occur,} \end{cases}$$

$$a_7 = \begin{cases} 1 & \text{occurs,} \\ & \text{if complication } C \\ 0 & \text{does not occur.} \end{cases}$$

In Section 5.2, we showed how one might find some number a_3''' so that $(a_1, a_2, a_3, a_4, a_5, a_6, a_7)$ was judged indifferent to $(0, 0, a_3''', 0, a_5, a_6, a_7)$. Let us assume that it is only possible to get one of the complications A, B, and C, and so (a_5, a_6, a_7) is only $(0, 0, 0)$, $(1, 0, 0)$, $(0, 1, 0)$, or $(0, 0, 1)$. Fix a_3''' and let $\alpha_1 = (0, 0, a_3''', 0)$. The vector α_1 represents a certain level of "discomfort." Suppose that complication A is the worst possible complication, complication B is next worst, and having no complication is of course the best. Thus, for fixed α_1, we have

$$(\alpha_1, 0, 0, 0) \succ (\alpha_1, 0, 0, 1) \succ (\alpha_1, 0, 1, 0) \succ (\alpha_1, 1, 0, 0),$$

where (α_1, x, y, z) is short for $(0, 0, a_3''', 0, x, y, z)$. Using $c^* = (\alpha_1, 0, 0, 0)$ and $c_* = (\alpha_1, 0, 0, 1)$, we can calculate $u(\alpha_1, 0, 1, 0)$ and $u(\alpha_1, 0, 0, 1)$, by the method of Theorem 7.1. Specifically, using c^* and c_*, we seek a number π_B so that the constant lottery $\ell(\alpha_1, 0, 1, 0)$ is judged indifferent to

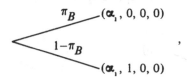

the lottery with a π_B chance of "discomfort" α_1 and no complications and a $(1 - \pi_B)$ chance of "discomfort" α_1 and complication A. We also seek a similar number π_C. Using these numbers we can calculate $u(\alpha_1, 0, 1, 0)$ and $u(\alpha_1, 0, 0, 1)$ for all α_1, knowing only $u(\alpha_1, 0, 0, 0)$ and $u(\alpha_1, 1, 0, 0)$. Thus, we can reduce calculation of $u(a_1, a_2, a_3, a_4, a_5, a_6, a_7)$ to calculation of $u(0, 0, a_3''', 0, 0, 0, 0)$ and $u(0, 0, a_3''', 0, 1, 0, 0)$, for all a_3'''. This has essentially reduced matters to a two-dimensional problem, using dimensions 3 and 5. For recent references on the subject of reducing the number of dimensions, see MacCrimmon and Siu [1974], MacCrimmon and Wehrung [1978], or Keeney [1971, 1972b].

7.3.2 Additive Utility Functions

In Section 5.4, we considered the problem of finding an additive utility function, that is, an order-preserving utility function u on a product $A_1 \times A_2 \times \cdots \times A_n$ such that for all $(a_1, a_2, \ldots, a_n)$ belonging to $A_1 \times A_2 \times \cdots \times A_n$,

$$u(a_1, a_2, \ldots, a_n) = u_1(a_1) + u_2(a_2) + \cdots + u_n(a_n), \qquad (7.13)$$

where $u_i (i = 1, 2, \ldots, n)$ is a real-valued function on A_i. Let us call such a function u an *additive order-preserving utility function*. In case $n = 2$, Eq. (7.13) takes the form

$$u(a, b) = u_1(a) + u_2(b). \tag{7.14}$$

In Section 5.4.3 we stated very general conditions on a set $A_1 \times A_2$ sufficient to guarantee the existence of an order-preserving utility function satisfying Eq. (7.14).

In this subsection, we shall assume that the set K of consequences has a product structure $A_1 \times A_2$. Assuming the existence of lotteries and the EU Hypothesis, we state rather simple assumptions which give rise to an order-preserving utility function u on K satisfying Eq. (7.14). Our approach follows that of Raiffa [1969], though the details are different. Let us say that the structure (K, L, R) satisfies the *marginality assumption* (relative to a_* and b_*) if, for any $a \in A_1$ and $b \in A_2$, the decisionmaker is indifferent between the following lotteries ℓ_1 and ℓ_2:

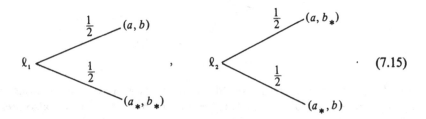

$$\tag{7.15}$$

In each of ℓ_1 and ℓ_2, there is a 50–50 chance between a and a_* and a 50–50 chance between b and b_*, and so it is reasonable to be indifferent between them.* We say (K, L, R) satisfies the *marginality assumption* if it satisfies the marginality assumption relative to some a_* and b_*. If there is an additive order-preserving utility function u on K and if u satisfies the EU Hypothesis, then the marginality assumption (relative to any a_* and b_*) follows easily. Indeed, the marginality assumption follows from the existence of a value function u on K which is additive [satisfies Eq. (7.14)] and satisfies the EV Hypothesis. We shall show that the marginality assumption is essentially sufficient for an additive value function. Suppose the value function $u : K \to Re$ satisfies the EV Hypothesis. Then by the

---

*This assumption is sometimes called *marginal independence* or *value independence*. The reason for the term "marginality" comes from probability theory: each lottery has the same marginal probability distributions for a outcomes and b outcomes, though the joint distributions differ.

marginality assumption,

$$\tfrac{1}{2}u(a, b) + \tfrac{1}{2}u(a_*, b_*) = \tfrac{1}{2}u(a, b_*) + \tfrac{1}{2}u(a_*, b),$$

for all $a \in A_1$ and $b \in A_2$. Thus,

$$u(a, b) + u(a_*, b_*) = u(a, b_*) + u(a_*, b). \tag{7.16}$$

Define u_i on A_i $(i = 1, 2)$ by

$$u_1(a) = u(a, b_*) - u(a_*, b_*), \qquad u_2(b) = u(a_*, b). \tag{7.17}$$

By Eq. (7.16), we find immediately that

$$u(a, b) = u_1(a) + u_2(b). \tag{7.18}$$

If (L, R) extends $(K, \succ)$, it follows that u is an order-preserving utility function for $(K, \succ)$, and hence by (7.18) an additive order-preserving utility function. This proof follows one of Raiffa [1969, p. 36], though the result is due to Fishburn [1965]. (For references, see also Fishburn [1967, 1969a,b, 1970, Chapter 11], Fishburn and Keeney [1974].) We summarize the results as a theorem.

THEOREM 7.2 (Fishburn.) *Suppose $K = A_1 \times A_2$ is a set, $\succ$ is a binary relation on K, L is a collection of lotteries with consequences in K, and R is a binary relation on L. If (K, L, R) satisfies the EV Hypothesis and if the marginality assumption holds for (K, L, R), then any value function u on K satisfying the EV Hypothesis is additive. Moreover, if (L, R) extends $(K, \succ)$, then u is an additive order-preserving utility function for $(K, \succ)$.*

Remark: This theorem generalizes easily to the situation where we have $K = A_1 \times A_2 \times \cdots \times A_n$, $n > 2$. See Raiffa [1969, p. 37] or Farquhar [1974] for a formulation.

Let us ask whether or not the marginality assumption is satisfied in various examples. In the public health problem, we can consider a choice between the following lotteries, with $n \neq 0$ or $a \neq 0$:

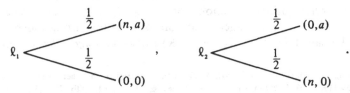

Certainly ℓ_1 would usually be preferred; in ℓ_2 both consequences are as bad as can be. This violates the marginality assumption (relative to 0 and 0). Thus, either the EV Hypothesis is false or the existence of an additive value (utility) function is false, for we have seen that the marginality assumption (relative to any a_* and b_*) follows from these two assumptions. Of course, here additivity seems to fail, at least for the second hypothetical society of Sec. 7.2.6. For suppose $u(n, a) = u_1(n) + u_2(a)$, some u_1, u_2. We have $u(n, 0) = u(m, 0)$, all n, m. Let α be the common value. Then $u_1(n) = \alpha - u_2(0)$, all n. Hence, $u(n, a) = \alpha - u_2(0) + u_2(a) = u(m, a)$, all n, m. Note that if additivity (or the marginality assumption) fails, this does not mean that there is no value (utility) function. It only means that value (utility) cannot be measured by separating out components and adding.

If the set K of consequences consists of market baskets, the marginality assumption might be reasonable. However, this is not reasonable if there is an interaction between components. If the first component is eggs and the second salt, you might consider 2 eggs and no salt very bad, and also no eggs and 1/2 a teaspoon of salt. But you would be happy with 2 eggs and 1/2 a teaspoon of salt. Thus, you might prefer ℓ_1 to ℓ_2 in (7.15). Naturally, if there is such an interaction, we do not expect an additive order-preserving utility function u.

7.3.3 Quasi-additive Utility Functions*

Since the marginality assumption might be too strong, let us consider some weaker conditions.

Let us recall that one of the necessary conditions on $(A_1 \times A_2, \succ)$ for the existence of an additive order-preserving utility function (Section 5.4.3) was *independence*: for all $a, a' \in A_1$ and $b_0, b_0' \in A_2$,

$$(a, b_0) \succ (a', b_0) \Leftrightarrow (a, b_0') \succ (a', b_0'),$$

and for all $a_0, a_0' \in A_1$ and $b, b' \in A_2$,

$$(a_0, b) \succ (a_0, b') \Leftrightarrow (a_0', b) \succ (a_0', b').$$

Suppose again that there is an *additive* value function u which satisfies the EV Hypothesis for $(A_1 \times A_2, L, R)$. Such a function u is an additive order-preserving utility function for $(A_1 \times A_2, \succ)$ if (L, R) extends $(K, \succ)$. If u exists, a stronger condition to be called strong independence holds. We say that $(A_1 \times A_2, \succ, L, R)$ satisfies *strong independence*[†] (on

---

*Although we follow Raiffa [1969], much of the material in this section is based on work of Keeney [1968a,b]. For more recent references, see Keeney [1972b, 1973c] and Farquhar [1974, 1977].

[†]This assumption is sometimes called *utility independence*.

the first component) if, for all $b_0 \in A_2$, whenever ℓ and ℓ' are lotteries all of whose consequences have the form (a, b_0), then preferences for ℓ versus ℓ' do not change if the common value b_0 is changed *in every consequence* to the same b_0' in A_2, and the probabilities $p(a, b_0)$ and $p(a, b_0')$ are the same. A similar definition applies to the second component, and we say that *strong independence* holds if strong independence holds on both components. To give an example, we observe that strong independence says that one prefers lottery ℓ_1 to lottery ℓ_1' if and only if one prefers lottery ℓ_2 to lottery ℓ_2', where the lotteries ℓ_1, ℓ_1', ℓ_2, and ℓ_2' are given by

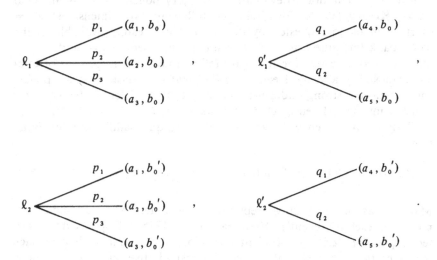

The verification that strong independence follows from the existence of an additive value function u on $A_1 \times A_2$ satisfying the EV Hypothesis is left to the reader. Incidentally, trivially, strong independence implies independence, if we assume that (L, R) extends $(K, \succ)$.

Under the EV Hypothesis, the condition of strong independence does not imply that there is an additive value function u on $A_1 \times A_2$ satisfying the EV Hypothesis or that there is an additive order-preserving utility function u on $(A_1 \times A_2, \succ)$. However, under certain simple assumptions, we shall conclude the existence of a u that is almost additive. Let us say a real-valued function u on $A_1 \times A_2$ is *quasi-additive* (*bilinear, multilinear*) if there are real-valued functions u_i on A_i $(i = 1, 2)$ and a real number λ so that for all $(a, b) \in A_1 \times A_2$,

$$u(a, b) = u_1(a) + u_2(b) + \lambda u_1(a)u_2(b). \tag{7.19}$$

The third term on the right-hand side represents an interaction effect. We have encountered a variant (a generalization) of this representation, which we also called quasi-additive, in Section 5.5.2. We shall see that under certain additional assumptions, strong independence implies the existence of a quasi-additive value or utility function.

Nonadditive representations for utility, in particular the quasi-additive representation, have had a wide variety of applications, for example, to transportation (Keeney [1973a], de Neufville and Keeney [1972, 1973]), to medical decisionmaking (Keeney [1972a]), to management (Huber [1974], Ting [1971]), to solid waste disposal (Collins [1974]), to air pollution (Ellis [1970]), and to urban services (Keeney [1973b]). In applying the quasi-additive representation to evaluating inventory policies of hospital blood banks, Keeney [1972a] identified the following two dimensions: $A_1 =$ shortage of blood requested by doctors but not readily available in the blood bank (measured in percent of units demanded); $A_2 =$ outdating of blood caused by exceeding its legal lifetime (measured in percent of total units stocked in a year). Keeney verified that his decisionmaker's preferences satisfied strong independence on $A_1 \times A_2$. Her preferences for lotteries involving fixed $b_0 \in A_2$ did not depend on the level of b_0, and similarly for A_1. Keeney derived the following quasi-additive utility function:

$$u(a, b) = 0.32(1 - e^{.13a}) + 0.57(1 - e^{.40b}) + 0.11(1 - e^{.13a})(1 - e^{.40b}).$$

Many kinds of nonadditive representations other than quasi-additivity might be useful. Recently, Farquhar [1974, 1975, 1976] has developed techniques for deriving sufficient conditions for a wide variety of such representations. We concentrate on the quasi-additive representation here.

We say that the component A_1 is *bounded* if there are $a_*, a^* \in A_1$, such that for all $a \in A_1$,

$$a^* \succsim_1 a \succsim_1 a_*,$$

where

$$a \succsim_1 a' \Leftrightarrow (\exists b \in A_2)\left[(a, b) \succsim (a', b)\right].$$

A similar definition applies on the second component. If strong independence holds, indeed even if only independence holds, and (L, R) extends $(K, \succ)$, then

$$a \succsim_1 a' \Leftrightarrow (\forall b \in A_2)\left[(a, b) \succsim (a', b)\right],$$

and similarly for $\succsim_2$.

THEOREM 7.3. *Suppose* $(A_1 \times A_2, \succ, L, R)$ *satisfies the following conditions:*

(a) *The EV Hypothesis.*
(b) *Strong independence.*
(c) *Each component A_i is bounded.*
(d) *Continuity*.*

Then there is a quasi-additive value function u on $A_1 \times A_2$ which satisfies the EV Hypothesis. If in addition (L, R) extends $(K, \succ)$, then u is a quasi-additive order-preserving utility function for $(A_1 \times A_2, \succ)$.

Proof. Let u be a value function on $A_1 \times A_2$ satisfying the EV Hypothesis. Let a_* and a^* be bounds for A_1, and let b_* and b^* be bounds for A_2; that is, assume

$$a^* \succsim_1 a \succsim_1 a_*, \quad \text{all } a \in A_1$$

and

$$b^* \succsim_2 b \succsim_2 b_*, \quad \text{all } b \in A_2.$$

We may assume that $u(a_*, b_*) = 0$. For if u is a value function satisfying the EV Hypothesis, so is $u + \lambda$, any real constant λ. By continuity, we can define a function $v_1(a)$ on A_1 such that $\ell(a, b)$, the lottery that gives (a, b) with certainty[†], is indifferent to the lottery

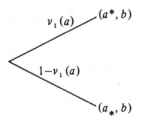

This lottery can be thought of as a $v_1(a)$-brlt (Section 7.2.3). By strong independence, v_1 is a function of a alone. Similarly, we can define a function $v_2(b)$ on A_2, which is a function of b alone, such that $\ell(a, b)$ is

---

*See Section 7.2.2 for definition.
[†]We are assuming throughout this chapter that all such certain lotteries belong to the set L.

indifferent to the lottery

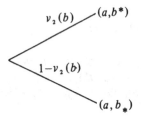

Let $u(a^*, b_*) = \alpha_1$, $u(a_*, b^*) = \alpha_2$, and $u(a^*, b^*) = \alpha_3$. Recall that $u(a_*, b_*) = 0$. Now for all a, b, by the EV Hypothesis,

$$
\begin{aligned}
u(a, b) &= v_1(a)u(a^*, b) + [1 - v_1(a)]u(a_*, b) \\
&= v_1(a)\{v_2(b)u(a^*, b^*) + [1 - v_2(b)]u(a^*, b_*)\} \\
&\quad + [1 - v_1(a)]\{v_2(b)u(a_*, b^*) + [1 - v_2(b)]u(a_*, b_*)\} \\
&= v_1(a)\{v_2(b)\alpha_3 + [1 - v_2(b)]\alpha_1\} + [1 - v_1(a)]v_2(b)\alpha_2 \\
&= \alpha_1 v_1(a) + \alpha_2 v_2(b) + (\alpha_3 - \alpha_1 - \alpha_2)v_1(a)v_2(b).
\end{aligned}
$$

Thus, letting $u_1(a) = \alpha_1 v_1(a)$ and $u_2(b) = \alpha_2 v_2(b)$, one obtains a quasi-additive representation for u.   ∎

Remark: Under the EV Hypothesis, strong independence is a necessary condition for quasi-additivity. The proof is left to the reader.

COROLLARY (Keeney and Raiffa [1976]). *Under the hypotheses of Theorem 7.3, either there is an additive value function u on $A_1 \times A_2$ which satisfies the EV Hypothesis, or there is a multiplicative value function u on $A_1 \times A_2$ which satisfies the EV Hypothesis.*

Proof. Suppose u is a value function on $A_1 \times A_2$ which satisfies the EV Hypothesis and also Eq. (7.19). If $\lambda = 0$, then (7.19) reduces to an additive representation for u. Thus, suppose $\lambda \neq 0$. Suppose first that $\lambda > 0$. Then clearly

$$
v(a, b) = \lambda u(a, b) + 1
$$

is again a value function on $A_1 \times A_2$ which satisfies the EV Hypothesis.

Moreover, we have

$$v(a, b) = \lambda u_1(a) + \lambda u_2(b) + \lambda^2 u_1(a) u_2(b) + 1$$

$$= [\lambda u_1(a) + 1][\lambda u_2(b) + 1]$$

$$= v_1(a) v_2(b),$$

where $v_i(x) = \lambda u_i(x) + 1$. Thus, v is multiplicative. If $\lambda < 0$, let

$$v(a, b) = -\lambda u(a, b) - 1,$$

and use

$$v_1(a) = -\lambda u_1(a) - 1$$

and

$$v_2(b) = \lambda u_2(b) + 1. \qquad ■$$

Let us ask if the axioms of Theorem 7.3 seem reasonable. In both market baskets and the public health problem, continuity seems reasonable. Let us accept the EV Hypothesis. There is a lower bound a_* for each component in the market baskets example, namely 0. We could create an upper bound a^* for each component by using a reasonable upper limit to consumption, say 100 times the GNP for safety's sake. In the public health problem, suppose average assets are used for the utility function. For fixed $a \neq 0$, if $n, m \neq 0$, $(n, a) \succ (m, a)$ if and only if $n < m$. Thus 1 could serve as an upper bound (best alternative) on the population dimension. The lower bound would be 0. On the assets dimension, 0 would be a lower bound. The maximum conceivable asset level could serve as an upper bound. Suppose next that we are in the maximum comfort level society. Then $\succsim_1$ and $\succsim_2$ are strange relations. For every n and m, there is a such that $(n, a) \succsim (m, a)$. Namely, choose a such that $a/n = C$ and $a/m \neq C$. Thus, $n \succsim_1 m$. It follows that every n qualifies as an upper bound for the first component. But of course this is not what we had in mind. What is happening is that, as we shall see, independence is violated, in which case $\succsim_1$ is not an interesting relation. The same is true for $\succsim_2$.

Let us consider next strong independence. For an extensive discussion of empirical tests of this assumption, see Keeney and Raiffa [1976]. See in particular the references to applications of utility functions over lotteries with multidimensional consequences given at the beginning of Section 7.3.1.

Even independence might not hold for market baskets, as we observed in Section 5.4.3. For suppose the first component is coffee, and the second sugar. You might prefer $(1,1)$ to $(0,1)$, but $(0,0)$ to $(1,0)$, having a violent dislike for coffee without sugar and so having to find a place to dispose of it.

Independence seems to hold in the public health problem for the society that uses average assets to measure utility, if we do not allow assets to be 0. For this society, if $n, m > 0$,

$$(n, a) \succ (n, b) \;\Rightarrow\; a/n > b/n$$
$$\Rightarrow\; a/m > b/m$$
$$\Rightarrow\; (m, a) \succ (m, b).$$

Similarly, if $m, n > 0$ and $b > 0$, then $(n, a) \succ (m, a) \;\Rightarrow\; (n, b) \succ (m, b)$. However, this fails if $b = 0$. Strong independence also seems basically to be satisfied for this society, at least if we do not allow population or assets to be 0. For consider lotteries

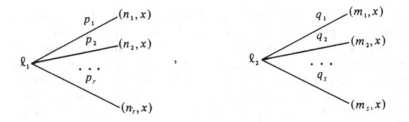

If the EV Hypothesis holds, then if all n_i, m_j, x, and x' are positive,

$$\ell_1 R \ell_2 \Leftrightarrow \sum_i p_i \cdot (x/n_i) > \sum_j q_j \cdot (x/m_j)$$

$$\Leftrightarrow x \sum_i (p_i/n_i) > x \sum_j (q_j/m_j)$$

$$\Leftrightarrow x' \sum_i (p_i/n_i) > x' \sum_j (q_j/m_j)$$

$$\Leftrightarrow \sum_i p_i \cdot (x'/n_i) > \sum_j q_j \cdot (x'/m_j)$$

$$\Leftrightarrow \ell_1' R \ell_2',$$

where ℓ_i' is obtained from ℓ_i by substituting x' for x. A similar argument works on the second component.

Turning finally to the public health problem in the maximum comfort level society, we see that, as suggested earlier, even independence is violated here. Specifically,

$$(n, a) \succ (n, b) \Leftrightarrow (m, a) \succ (m, b)$$

fails. For given population level n, there is an optimal total wealth nC. Suppose $m \neq n$. Then

$$(n, nC) \succ (n, mC),$$

but

$$(m, mC) \succ (m, nC).$$

Strong independence is violated here in an interesting way. Consider the two lotteries

Suppose a is small, a' is large, and b is moderate in size. Suppose a'/n is close to the ideal comfort level. If a/n is above a level of subsistence, it might be worth a $7/8$ chance of reaching ideal comfort level, rather than taking a moderate level b/n for certain. Thus, ℓ_1 might be preferred to ℓ_2. However, if a/n is below the subsistence level, ℓ_2 might be preferred to ℓ_1. Thus, by changing n, we might change preference between these lotteries. The choice between ℓ_1 and ℓ_2 could arise for a country that is considering a radical new trade policy or foreign policy.

To close this section, we prove a theorem that tells when a quasi-additive representation can be made into an additive representation. We use the notation $a_1 \sim_i a_2$ to stand for $a_1 \succeq_i a_2$ and $a_2 \succeq_i a_1$.

Suppose that a_1, $a_2 \in A_1$, b_1, $b_2 \in A_2$ and not $(a_1 \sim_1 a_2)$ and not $(b_1 \sim_2 b_2)$. Suppose we are indifferent between the lotteries

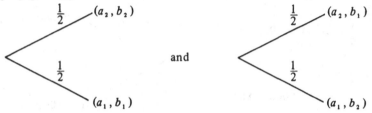

In this case, we shall say that the *marginality assumption holds for* a_1, a_2, b_1, b_2.

THEOREM 7.4 (Raiffa [1969]). *Given* $(A_1 \times A_2, \succ, L, R)$, *suppose there is a quasi-additive value function* u *on* K *satisfying the EV Hypothesis and suppose* (L, R) *extends* $(K, \succ)$. *Suppose also that for some* $a_1, a_2 \in A_1$ *and* $b_1, b_2 \in A_2$ *such that not* $a_1 \sim_1 a_2$ *and not* $b_1 \sim_2 b_2$, *the marginality assumption holds for* a_1, a_2, b_1, b_2. *Then* u *is additive.*

Proof. By the marginality assumption for a_1, a_2, b_1, b_2 and the EV Hypothesis, we have

$$\tfrac{1}{2}u(a_2, b_2) + \tfrac{1}{2}u(a_1, b_1) = \tfrac{1}{2}u(a_2, b_1) + \tfrac{1}{2}u(a_1, b_2).$$

Canceling $\tfrac{1}{2}$ and using quasi-additivity (7.19), we obtain

$$u_1(a_2) + u_2(b_2) + u_1(a_1) + u_2(b_1) + \lambda u_1(a_2)u_2(b_2) + \lambda u_1(a_1)u_2(b_1)$$
$$= u_1(a_2) + u_2(b_1) + u_1(a_1) + u_2(b_2) + \lambda u_1(a_2)u_2(b_1) + \lambda u_1(a_1)u_2(b_2).$$

It follows that

$$\lambda u_1(a_2)u_2(b_2) + \lambda u_1(a_1)u_2(b_1) = \lambda u_1(a_2)u_2(b_1) + \lambda u_1(a_1)u_2(b_2).$$

Then we have

$$\lambda[u_1(a_1) - u_1(a_2)][u_2(b_1) - u_2(b_2)] = 0.$$

But not $a_1 \sim_1 a_2$ implies that $u_1(a_1) - u_1(a_2) \neq 0$, and not $b_1 \sim_2 b_2$ implies that $u_2(b_1) - u_2(b_2) \neq 0$. (The proof, which uses the fact that $(L; R)$ extends $(K, \succ)$, is left to the reader.) We conclude that $\lambda = 0$, and so additivity follows. ∎

Exercises

1. Let (a, b) represent consumption in two time periods. Consider the following lotteries:

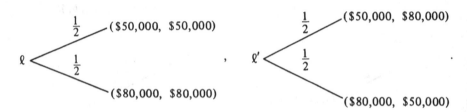

If an individual prefers ℓ' to ℓ, conclude that there can be for him no additive value function over K satisfying the EV Hypothesis.

2. (Farquhar [1974, p. 16]) Consider a two-element system with identical components operating in parallel, for example, such human organ systems as eyes, lungs, limbs, or kidneys. Let A_1 be an index of relative performance of one component, say the left kidney, and A_2 an index of performance of the second component, say the right kidney. Is it reasonable to argue that preferences for removal or nonremoval of either component (kidney) do not satisfy strong independence?

3. (Farquhar [1974, p. 16]) In the study of painkillers, suppose A_1 is the amount of pain endured during a given time period, and A_2 measures the distribution of pain. Is it reasonable to argue that strong independence holds for one component, but not for the other?

4. Fishburn and Keeney [1975] study the case where A_1 is the number of days until death and A_2 is the quality of a patient's life.
   (a) Consider whether marginality holds.
   (b) Consider whether strong independence holds.

5. Show that if there is an additive value function u satisfying the EV Hypothesis, then strong independence holds.

6. Show that if there is a quasi-additive value function satisfying the EV Hypothesis, then strong independence holds.

7. Show that under the hypotheses of Theorem 7.4, not $a_1 \sim_1 a_2$ implies $u_1(a_1) - u_1(a_2) \neq 0$, and not $b_1 \sim_2 b_2$ implies $u_2(b_1) - u_2(b_2) \neq 0$.

8. Show that the proof of Theorem 7.3 goes through if we replace the assumption of strong independence by the following weaker assumption and its analogue for the second component: if $\ell(a, b)I\ell'$, where ℓ' is as follows, then $\ell(a, b')I\ell''$ for all b', where I is the tying relation defined by

$$kIk' \Leftrightarrow \, \sim kRk' \, \& \sim k'Rk$$

and ℓ'' is as follows:

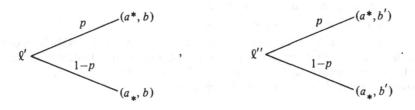

(I is the relation we have usually denoted E; however, in this chapter it will be useful to reserve the letter E for another purpose.)

9. Suppose $A_1 = A_2 = \{1, 2, \ldots, n\}$ and

$$(a_1, a_2) \succ (b_1, b_2) \Leftrightarrow u(a_1, a_2) > u(b_1, b_2).$$

Suppose (L, R) extends $(K, \succ)$, and v is a value function satisfying the EV Hypothesis. Suppose $u(a_1, a_2) = a_1 a_2$. Show the following:

    (a) Strong independence holds.

    (b) Continuity holds.

    (c) Each component is bounded.

    (d) There is a quasi-additive representation.

    (e) The quasi-additive representation given by the proof of Theorem 7.3 uses

$$v_1(a) = \frac{a - 1}{n - 1}, \qquad v_2(b) = \frac{b - 1}{n - 1},$$

$$u_1(a) = \frac{n}{n - 1}(a - 1), \qquad u_2(a) = \frac{n}{n - 1}(b - 1),$$

and

$$\lambda = \frac{n^2 - 2n}{n^2}.$$

    (f) The marginality assumption fails.

    (g) There is no additive representation.

10. Consider the assertions of parts (a), (b), (c), (d), (f), and (g) of Exer. 9 for the following functions u:

    (a) $u(a_1, a_2) = a_1^2 + a_1 a_2$.

    (b) $u(a_1, a_2) = \max\{a_1, a_2\}$.

    (c) $u(a_1, a_2) = |a_1 - a_2|$.

    (d) $u(a_1, a_2) = 1/a_1 a_2$.

    (e) $u(a_1, a_2) = a_1/(a_1 + a_2)$.

    (f) $u(a_1, a_2) = a_1^\alpha a_2^\beta/(a_1 + a_2)$.

    (g) $u(a_1, a_2) = f(a_2) + g(a_2)h(a_1)$.

    (h) $u(a_1, a_2) = \alpha f(a_1) + \beta g(a_2) + \gamma f(a_1)g(a_2)$.

    (i) $u(a_1, a_2) = [\alpha + \beta f(a_1)][\gamma + \delta g(a_2)]$.

    (j) $u(a_1, a_2) = \alpha f(a_1) + \beta g(a_2)$.

11. Repeat Exer. 9 if $A_1 = A_2 = Re$.

12. Repeat Exer. 10 if $A_1 = A_2 = Re$.

13. (Keeney [1971], Keeney and Raiffa [1976, pp. 226, 243]) Suppose strong independence holds on the second component. Identify what hypotheses are needed to show that there is a value function u on $A_1 \times A_2$ satisfying the EV Hypothesis and such that

$$u(a, b) = u_1(a) + u_1'(a)u_2(b).$$

14. (Keeney [1974], Keeney and Raiffa [1976]) Given the structure $(A_1 \times A_2 \times \cdots \times A_n, \succ, L, R)$, one generalization of the quasi-additive representation is that there is a value function u on $A_1 \times A_2 \times \cdots \times A_n$ that satisfies the EV Hypothesis, and there are real-valued functions u_i on

A_i, $i = 1, 2, \ldots, n$, and there is a constant λ, such that

$$
\left.
\begin{aligned}
u(a_1, a_2, \ldots, a_n) = &\sum_{i=1}^{n} u_i(a_i) + \lambda \sum_{\substack{i=1 \\ j>i}}^{n} u_i(a_i)u_j(a_j) \\
&+ \lambda^2 \sum_{\substack{i=1 \\ j>i \\ l>j}}^{n} u_i(a_i)u_j(a_j)u_l(a_l) \\
&+ \cdots + \lambda^{n-1} u_1(a_1)u_2(a_2) \ldots u_n(a_n).
\end{aligned}
\right\} \quad (7.20)
$$

One hypothesis required for the representation (7.20) is the variant of strong independence which says that if consequences in all components but the ith are held fixed, then preferences among lotteries do not depend on the level of the fixed components. Formalize this and other hypotheses that allow the derivation of the representation (7.20). *Hint:* Repeatedly use

$$
u(a_1, a_2, \ldots, a_n) = u_i(a_i) + c_i(a_i)w_i(a_1, a_2, \ldots, a_{i-1}, a_{i+1}, \ldots, a_n).
$$

15. (Keeney and Raiffa [1976]) Show that if the hypotheses you introduced for the representation (7.20) in the previous exercise are satisfied, then there is a value function on $A_1 \times A_2 \times \cdots \times A_n$ satisfying either an additive or a multiplicative representation.

16. (Keeney and Raiffa [1976]) A variant of the representation (7.20) is the following representation for a value function u satisfying the EV Hypothesis, where the $\lambda_{ijk\ldots}$ are constants.

$$
\begin{aligned}
u(a_1, a_2, \ldots, a_n) = &\sum_{i=1}^{n} \lambda_i u_i(a_i) + \sum_{i=1}^{n} \sum_{j>i} \lambda_{ij} u_i(a_i)u_j(a_j) \\
&+ \sum_{i=1}^{n} \sum_{j>i} \sum_{l>j} \lambda_{ijl} u_i(a_i)u_j(a_j)u_l(a_l) \\
&+ \cdots + \lambda_{123\ldots n} u_1(a_1)u_2(a_2) \ldots u_n(a_n).
\end{aligned}
$$

One hypothesis required for this representation is that if consequences in all components in any set X of components are held fixed, then preferences among lotteries do not depend on the level of the fixed components. Formalize this and other hypotheses which allow the derivation of such a representation. *Hint:* Use

$$
u(a_1, a_2, \ldots, a_n) = v(a_{i_1}, a_{i_2}, \ldots, a_{i_p}) + c(a_{i_1}, a_{i_2}, \ldots, a_{i_p})w(a_{j_1}, a_{j_2}, \ldots, a_{j_q}),
$$

where

$$
\{i_1, i_2, \ldots, i_p\} \quad \text{and} \quad \{j_1, j_2, \ldots, j_q\}
$$

form a partition of $\{1, 2, \ldots, n\}$. Let X be any subset of $\{1, 2, \ldots, n\}$.

Show that

$$u(a_1, a_2, \ldots, a_n) = u_i(a_i)u(a_1', a_2', \ldots, a_n') + [1 - u_i(a_i)]u(a_1'', a_2'', \ldots, a_n''),$$

where

$$a_i' = \begin{cases} a_i & \text{if } i \notin X \\ a_i^* & \text{if } i \in X, \end{cases}$$

$$a_i'' = \begin{cases} a_i & \text{if } i \notin X \\ a_{i*} & \text{if } i \in X, \end{cases}$$

and the a_i^* determine a maximum element using the components of X, and the a_{i*} determine a minimum element using the components of X.

7.4 Mixture Spaces

In this section, we shall seek a representation theorem for the Expected Utility (Value) Hypothesis. That is, given a preference relation R on a set L of lotteries, we shall ask for conditions on (L, R) sufficient to guarantee the existence of a function E on L satisfying, for all $\ell, \ell' \in L$,

$$\ell R \ell' \Leftrightarrow E(\ell) > E(\ell'). \tag{7.21}$$

Of course, we already know conditions on (L, R) sufficient for the existence of such an E: (L, R) must be a strict weak order, and (L^*, R^*) must have a countable order-dense subset. We really want E to have more properties than (7.21): we want it to act like an expected utility. A key property of expected utility is the following. Suppose ℓ and ℓ' are lotteries, p is a real number in $[0, 1]$, and ℓ'' is the complex lottery

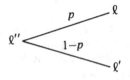

Then

$$E(\ell'') = pE(\ell) + (1 - p)E(\ell'). \tag{7.22}$$

We shall seek conditions on (L, R) sufficient to guarantee the existence of a function E on L satisfying Eqs. (7.21) and (7.22). If we can find such a function E, and if (L, R) extends $(K, \succ)$, then we can define $u{:}K \to Re$ by

$$u(c) = E[\ell(c)].$$

It is easy to see that u is an order-preserving utility function over $(K, \succ)$ and E is expected utility relative to u. The problem we study has also been considered by, among others, von Neumann and Morgenstern [1944], Friedman and Savage [1948, 1952], Marschak [1950], Herstein and Milnor [1953], Cramer [1956], Luce and Raiffa [1957], Blackwell and Girshick [1954], and Huang [1971]. Our approach follows that of Fishburn [1970, Chapter 8], Luce and Suppes [1965], and Fishburn and Roberts [1978].

The conditions we shall present hold in a more general setting than that of lotteries. Suppose A is a set of alternatives and R is a binary relation of (strict) preference on A. For every real number p in $[0, 1]$ and for every a, b in A, we shall speak of the *mixture apb*. This is a new alternative, an element of A, which is *interpreted* as a lottery with probability p of obtaining a and probability $1 - p$ of obtaining b. It will be convenient to think of A as a collection of lotteries with consequences in a set K and apb as the new complex lottery

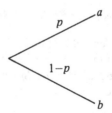

But these interpretations are merely for motivation purposes. Formally, one thinks of a function $\theta{:}A \times [0, 1] \times A \to A$, with $\theta(a, p, b) = apb$. We shall study the triple (A, R, θ), and we shall call it a *mixture space*.

As Gudder [1977] points out, mixture spaces arise in a wide variety of contexts. One example is in the making of bread. If F is flour and B is butter, we can speak of the mixture FpB: this means p cups of flour for each $1 - p$ cups of butter. In color vision, we can mix colored lights by superposition of light beams. If a and b are two colored lights and $p \in [0, 1]$, then apb is the light obtained by superposing a and b in the proportion $p{:}1 - p$. In quantum mechanics, if a and b are states of a system and $p \in [0, 1]$, apb represents the state in which the system is in state a with probability p and in state b with probability $1 - p$. (Axiomatic developments of quantum mechanics using ideas like mixtures can be found in Gudder [1973] and Mielnik [1969].)

Let us return to the mixture space (A, R, θ) in the special case where A is a collection of lotteries and the EU Hypothesis holds. Then there is a real-valued function E on A, the expected utility, such that for all $a, b \in A$ and $p \in [0, 1]$,

$$aRb \Leftrightarrow E(a) > E(b) \qquad (7.23)$$

and

$$E(apb) = pE(a) + (1 - p)E(b). \qquad (7.24)$$

Speaking abstractly again, we now seek conditions on the triple (A, R, θ) sufficient to guarantee the existence of a real-valued function E on A such that Eqs. (7.23) and (7.24) are satisfied. This abstract formulation of the problem of axiomatizing expected utility goes back to von Neumann and Morgenstern [1944]. They gave axioms sufficient for the representation (7.23), (7.24). We shall present a set of axioms based on some of Fishburn and Roberts [1978], which are all necessary as well as being jointly sufficient. These axioms are closely related to a set given in Fishburn [1970].

To present these axioms, we define the "tying" relation I from R by

$$aIb \Leftrightarrow \sim aRb \ \& \sim bRa. \qquad (7.25)$$

(This is the relation we have usually denoted E; however, here we use the notation I to avoid confusion with the function E). A mixture space (A, R, θ) is defined to be an *EU Mixture Space* if, for all $a, b, c \in A$, the following axioms are satisfied:

AXIOM M1. (A, R) *is a strict weak order.*

AXIOM M2. $(apb)I[b(1 - p)a], \quad all \ p \in [0, 1].$

AXIOM M3. $[(apb)qb]I[a(pq)b], \quad all \ p, q \in [0, 1].$

AXIOM M4. *If* aRb, *then* $(apc)R(bpc), \quad all \ p \in (0, 1).$

AXIOM M5. *If* $aRbRc$, *then there are* $p, q \in (0, 1)$ *such that*

$$(apc)Rb \ and \ bR(aqc).$$

Axiom M1 is a standard axiom which clearly follows from Eq. (7.23). Axioms M2 and M3 follow simply in the case of lotteries. For instance, to verify Axiom M3 here, the lottery on the left side can be replaced by the

tree diagram

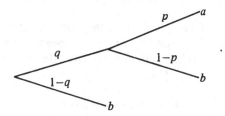

Multiplying along the branches, we obtain the lottery

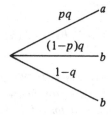

Combining different branches leading to the consequence b, we have

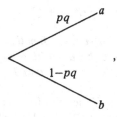

which is the lottery on the right side of Axiom M3. Axioms M2 and M3 also follow easily from Eqs. (7.23) and (7.24), as the reader can readily verify.

Axiom M4 also directly follows from the representation (7.23), (7.24). For if aRb, then $E(a) > E(b)$, so

$$pE(a) + (1 - p)E(c) > pE(b) + (1 - p)E(c),$$

so

$$E(apc) > E(bpc),$$

so

$$(apc)R(bpc).$$

Axiom M4 is a kind of independence condition, and it is also reminiscent of the monotonicity condition of extensive measurement (Section 3.2). Finally, to show that Axiom M5 follows from the representation (7.23), (7.24), suppose that $aRbRc$. Then

$$E(a) > E(b) > E(c),$$

so

$$1 \cdot E(a) + 0 \cdot E(c) > E(b) > 0 \cdot E(a) + 1 \cdot E(c).$$

There must be $p, q \in (0, 1)$ so that

$$pE(a) + (1 - p)E(c) > E(b) > qE(a) + (1 - q)E(c).$$

THEOREM 7.5 (Fishburn and Roberts [1978])*. *If (A, R, θ) is a mixture space, then (A, R, θ) is an EU Mixture Space if and only if there is a real-valued function E on A such that for all $a, b \in A$ and $p \in [0, 1]$,*

$$aRb \Leftrightarrow E(a) > E(b) \tag{7.23}$$

and

$$E(apb) = pE(a) + (1 - p)E(b). \tag{7.24}$$

Moreover, if E' is another real-valued function on A satisfying (7.23) and (7.24), then E' is related to E by a positive linear transformation; that is, there are real numbers $\lambda > 0$ and λ' such that $E' = \lambda E + \lambda'$.

We omit the proof of the sufficiency of the conditions of an EU Mixture Space for the representation (7.23) and (7.24), and of the uniqueness statement. But see Exers. 8 through 12.

The following is an interesting corollary of Theorem 7.5.

COROLLARY. *Suppose L consists of all lotteries with consequences in a set K, and R is a binary relation on L. Then there is a real-valued function u on K satisfying the EV Hypothesis if and only if the following axioms hold:*
(a) (L, R) is a strict weak order.

---

*The theorem in Fishburn and Roberts states explicitly the axiom that for all $a, b \in A$ and $p \in [0, 1]$, $apb \in A$. This is contained in our definition of a mixture space.

(b) If $\ell_1 R \ell_2$, then for all $p \in (0, 1)$ and $\ell_3 \in L$, $\ell R \ell'$, where ℓ and ℓ' are the following complex lotteries:

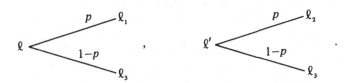

(c) If $\ell_1 R \ell_2 R \ell_3$, then there are $p, q \in (0, 1)$ such that $\ell R \ell_2$ and $\ell_2 R \ell'$, where ℓ and ℓ' are the following complex lotteries:

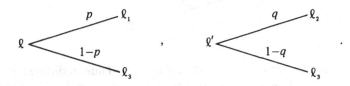

Proof. Necessity of conditions (a), (b), and (c) is clear. To show sufficiency, note that the set L forms a mixture space in the obvious way and that (a), (b), and (c) are Axioms M1, M4, and M5 of an EU Mixture Space. The other axioms of an EU Mixture Space hold trivially by conventions for lotteries. Thus, there is a real-valued function E on L satisfying Eqs. (7.23) and (7.24). Using the convention that c_i and $\ell(c_i)$ are interchangeable, one easily verifies from (7.24) that if ℓ is the lottery

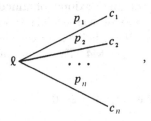

then

$$E(\ell) = \sum_{i=1}^{n} p_i E(\ell(c_i)).$$

The EV Hypothesis follows by taking $u(c_i) = E(\ell(c_i))$. ∎

The axioms for an EU Mixture Space seem intuitively pleasing. In particular, Axioms M2 and M3 seem very plausible. However, we have previously mentioned evidence against Axiom M1. As another argument against Axiom M1, consider the following lotteries:

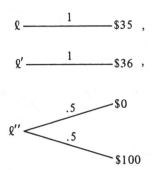

It is possible that $\ell''I\ell$ and $\ell''I\ell'$, while $\ell'R\ell$. Thus, indifference is not transitive, and Axiom M1 is violated.[*][†] The lotteries in the Allais Paradox (Section 7.2.5) can be used to make an argument against Axiom M4 (see Exer. 6). It is also easy to make arguments against Axiom M5. The following example is due to Gudder [1977]. Suppose a = "be given two candy bars," b = "be given one candy bar," and c = "be hanged at dawn." Then $aRbRc$, but in the lottery interpretation, there is no $p < 1$ such that $(apc)Rb$. For more serious examples, see Exers. 4 and 5 below.

The reader might wish to think about whether Axioms M1 through M5 hold in the various other interpretations of mixtures given above (bread and flour, colored lights, quantum mechanics). See Gudder [1977] for a discussion of the variants of these axioms obtained by replacing I by = .

An interesting consequence of Axioms M1 through M4 has been subjected to experimental test. If we believe Axioms M1 through M3, we can regard this as a test of Axiom M4. The consequence in question is the following condition:

$$\text{If } aRb, \text{ then for all } p \in (0, 1),\ aR(apb)Rb. \tag{7.26}$$

Coombs and Huang [1976] have subjected condition (7.26) to experimental test by providing subjects choices of gambles. In both of their experiments, there were, surprisingly, a large number of violations of this condition.

It is interesting to see how condition (7.26) follows from the axioms. The proof should provide a little feeling for how one reasons about mixture spaces. We first prove the following lemmas.

LEMMA 1. *In an EU Mixture Space, $(a1b)Ia$.*

Proof. Suppose first that $(a1b)Ra$. Then by Axiom M4,

$$\left[(a1b)\tfrac{1}{2}b\right]R\left(a\tfrac{1}{2}b\right).$$

But by Axiom M3,

$$\left[(a1b)\tfrac{1}{2}b\right]I\left(a\tfrac{1}{2}b\right).$$

This contradicts the definition of I. A similar proof shows that $aR(a1b)$ is impossible.   ∎

It should be noted that in the mixture space axioms given by Fishburn [1970], the result $a1b = a$ is an axiom, and Axioms M2 and M3 are replaced by axioms in which I is replaced by equality. When equality is replaced by I to obtain Axioms M2 and M3, the weaker statement $(a1b)Ia$ can be proved from the remaining axioms, as we have seen.

LEMMA 2. *In an EU Mixture Space, if $p \in [0, 1]$, then $(apa)Ia$.*

Proof.[*] If $p = 0$ or 1, $(apa)Ia$ follows by Lemma 1 and M2 and an application of M1. Suppose $p \in (0, 1)$, and suppose $aR(apa)$. Let $q = (1 + p)^{-1}$. By M4,

$$(aqa)R[(apa)qa].$$

Hence, by M3 and M1,

$$(aqa)R[a(pq)a].$$

Now $pq = 1 - q$, since $q = (1 + p)^{-1}$. Thus we have

$$(aqa)R[a(1 - q)a].$$

---

[*]The author thanks Peter Fishburn for this proof.

By definition of I this violates Axiom M2. A similar proof shows that $(apa)Ra$ is impossible.   ∎

We now show that condition (7.26) holds. Assuming aRb and taking $c = b$ in Axiom M4 yields

$$(apb)R(bpb).$$

Using $(bpb)Ib$ (which follows by Lemma 2) and Axiom M1, we find

$$(apb)Rb. \tag{7.27}$$

Next, using $c = a$ and using $1 - p$ instead of p in Axiom M4, we obtain

$$[a(1 - p)a]R[b(1 - p)a]. \tag{7.28}$$

By Lemma 2 with $1 - p$, by Axiom M2, and by Axiom M1, (7.28) implies that

$$aR(apb). \tag{7.29}$$

Equations (7.27) and (7.29) give us condition (7.26).

Exercises

1. Fishburn [1970] Suppose X is a set. A *simple probability measure* on X is a function P, which assigns a real number $P(A)$ to each subset A of X, and satisfies the following conditions:
      (i) $P(A) \geqq 0$, for all $A \subseteqq X$,
      (ii) $P(X) = 1$,
      (iii) $P(A \cup B) = P(A) + P(B)$ when $A, B \subseteqq X$ and $A \cap B = \varnothing$,
      (iv) $P(A) = 1$, for some finite $A \subseteqq X$.
If P and Q are simple probability measures on X, and $\alpha \in [0, 1]$, then $P\alpha Q$ is defined to be the function $\alpha P + (1 - \alpha)Q$.
    (a) Show that $P\alpha Q$ is again a simple probability measure.
    (b) Suppose R is a binary relation on the set of all simple probability measures on X, and suppose $P = Q$ implies PIQ. Show the following:
        (i) The set of simple probability measures on X under the relation R satisfies Axiom M2 of an EU Mixture Space.
        (ii) Axiom M1 might fail. (An example analogous to the one in the text showing that indifference between lotteries may not be transitive suffices to show this.)
        (iii) What about Axiom M3?
        (iv) What about Axiom M4?
        (v) What about Axiom M5?
        (vi) What about the conclusion of Lemma 1?
        (vii) What about the conclusion of Lemma 2?
        (viii) What about condition (7.26)?

2. (Fishburn [1970, p. 109]) Suppose ℓ and ℓ_p are the following lotteries:

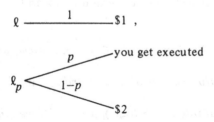

$$\ell \;\underline{\qquad 1 \qquad}\; \$1 \;,$$

Use these lotteries to argue against one of the axioms for an EU Mixture Space.

3. (Gudder [1977]) A recipe for bread calls for 8 cups of flour, 1 cup of butter, and 2 cups of water. We can mix flour and butter first and then add water, or mix butter and water first and then add flour. The results are equivalent. From this, conclude that $[(F\frac{8}{9}B)\frac{9}{11}W]I[F\frac{8}{11}(B\frac{1}{3}W)]$.

4. (Gudder [1977]) In a war, suppose a = losing 100 tanks, b = losing 100 men, and c = losing the war. Observe that under the preferences of most military leaders, Axiom M5 fails in the lottery interpretation.

5. (Gudder [1977]) In a society, action a is considered moderately good, action b neither good nor bad, and action c taboo. Observe that Axiom M5 fails in a natural mixture of actions interpretation.

6. (Fishburn [1970, p. 109]) (a) Show that the following condition is necessary for the representation (7.23), (7.24): if $p \in (0, 1)$ and $(apc)R(bpc)$, then aRb.

(b) Suppose lotteries $\ell^{(1)}$, $\ell^{(2)}$, $\ell^{(3)}$, and $\ell^{(4)}$ are as in the Allais Paradox (Section 7.2.5). Show that if $\ell^{(1)}$ is preferred to $\ell^{(2)}$, and $\ell^{(3)}$ to $\ell^{(4)}$, then the condition in (a) is violated. To show this, use the lotteries

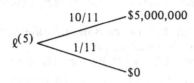

and

$$\varrho^{(6)} \;\underline{\qquad 1 \qquad}\; \$0$$

and write $\ell^{(1)}$, $\ell^{(2)}$, $\ell^{(3)}$, and $\ell^{(4)}$ as mixtures making use of these lotteries.

(c) Show that in the presence of the other axioms, Axiom M4 implies the condition in (a). (Thus, the Allais Paradox is an argument against Axiom M4, as is claimed by Allais.)

7. Huang [1971] has replaced Axioms M4 and M5 by the following three axioms:

AXIOM H1. *If aRb, then for all $p \in (0, 1)$, aR(bpa) and (bpa)Rb.*

AXIOM H2. *If aIb, then for all $c \in A$ and $p \in [0, 1]$, (apc)I(bpc).*

AXIOM H3. *If aRbRc and bR(apc), $p \in (0, 1)$, then there is $q \in (0, 1)$ so that $q > p$ and*

$$bR(aqc).$$

Similarly, if aRbRc and (apc)Rb, $p \in (0, 1)$, then there is $q \in (0, 1)$ so that $q < p$ and

$$(aqc)Rb.$$

She also uses the following axiom from the original list in Fishburn [1970]. This axiom is the result of Lemma 1.

AXIOM H4. *(a1b)Ia.*

(a) Show that Axioms H1 through H4 are necessary for the representation (7.23), (7.24), and hence follow from the axioms M1 through M5.

(b) Axioms M1 through M3 and H1 through H4 together are sufficient for the representation (7.23), (7.24). To show this, first show that if *aRb*, then

$$p > q \Rightarrow (apb)R(aqb).$$

(c) Next, using the result of (b), show that

$$aRcRb \Rightarrow cI(apb) \quad \text{for some } p \in (0, 1).$$

(d) Verify Axiom M4 by considering cases such as *bRcRa* and *bRaRc*, etc., and using the result of (c).

(e) Finally, verify Axiom M5 by applying the result of (c).

8. Exercises 8 through 12 sketch a proof of Theorem 7.5. As a first step, one uses the axioms (and Lemmas 1 and 2) to prove the following lemmas. Provide a proof of each. You may assume all of the previously proved statements in verifying these lemmas.

(a) If *aRb* and $0 \le p < q \le 1$, then (aqb)R(apb).

(b) Suppose *aSbSc* and *aRc*, where

$$xSy \Leftrightarrow \sim yRx. \tag{7.30}$$

Then *bI(apc)* for exactly one $p \in [0, 1]$.

(c) If aRb and cRd and $p \in [0, 1]$, then

$$(apc)R(bpd).$$

(d) If aIb and $p \in [0, 1]$, then

$$(apb)Ia.$$

(e) If aIb and $p \in [0, 1]$, then for all c,

$$(apc)I(bpc).$$

9. Continuing the proof of Theorem 7.5, prove the following:

$$[(aqb)p(arb)]I\{a[\,pq + (1 - p)r\,]b\}.$$

10. Continuing the proof of Theorem 7.5, fix c and d in A with cRd and consider

$$cd = \{a \in A : cSaSd\},$$

where S is defined in (7.30). By (b) of Exer. 8, there is a unique number $f(a) \in [0, 1]$ for each $a \in cd$, such that

$$a\,I\{c[\,f(a)\,]d\},$$

and such that $f(c) = 1$ and $f(d) = 0$.

(a) Prove that for all $a, b \in cd$,

$$aRb \Leftrightarrow f(a) > f(b). \tag{7.31}$$

(b) Prove that if $p \in [0, 1]$ and $a, b \in cd$,

$$f(apb) = pf(a) + (1 - p)f(b). \tag{7.32}$$

11. As the last step in the sufficiency proof of Theorem 7.5, do the following.

(a) Given $a, b \in A$, show that there are $a_i, b_i \in A$ so that a and b are both in the set $a_ib_i = \{e \in A : a_iSeSb_i\}$ and so that $a_ib_i \supseteq cd$, where c, d are as in Exer. 10

(b) Show that there is a function f_i^* on a_ib_i satisfying (7.31) and (7.32) on a_ib_i.

(c) Let f_i be obtained from f_i^* by a positive linear transformation, in such a way that $f_i(c) = 1$ and $f_i(d) = 0$. Show that f_i also satisfies (7.31) and (7.32) on a_ib_i.

(d) Show that if $a \in a_ib_i$ and $a \in a_jb_j$, then $f_i(a) = f_j(a)$.

(e) Let $E(a)$ be the common value of $f_i(a)$ over all i so that $a \in a_ib_i$. Show that E satisfies (7.23) and (7.24) for all $a, b \in A$ and $p \in [0, 1]$.

12. To prove the uniqueness statement in Theorem 7.5, fix c and d in A and consider the functions

$$F(a) = \frac{E(a) - E(d)}{E(c) - E(d)},$$

$$F'(a) = \frac{E'(a) - E'(d)}{E'(c) - E'(d)}.$$

Show that $F(a) = F'(a)$, all a, and derive the uniqueness statement.

13. (Pfanzagl [1968, pp. 217-218]) Suppose (A, R, θ) is an EU Mixture Space and the function E on A is non-constant and satisfies (7.23) and

$$E(apb) = s(p)E(a) + [1 - s(p)]E(b),$$

for all a, b in A and $p \in [0, 1]$. The function $s(p)$ can be thought of as a measure of the subjective probability of p (see Chapter 8). Show that the function s has the following properties.
  (a) $s(1 - p) = 1 - s(p)$, for all $p \in [0, 1]$.
  (b) $s(pq) = s(p)s(q)$, for all $p, q \in [0, 1]$.
  (c) $s(1/2) = 1/2$.
  (d) s is monotone increasing on $[0, 1]$.
  (e) $s(p) = p$, all p. (This is shown using the methods of Chapter 4.)

7.5 Subjective Probability

In Section 7.1, we associated with an act or choice a set of possible events $A_1, A_2, \ldots, A_n$, and with each event A we associated a consequence c_i. The expected utility of the act was defined as

$$\Sigma p(A_i)u(c_i),$$

where $p(A_i)$ is the probability that event A_i occurs and $u(c_i)$ is the utility of consequence c_i. In passing to the notion of a lottery in Section 7.2, we suppressed the distinction between events and consequences. We also assumed that the probabilities $p(A_i)$, or p_i in our lotteries, were known beforehand. Often this is not a reasonable assumption. For example, in buying fire insurance, we may not know exactly the probability of our house catching fire over the period for which we want to buy insurance (though our broker may have good estimates). In betting on a horse race, we do not know exactly the probability that a particular horse will win. In making decisions about alternative sources of energy, we do not know the probabilities that various significant events will take place. We often have some ideas of how probable different outcomes are, or at least that one outcome is more probable than another. In the next chapter, we shall discuss how to use this information to make subjective or qualitative assessment of probabilities and we shall mention various applications of subjective assessments of probabilities. We shall then return to the EU Hypothesis in the situation where the probabilities are subjective. We shall study the representation (7.23), (7.24) in the situation where probabilities are not known, and mention some of the foundational work of de Finetti [1931, 1937], Ramsey [1931], Savage [1954], and von Neumann and Morgenstern [1944]. Finally, we shall discuss recent results on the measurement of subjective probability, and modern applications of such measurement.

References

Aitchison, J., "Decision-Making in Clinical Medicine," *J Roy. Coll. Physicians, London*, **4** (1970), 195–202.

Allais, M., "Le comportement de l'homme rationnel devant le risque: Critique des postulats de l'école Americaine," *Econometrica*, **21** (1953), 503–546.

Baldwin, M. M. (ed.), "Portraits of Complexity: Applications of Systems Methodologies to Social Problems," Battelle Monograph, Battelle Memorial Institute, Columbus, Ohio, 1975.

Becker, G. M., and McClintock, C. G., "Value: Behavioral Decision Theory,", *Ann. Rev. Psychol.*, **18** (1967), 239–286.

Bell, D. E., "Evaluating Time Streams of Income," *Omega*, **2** (1974), 691–699.

Bell, D. E., "A Decision Analysis of Objectives for a Forest Pest Problem," RR-75-43, International Institute for Applied Systems Analysis, Laxenburg, Austria, 1975.

Bernoulli, D., "Specimen Theoriae Novae de Mensura Sortis," *Comentarii Academiae Scientiarum Imperiales Petropolitanae*, **5** (1738), 175–192, Translated by L. Sommer in *Econometrica*, **22** (1954), 23–36.

Blackwell, D., and Girshick, M. A., *Theory of Games and Statistical Decisions*, Wiley, New York, 1954.

Busacker, R. G., and Saaty, T. L., *Finite Graphs and Networks*, McGraw-Hill, New York,1965.

Collins, J. P., "The Development of a Solid Waste Environmental Evaluation Procedure," Doctoral Dissertation, Department of Civil Engineering, University of Michigan, Ann Arbor, Michigan, 1974.

Coombs, C. H., Dawes, R. M., and Tversky, A., *Mathematical Psychology: An Elementary Introduction*, Prentice-Hall, Englewood Cliffs, New Jersey, 1970.

Coombs, C. H., and Huang, L. C., "Tests of the Betweenness Property of Expected Utility," *J. Math. Psychol.*, **13** (1976), 323–337.

Coombs, C. H., and Komorita, S. S., "Measuring Utility of Money Through Decisions," *Amer. J. Psychol.*, **71** (1958), 383–389.

Cramer, H., "A Theorem on Ordered Sets of Probability Distributions," *Theory of Probability and Its Applications*, **1** (1956), 16–21.

Davidson, D., Suppes, P., and Siegel, S., *Decision-Making: An Experimental Approach*, Stanford University Press, Stanford, California, 1957.

de Finetti, B., "Sul Significato Soggettivo della Probabilita," *Fund. Math.*, **31** (1931), 298–329.

de Finetti, B., "La Prevision: Ses Lois Logiques, Ses Sources Subjectives," *Ann. Inst. H. Poincaré*, **7** (1937), 1–68. Translated into English in H. E. Kyburg and H. E. Smokler (eds.), *Studies in Subjective Probability*, Wiley, New York, 1964, pp. 93–158.

de Neufville, R., and Keeney, R. L., "Use of Decision Analysis in Airport Development for Mexico City," in A. W. Drake, R. L. Keeney, and P. M. Morse (eds.), *Analysis of Public Systems*, M.I.T. Press, Cambridge, Massachusetts, 1972.

de Neufville, R., and Keeney, R. L., "Multiattribute Preference Analysis for Transportation Systems Evaluation," *Transportation Res.*, **7** (1973), 63–76.

Edwards, W., "The Theory of Decision Making," *Psychol. Bull.*, **51** (1954), 380–417.

Edwards, W., "Behavioral Decision Theory," *Ann. Rev. Psychol.*, **12** (1961), 473–498.

Ellis, H. M., "The Application of Decision Analysis to the Problem of Choosing an Air Pollution Control Program for New York City," Doctoral Dissertation, Graduate School of Business Administration, Harvard University, Cambridge, Massachusetts, 1970.

Ellis, H. M., and Keeney, R. L., "A Rational Approach for Government Decisions Concerning Air Pollution," in A. W. Drake, R. L. Keeney, and P. M. Morse (eds.), *Analysis of Public Systems*, M.I.T. Press, Cambridge, Massachusetts, 1972.

Farquhar, P. H., "Fractional Hypercube Decompositions of Multiattribute Utility Functions," Tech. Rept. 222, Department of Operations Research, Cornell University, Ithaca, New York, August 1974.

Farquhar, P. H., "A Fractional Hypercube Decomposition Theorem for Multiattribute Utility Functions," *Operations Research*, **23** (1975), 941–967

Farquhar, P. H., "Pyramid and Semicube Decompositions of Multiattribute Utility Functions," *Operations Research*, **24** (1976), 256–271.

Farquhar, P. H., "A Survey of Multiattribute Utility Theory and Applications," in M. Starr and M. Zeleny (eds.), *TIMS / North Holland Studies in the Management Sciences: Multiple Criteria Decisionmaking*, **6** (1977), 59–89.

Fishburn, P. C., "Independence in Utility Theory with Whole Product Sets," *Operations Research*, **13** (1965), 28–45.

Fishburn, P. C., "Methods for Estimating Additive Utilities, " *Management Sci.*, **13** (1967), 435–453.

Fishburn, P. C., "Utility Theory," *Management Sci.*, **14** (1969a), 335–378.

Fishburn, P. C., "A Study of Independence in Multivariate Utility Theory," *Econometrica*, **37** (1969b), 107–121.

Fishburn, P. C., *Utility Theory for Decision Making*, Wiley, New York, 1970.

Fishburn, P. C., "Bernoullian Utilities for Multiple-Factor Situations," in J. L. Cochrane and M. Zeleny (eds.), *Multiple Criteria Decision Making*, University of South Carolina Press, Columbia, South Carolina, 1973, pp. 47–61.

Fishburn, P. C., and Keeney, R. L., "Seven Independence Concepts and Continuous Multiattribute Utility Functions," *J. Math. Psychol.*, **11** (1974), 294–327.

Fishburn, P. C., and Keeney, R. L., "Generalized Utility Independence and Some Implications," *Operations Research*, **23** (1975), 928–940.

Fishburn, P. C., and Roberts, F. S., "Mixture Axioms in Linear and Multilinear Utility Theories," *Theory and Decision*, **9** (1978), 161–171.

Ford, L. R., Jr., and Johnson, S. M., "A Tournament Problem," *Am. Math. Monthly*, **66** (1959), 387–389.

Forst, B. E., "Decision Analysis and Medical Malpractice," Center for Naval Analyses, Arlington, Virginia, Professional Paper No. 92, May 1972.

Friedman, M., and Savage, L. J., "The Utility Analysis of Choices Involving Risk," *J. Polit. Econ.*, **56** (1948), 279–304.

Friedman, M., and Savage, L. J., "The Expected-Utility Hypothesis and the Measurability of Utility," *J. Polit. Econ.*, **60** (1952), 463–474.

Fryback, D. G., "Use of Radiologists' Subjective Probability Estimates in a Medical Decision Making Problem," Tech. Rept. MMPP 74-14, Department of Psychology, University of Michigan, Ann Arbor, Michigan, 1974.

Ginsberg, A. S., "Decision Analysis in Clinical Patient Management with an Application to the Pleural-Effusion Syndrome," Rept. R-751-RC/NLM, The RAND Corporation, Santa Monica, California, July 1971.

Grayson, C. J., *Decisions under Uncertainty: Drilling Decisions by Oil and Gas Operators*, Division of Research, Harvard Business School, Boston, Massachusetts, 1960.

Gudder, S. P., "Convex Structures and Operational Quantum Mechanics," *Comm. Mathematical Physics*, **29** (1973), 249–264.

Gudder, S. P, "Convexity and Mixtures," *SIAM Rev.*, **19** (1977), 221–240.

Herstein, I. N., and Milnor, J., "An Axiomatic Approach to Measurable Utility," *Econometrica*, **21** (1953), 291–297.

Horgan, D. N., Jr., "A Decision Analysis of Sewage Sludge Disposal Alternatives for Boston Harbor," Masters Thesis, Department of Electrical Engineering, Massachusetts Institute of Technology, Cambridge, Massachusetts, 1972.

Huang, L. C., "The Expected Risk Function," Tech. Rept. MMPP 71-6, Department of Psychology, University of Michigan, Ann Arbor, Michigan, July 1971.

Huber, G. P., "Multi-Attribute Utility Models: A Review of Field and Field-Like Studies," *Management Sci.*, **20** (1974), 1393–1402.

Kalelkar, A. S., Partridge, L. J., and Brooks, R. E., "Decision Analysis in Hazardous Material Transportation," *Proceedings of the 1974 National Conference on Control of Hazardous Material Spills*, American Institute of Chemical Engineers, San Francisco, California, 1974.

Keeney, R. L., "Evaluating Multidimensional Situations using a Quasi-Separable Utility Function," *IEEE Transactions on Man-Machine Systems*, MMS-9 (1968a), 25–28.

Keeney, R. L., "Quasi-Separable Utility Functions," *Naval Research Logistics Quarterly*, **15** (1968b), 551–565.

Keeney, R. L., "Utility Independence and Preferences for Multiatfributed Consequences," *Operations Res.*, **19** (1971), 875–893.

Keeney, R. L., "An Illustrated Procedure for Assessing Multiattributed Utility Functions," *Sloan Management Review*, 1972a, 38–50.

Keeney, R. L., "Utility Functions for Multiattribute Consequences," *Management Sci.*, **18** (1972b), 276–287.

Keeney, R. L., "A Decision Analysis with Multiple Objectives: The Mexico City Airport," *Bell J. Econ. Management Sci.*, **4** (1973a), 101–117.

Keeney, R. L., "A Utility Function for Response Times of Engines and Ladders to Fires," *Urban Analysis*, **1** (1973b), 209–222.

Keeney, R. L., "Concepts of Independence in Multiattribute Utility Theory," in J. L. Cochrane and M. Zeleny (eds.), *Multiple Criteria Decision Making*, University of South Carolina Press, Columbia, South Carolina, 1973c.

Keeney, R. L., "Multiplicative Utility Functions," *Operations Research*, **22** (1974), 22–34.

Keeney, R. L., and Raiffa, H., *Decisions with Multiple Objectives: Preferences and Value Tradeoffs*, Wiley, New York, 1976.

Knuth, D. E., *The Art of Computer Programing*, Vol. 3, Addison-Wesley, Reading, Massachusetts, 1973.

Ledley, R. S., and Lusted, L. B., "Reasoning Foundations of Medical Diagnosis," *Science*, **130** (1959), 9–21.

Lighthill, J. (ed.), *Newer Uses of Mathematics*, Penguin Books, Harmondsworth, Middlesex, England, 1978.

Luce, R. D., and Raiffa, H., *Games and Decisions*, Wiley, New York, 1957.

Luce, R. D., and Suppes, P., "Preference, Utility, and Subjective Probability," in R. D. Luce, R. R. Bush, and E. Galanter (eds.), *Handbook of Mathematical Psychology*, Vol. III, Wiley, New York, 1965, pp. 249–410.

Lusted, L. B., *Introduction to Medical Decision Making*, Thomas, Springfield, Illinois, 1968.

MacCrimmon, K. R., and Siu, J. K., "Making Tradeoffs," *Decision Sci.*, **5** (1974), 680–704.

MacCrimmon, K. R., and Wehrung, D. A., "Trade-off Analysis: Indifference and Preferred Proportion," in D. E. Bell (ed.), *Conflicting Objectives in Decisions*, Wiley, New York, 1978.

McGuire, C. B., and Radner, R., *Decision and Organization*, North-Holland, Amsterdam, 1972.

Marschak, J., "Rational Behavior, Uncertain Prospects, and Measurable Utility," *Econometrica*, **18** (1950), 111–141.

Mielnik, B., "Theory of Filters," *Comm. Mathematical Physics*, **15** (1969), 1–46.

Morrison, D. G., "On the Consistency of Preference in Allais' Paradox," *Behavioral Sci.*, **12** (1967).

Mosteller, F., and Nogee, P., "An Experimental Measurement of Utility," *J. Polit. Econ.*, **59** (1951), 371–404.

Pfanzagl, J. *Theory of Measurement*, Wiley, New York, 1968.

Raiffa, H., *Decision Analysis*, Addison-Wesley, Reading, Massachusetts, 1968.

Raiffa, H., *Preferences for Multi-Attributed Alternatives*, Memorandum RM-5868-DOT/RC, The RAND Corporation, Santa Monica, California, April 1969.

Ramsey, F. P., "Truth and Probability," in F. P. Ramsey, *The Foundations of Mathematics and Other Logical Essays*, Harcourt, Brace, New York, 1931. [Reprinted in H. E. Kyburg, and H. E. Smokler (eds.), *Studies in Subjective Probability*, Wiley, New York, 1964, pp. 61–92.]

Roche, J. G., "Preference Tradeoffs among Instructional Programs: An Investigation of Cost-benefit and Decision Analysis Techniques in Local Educational Decisionmaking," Doctoral Dissertation, Graduate School of Business Administration, Harvard University, Cambridge, Massachusetts, 1971.

Savage, L. J., *The Foundations of Statistics*, Wiley, New York, 1954.

Schelling, T., "The Life You Save May Be Your Own," in S. B. Chase, Jr. (ed.), *Problems in Public Expenditures*, Brookings Institution, 1968.

Schwartz, W. B., Gorry, G. A., Kassirer, J. P., and Essig, A., "Decision Analysis and Clinical Judgment," *Amer. J. Med.*, **55** (1973), 459–472.

Slack, W. V., "Patient Power: A Patient-Oriented Value System," in J. Jacquez (ed.), *Computer Diagnosis and Diagnostic Methods*, Thomas, Springfield, Illinois, 1972.

Spetzler, C. S., "The Development of a Corporate Risk Policy for Capital Investment Decisions," *IEEE Transactions on Systems Science and Cybernetics*, SSC-4 (1968), 279–300.

Steinhaus, H., *Mathematical Snapshots*, Oxford University Press, New York, 1950.

Stevens, S. S., "Measurement, Psychophysics, and Utility," in C. W. Churchman and P. Ratoosh (eds.), *Measurement: Definitions and Theories*, Wiley, New York, 1959, pp. 18–63.

Swalm, R. O., "Utility Theory—Insights into Risk Taking," *Harvard Business Review*, **44** (1966), 123–136.

Ting, H. M., "Aggregation of Attributes for Multiattributed Utility Assessment," Tech. Rept. 66, Operations Research Center, Massachusetts Institute of Technology, Cambridge, Massachusetts, August 1971.

von Neumann, J., and Morgenstern, O., *The Theory of Games and Economic Behavior*, Princeton University Press, Princeton, New Jersey, 1944, 1947, 1953.

Warfield, J. N., "An Assault on Complexity," Battelle Monograph, Battelle Memorial Institute, Columbus, Ohio, 1973.

Warfield, J. N., "Structuring Complex Systems," Battelle Monograph, Battelle Memorial Institute, Columbus, Ohio, 1974a.

Warfield, J. N., "Toward Interpretation of Complex Structural Models," *IEEE Trans. on Systems, Man, and Cybernetics*, SMC-4 (1974b), 441–449.

Warfield, J. N., *Societal Systems*, Wiley, New York, 1976.

Welford, B. P., "Statistical Decision Theory," *Bulletin IMA*, **6** (1970).

Wells, M. B., *Elements of Combinatorial Computing*, Pergamon Press, Oxford, 1971.

Yntema, D. B., and Klem, L., "Telling a Computer how to Evaluate Multidimensional Situations," *IEEE Transactions on Human Factors in Electronics*, HFE-6 (1965), 3–13.

CHAPTER 8

Subjective Probability

8.1 Objective Probability

At the end of the last chapter, we observed that sometimes we have some information or intuition about probabilities, but do not know them explicitly or have no way to calculate them. In this chapter we shall discuss how to "measure" our subjective or qualitative notions of probability. For comparison's sake, we shall briefly review in this section the classical definition of probability, which we call *objective probability*. (The reader may skip this section if desired.) In the next section, we constrast the situation where this notion of probability applies with the situation where subjective probability is appropriate.

Often in science we are concerned with laws of the form: if a certain set of conditions $\mathcal{C}$ is satisfied, for example if a certain experiment is performed, then an event A will take place. For example, we assert that if water is heated to a certain temperature, it will turn to steam; if supply is greater than demand, then prices will decrease; etc. Similarly, we assert laws to the effect that if conditions $\mathcal{C}$ are realized, an event A will not take place. If the occurrence of event A is inevitable whenever $\mathcal{C}$ is realized, we shall speak of A as *certain* (relative to $\mathcal{C}$). If it inevitably will not occur, we speak of A as *impossible*.

Often we can only assert that event A is likely (or unlikely) to occur if $\mathcal{C}$ is realized. Thus, for example, we might say that if two sounds are sufficiently different in intensity, then it is likely that people will be able to distinguish them. We are speaking of an event that is neither certain nor impossible. Sometimes conditions $\mathcal{C}$ can be realized very often (as in the case of heating water), and we can observe that, as a rule, event A occurs a certain proportion of the time. For example, we cannot predict whether a

ENCYCLOPEDIA OF MATHEMATICS and Its Applications, Gian-Carlo Rota (ed.).
Vol. 7: Fred S. Roberts, Measurement Theory

given atom of radium will disintegrate during a given interval of time of length t. But we can say that if we observe enough time periods, and before each one we specify a particular atom, then the proportion of such periods in which the specified atom will disintegrate is approximately $p = 1 - e^{-\alpha t}$, where α is the rate of decay. In situations such as these, it makes sense to speak of the (objective) *probability* of the event A, thinking of the probability as the relative frequency or proportion of occurrences of A if conditions $\mathcal{C}$ are realized a large number of times.

This notion of *objective probability* goes back to Pascal and Fermat, who corresponded about gambling in the seventeenth century. It has been well-developed and formalized in much the following way. Let X be a set of *outcomes* $a, b, c, \ldots$, each of which either occurs or fails to occur whenever the fixed set of conditions $\mathcal{C}$ occurs. We call certain subsets of outcomes *events* and let $\mathcal{E}$ denote the set of events. A set consisting of a single outcome can qualify as an event. So can a more complicated set. We assume that if A and B are in $\mathcal{E}$, then

$$A \cup B \in \mathcal{E} \tag{8.1}$$

$$A \cap B \in \mathcal{E} \tag{8.2}$$

$$A - B \in \mathcal{E} \tag{8.3}$$

$$A^c \in \mathcal{E}. \tag{8.4}$$

$A \cup B$ is the event that either A or B occurs, $A \cap B$ is the event that both A and B occur, $A - B$ is the event that A occurs but B does not, and A^c is the event that A does not occur.

A collection $\mathcal{E}$ will be called an *algebra of subsets* if it is nonempty and satisfies Eqs. (8.1) and (8.4). It is easy to see that an algebra of subsets must also satisfy (8.2) and (8.3). Moreover, since $\mathcal{E}$ is not empty, there is some A in $\mathcal{E}$. It follows that A^c is in $\mathcal{E}$, and hence so is $X = A \cup A^c$. X corresponds to the event that some outcome will occur. We shall assume that on any given realization of the conditions $\mathcal{C}$, one and only one of the outcomes in X occurs. The empty set $\varnothing$ is an event, since $\varnothing = A \cap A^c$ for any event A. The empty event $\varnothing$ is the event of none of the outcomes in X occurring. The empty event is impossible: we assume that on each realization of the conditions $\mathcal{C}$, one of the outcomes in X occurs.

We measure the probability of an event $A \in \mathcal{E}$ (the probability that one of the outcomes in A occurs) as a real number $p(A)$. These probabilities are assumed to have the following properties:

$$p(A) \geqq 0 \tag{8.5}$$

$$p(X) = 1 \tag{8.6}$$

$$A \cap B = \varnothing \Rightarrow p(A \cup B) = p(A) + p(B). \tag{8.7}$$

A function $p: \mathcal{E} \to Re$ satisfying (8.5), (8.6), and (8.7) is called a *probability measure*. The triple $(X, \mathcal{E}, p)$ is called a (*finitely*) *additive probability space* if X is a nonempty set, $\mathcal{E}$ is an algebra of subsets of X, and p is a probability measure on $\mathcal{E}$. This axiomatic definition of (objective) probability was first stated explicitly by Kolmogorov [1933]. The reader is referred to any standard treatment of probability (for example, Feller [1962]) for a more detailed discussion of the definition and its implications.

The important point to make about objective probability is that, in principle, it is universal and replicable. Namely, different individuals in different parts of the world computing the objective probability $p(A)$ of an event A should, if they agree on certain basics (the probability of certain elementary events), agree on $p(A)$.

Exercises

1. Show that the following are algebras of subsets under the usual notions of union and complementation:
   (a) $X = Re$, $\mathcal{E}$ = all subsets of X.
   (b) $X = \{1, 2, \ldots, n\}$, $\mathcal{E}$ = all subsets of X.
   (c) X = any infinite set, $\mathcal{E}$ = all subsets that are finite or whose complement is finite.
   (d) X = any uncountably infinite set, $\mathcal{E}$ = all subsets that are countable or whose complement is countable.
   (e) X = any nonempty set, $\mathcal{E}$ = all subsets of X that contain a given nonempty set A or are disjoint from A.

2. Show that the following define finitely additive probability spaces:
   (a) $(X, \mathcal{E})$ as in (b) of Exer. 1, and $p(A) = |A|/n$.
   (b) $(X, \mathcal{E})$ as in (c) of Exer. 1 and

$$p(A) = \begin{cases} 0 & \text{if } A \text{ is finite,} \\ 1 & \text{if } A^c \text{ is finite.} \end{cases}$$

1. (c) $(X, \mathcal{E})$ as in (d) of Exer. 1 and

$$p(A) = \begin{cases} 0 & \text{if } A \text{ is countable,} \\ 1 & \text{if } A^c \text{ is countable.} \end{cases}$$

3. Let X be a set of more than two elements and A a nonempty subset of X. Let $\mathcal{E}$ consist of $\varnothing$ and all subsets of X that intersect both A and A^c. Show that $(X, \mathcal{E})$ is not an algebra of subsets.

8.2 Subjective Probability

In a comprehensive study, Smil [1972, 1974a,b] asked a series of experts to estimate the probability of occurrence in the 1970's of certain "environmental episodes." For example, he asked them the probability of the event

A, a severe urban air pollution episode lasting several days with significant
consequences. And he asked them the probability of event B, a widespread
failure of power supply in a populated, industrial region, lasting several
hours. (Other events discussed by Smil are shown in Table 8.1.) Now for
such estimates of probability, the frequency interpretation of Section 8.1
does not make much sense. It hardly is reasonable to expect to realize the
conditions $\mathcal{C}$ (being in the 1970's) a large number of times. Rather, here,
the estimates of probability have a somewhat different interpretation. They
represent the "degree of certainty" or "degree of conviction" that the
expert has that an event will occur. They represent what is often called his
subjective probability of occurrence. Subjective probabilities do not have
the universal and replicable character that objective probabilities do.
Different people can have totally different subjective estimates of the
probability of an event, and an individual can from time to time change
his own estimate. For a further discussion on objective versus subjective
probability, see for example Carnap [1950], de Finetti [1937], Fishburn
[1964], Kemeny [1959], Keynes [1921], Nagel [1939], Savage [1954], or
Kyburg and Smokler [1964], where many basic papers have been collected.

Table 8.1. Probabilities of Environmental Episodes in the 1970's*
(Median and quartiles in percents.)

Episode	Lower Quartile	Median	Upper Quartile	Mode
Severe urban air pollution episode lasting several days with significant consequences	40	90	100	100
Widespread failure of power supply in populated, industrial region, lasting several hours	50	70	100	100
Catastrophe of fully loaded jumbo tanker (over 100,000 dwt) and spill of crude oil in the open sea	25	70	95	100
Serious oil spill from offshore drilling operation causing ecological disturbance over a large area	20	50	75	20
Radioactive contamination of environment outside of a reactor building caused by failure of nuclear plant protective systems	5	5	10	5

*Adapted from Smil [1974b] with permission of the author.

Our discussion is based on the observation that an individual often makes subjective estimates of the probabilities of certain events, or at least comparisons of the subjective probability of one event to that of another, even in situations where repetitions make sense and objective probability can be calculated. Subjective probability plays a role in a variety of applications—for example, in medical decisionmaking (Betaque and Gorry [1971], Fryback [1974]), in weather forecasting (Murphy and Winkler [1974], Peterson *et al.* [1972], Sanders [1963, 1967, 1973], Staël von Holstein [1971], Winkler and Murphy [1968a]), in assessing stock market trends (Bartos [1969], Staël von Holstein [1970a, 1972]), in production planning (of electricity and the like) (Kidd [1970]), and in predicting future sociopolitical events (Brown [1973]). Subjective probability judgments often have little connection with objective probability, as we shall see. On the other hand, since judgments of subjective probability are often made, especially in the social and policy sciences, it is incumbent upon us to understand the nature of subjective probability judgments, and if possible, to put these on a firm measurement-theoretic foundation much as judgments of preference were in earlier chapters.

We shall describe three approaches to the measurement of subjective probability. The first is simple direct estimation: Ask the expert (or an individual) to assign to each event a number that represents his subjective estimate of the probability of that event. An example of this procedure is the data gathered by Smil [1972] to which we have already referred. A variant of this approach involves direct estimation of odds, rather than of probabilities. Other variants ask for confidence intervals on probabilities, or ask for 50–50 points (points above which or below which the outcome is equally likely to lie), quartiles or fractiles, and so on. The second approach to the measurement of subjective probability uses preferences among complex choices or lotteries, where different prizes are given depending on the event which occurs after a certain experiment is performed. If utilities of the prizes are known, subjective probabilities of the occurrence of different events are calculated using the (Subjective) Expected Utility Hypothesis. If utilities are not known, subjective probabilities are estimated indirectly, by reducing to the third approach. This third approach is to ask an individual to make judgments comparing his subjective probabilities of two events. Thus, we deal with a binary relation $\succ$ on the set of events $\mathscr{E}$, with $A \succ B$ interpreted to mean that "A is judged more probable than B." The binary relation $\succ$ is called the *comparative probability relation* or the *qualitative probability relation*. Usually we assume that the set of events $\mathscr{E}$ forms an algebra of subsets of some set X of outcomes, that is, $(X, \mathscr{E})$ satisfies conditions (8.1) and (8.4). We try to assign numbers $p(A)$ to each event $A \in \mathscr{E}$ so that for all $A, B \in \mathscr{E}$,

$$A \succ B \Leftrightarrow p(A) > p(B). \qquad (8.8)$$

Usually we also ask that p satisfy Eqs. (8.5) to (8.7), that is, that $(X, \mathcal{E}, p)$ be a (finitely) additive probability space. If $(X, \mathcal{E}, p)$ satisfies Eqs. (8.8) and (8.5) to (8.7), then the function p is called a *measure of subjective probability* or a *measure of qualitative probability*. The important question, from a measurement point of view, is to find (necessary and) sufficient conditions on the structure $(X, \mathcal{E}, \succ)$ for the existence of a subjective probability measure. In the remainder of this chapter, we discuss these three approaches to the measurement of subjective probability. For a recent survey of the subjective probability literature, see Hogarth [1975]. Other surveys are Hampton *et al.* [1973], Huber [1974], Savage [1971], and Staël von Holstein [1970a,b].

8.3 Direct Estimation of Subjective Probability

Various studies are asking for direct estimates of (subjective) probabilities. For example, in the study of Smil [1972, 1974a,b], each expert referred to above was asked to state, for each of a number of "environmental episodes," his opinion of the probability of its occurrence during the 1970's. (The pooled data is shown in Table 8.1.) In another part of the same study, Smil listed a series of major scientific, technological, and management inventions, breakthroughs, and changes in the fields of energy systems and environmental protection and asked each member of his panel to indicate his opinion of the probability of practical implementation during certain given periods. (The pooled results of this experiment are listed in the last column of Table 8.2.)

Similarly, Fryback [1974] asked several radiologists to study excretory urograms of the urinary tract and to estimate the probability that there was a benign cyst, a malignant tumor, or a normal variant. Sanders [1963] asked forecasters to estimate the probability of various weather phenomena, such as wind reaching a certain speed, temperature changing by a certain amount, or rain falling within a prescribed period.

Direct subjective estimation of probabilities may be unavoidable in many studies, as, for example, in those of Smil or Fryback. However, the literature seems to show that when objective probability can be calculated, direct subjective probability estimates often differ significantly from objective probabilities.* There are well-known examples of the divergence between subjective and objective probability in most basic books on probability. The classic example is the *birthday problem*. In a group of 23 people, the probability that at least two people have the same birthday is greater than $1/2$, while most people guess it is much lower than $1/2$. Another common example is the so-called *gambler's fallacy*, namely, that

---

*Similar observations hold true for direct estimates of confidence intervals, means, variances, quartiles, and so on. For a discussion, see, for example, Hogarth [1975].

Table 8.2. Probability of Practical Implementation of Various Energy Systems Breakthroughs*

(The dates $a–b–c$ in the last column in the row corresponding to event A are interpreted as follows: 25% of the respondents (the optimists) felt that there was at least a 50% chance that event A would occur before year a; 25% of the respondents (the pessimists) believed that this was a possibility (of probability $\geqq$ 50%) only after the year c; and 50% of the respondents believed this was a possibility (of probability $\geqq$ 50%) first between the years a and c, with a median date of b. Items are ordered according to median date.)

Number	Item	Quartiles
1.	Fuel cells for small-scale power generation	1980–1980–1987
2.	Use of nuclear explosives in the production of natural gas and oil, geothermal heat, etc.	1980–1980–1993
3.	Coal gasification or liquefaction	1979–1982–1984
4.	"Fail-safe" nuclear power generation	1976–1983–1995
5.	High-temperature gas reactors (A-K cycle)	1979–1984–1990
6.	Extra-high-voltage transmission on very long distances (at least 1000 kV and 1000 km)	1979–1985–1990
7.	Fast breeder reactors	1981–1985–1990
8.	Cryogenic transmission systems using underground superconducting cables	1983–1985–1995
9.	Large-scale shale oil recovery	1983–1986–1996
10.	Fossil fuel-fired magnetohydrodynamics	1981–1988–1990
11.	Development of all practically feasible hydroelectric sites in populated regions	1982–1988–2000
12.	Techniques for economical recovery of additional 25% of crude oil from known resources	1983–1988–1998
13.	Fully automated underground coal mining	1983–1988–2000
14.	Cryogenic pipeline transportation of natural gas	1986–1988–2000
15.	Simple solar furnace for home power generation in tropical and subtropical regions	1986–1990–2000
16.	Low-cost high-voltage underground transmission	1988–1990–2000
17.	Microwave power transmission	1990–1993–2000
18.	"Fail-safe" systems for drilling and producing hydrocarbons at any water depth	1987–1995–2002
19.	Direct conversion—thermionics	1985–1998–2010
20.	Utilization of low thermal difference systems	1990–1999–Never
21.	Controlled thermonuclear power	1990–2000–2000
22.	Efficient storage of electric energy in large quantities	1990–2000–2010
23.	Laser power transmission	1990–2000–2010
24.	Large and efficient tidal power plants	1992–2000–Never
25.	High-temperature gas reactors with thermal cycle other than helium	2010–2010–2020
26.	Widespread use of geothermal power	1990–2020–Later
27.	Relay of solar energy via satellite collectors	2000–2020–Later
28.	Solar energy devices for bulk power generation	2000–Later–Never
29.	Cryogenic superfluid transportation of mechanical energy on long distances	2020–Later–Never
30.	Utilization of gravitational energy (anti-gravity)	Later–Later–Never

*Adapted from Smil [1974b] with permission of the author.

after a sequence of heads, a tail on the next trial becomes more likely. Kahneman and Tversky [1972] give many other examples, taken from a series of experiments. In one experiment, 75 out of 92 subjects said that in a survey of families with six children, the exact order of birth of boys and girls would be more likely to be GBGBBG than BGBBBB. (These sequences are about equally likely, though not exactly since the number of boys and girls born is not equal.) The subjects were told that 72 families in a given city had the first sequence of children and were asked to estimate how many families had the second sequence. The median estimate was 30. In another experiment of Kahneman and Tversky, subjects were asked the probability of finding more than 600 boys in a sample of 1000 babies, and the probability of finding more than 60 boys in a sample of 100. These probabilities were judged equal, though the latter is much more likely. In general, in comparable probability problems, subjects made estimates which were independent of sample size, violating properties of objective probability. In other studies, subjects have tended to overestimate the joint probability of independent events, and to underestimate the probabilities of disjunctive events such as "at least one . . . ." (See, for example, Bar-Hillel [1973].)

Studies of subjective probability which ask subjects to estimate subjective probability indirectly seem to confirm these observations. Preston and Baratta [1948] made the first attempt to measure subjective probability experimentally, using a simple auction game. They found that small objective probabilities (less than 0.20) were systematically overestimated and others (greater than 0.20) were systematically underestimated. Other studies using indirect estimates exhibit different types of differences between subjective and objective probabilities, though the Preston–Baratta results have been obtained in a variety of studies. For a summary of such studies, the reader is referred to Hogarth [1975, p. 274] and Luce and Suppes [1965, Section 4.3].

There are several serious problems with direct assessment of subjective probability. How one asks for the assessment can lead to significant discrepancies. For example, asking for odds can lead to different results than asking for numbers between 0 and 1. Offering payoffs can affect direct estimates (Phillips and Edwards [1966]). Betaque and Gorry [1971] discuss the effects of the likelihood of mistreatment on doctors' estimates of probabilities in diagnosis. In spite of these problems, direct assessment of probabilities is becoming an increasingly popular decisionmaking tool in government and industry, and so procedures for making these estimates have to be better understood in the future.

Before leaving this section, we should ask what kind of a scale direct estimates of subjective probability define. As we mentioned in Chapter 2, where there is no obvious representation, admissible transformations must be defined as functions preserving the empirical information depicted by a scale, and hence scale type is not formally defined, but depends on our interpretation of what the empirical information content of the scale is. There does not seem to be a treatment of scale type in the literature of direct estimates of subjective probability. However, a discussion of scale type will be important in analyzing studies such as Smil's. We begin some of this analysis here.

It is tempting to suggest that direct estimates of subjective probability define an absolute scale. Both the zero point and the unit are fixed. However, if the sum of judged probabilities by an individual is not equal to unity, it may only make sense to treat the data as a ratio scale to be properly normalized.* In Section 8.5 we shall obtain some representation theorems for subjective probability. The representations will, in some cases, give rise to absolute scales. However, in other cases, they will not even lead to ratio scales.

Let us apply these observations. Many investigators seek a pooled probability assessment from a group of assessors. Following Stone [1961], the most frequent method used for pooling probability assessments is to take a weighted average

$$G(A) = \Sigma w_i p_i(A),$$

where $G(A)$ is the group estimate of the probability of A, $p_i(A)$ is the estimate of the probability of A by the ith assessor, and w_i is a weighting factor, with $w_i \geq 0$ and $\Sigma w_i = 1$. If all the weights are equal, this is the mean estimate. Smil uses $M(A)$, the median of the $p_i(A)$, as his pooled estimate.

Smil makes statements to the effect that the group's estimate of the probability of A is greater than its estimate of the probability of B. This is the statement

$$M(A) > M(B).$$

It is not hard to see that this comparison is meaningful if each p_i is the same ratio scale. Comparison of means is meaningful in this case also; that is, the statement

$$G(A) > G(B)$$

is meaningful. However, if we treat direct estimates of subjective probability as defining a ratio scale, then perhaps it is reasonable to allow different transformations of each expert's scale, in which case comparison of medians or of means is not meaningful.

In pooling the data of Table 8.2, Smil makes the statement, "25% of the $p_i(A)$ are at least 1/2." This statement is meaningless, even if each p_i is the

*Amos Tversky (personal communication).

same ratio scale. However, it is meaningful if each p_i is considered an absolute scale, in which case no admissible transformations (but the identity) are allowed.

Exercises

1. (a) Show that the statement $M(A) > M(B)$ is meaningless if the p_i are possibly different ratio scales.
   (b) Do the same for the statement $G(A) > G(B)$.

2. Consider the meaningfulness of the following statements based on Table 8.1.
   (a) The median probability estimate of a severe air pollution episode was higher than the median estimate of a widespread power failure.
   (b) The former median was less than twice the latter median.
   (c) The lower quartile estimate of the probability of a serious oil spill was less than the median estimate of this probability.
   (d) The median estimate of the probability of a widespread power failure was equal to the median estimate of the probability of a catastrophe with a fully loaded tanker.

3. Consider the meaningfulness of the following statements based on Table 8.2.
   (a) The pessimistic date 1987 for fuel cells for small-scale power generation (item 1) is earlier than the pessimistic date 2000 for low-cost high-voltage underground transmission (item 16).
   (b) The median ("b") date for fail-safe nuclear power generation (item 4) is earlier than that for laser power transmission (item 23).

4. In their assessment of alternatives for Mexico City's future airport expansion, de Neufville and Keeney [1972] and Keeney [1973] obtained many judgments by the Ministry of Public Works. A typical statement was the following: If no new location is chosen for the airport, then by 1975, the probability that the number of people impacted by a noise level of at least 90 CNR (a measure of noise impact) will be less than 640,000 is one-half. Discuss the meaningfulness of this statement (assuming CNR defines an absolute scale).

5. If an individual makes many direct estimates of probability of (similar) events, one would like a way to "score" how well he does against the objective probability or real frequency of occurrence. One *scoring rule* due to Brier [1950] and Brier and Allen [1951], which has been used in weather forecasting and elsewhere (e.g., Sanders [1963]), goes as follows. Let

$$B^{(j)} = \frac{1}{N} \sum_{i=1}^{N} \left(f_i^{(j)} - O_i \right)^2,$$

where $f_i^{(j)}$ is the forecast probability by the jth forecaster that event i will occur (or the event in question will occur on the ith occasion), O_i is 1 if the ith event occurs (or the event in question occurs on the ith occasion), and 0 otherwise, and N is the number of forecasts made by the jth forecaster.

The *Brier Score* $B^{(j)}$ is 0 if the forecaster is perfectly correct, and 1 if the forecaster is as wrong as can be. (For a discussion of other scoring rules, see Hogarth [1975], Winkler [1969], Winkler and Murphy [1968b], and de Finetti [1962].*)

The Brier Score has been used to measure performance of weather forecasters who deal with such problems as the probability that the wind speed will reach a certain critical level within a certain time, that the temperature will change by at least a certain amount within a certain time, that it will rain within a certain time, etc.

(a) Is it meaningful to say that one individual forecaster scores better than another, that is, that $B^{(j)} < B^{(k)}$?

(b) Is it meaningful to say that the (arithmetic) mean score of one group of forecasters is better than the (arithmetic) mean score of another group?

(c) Let $\bar{f_i}$ be the arithmetic mean of the group estimates $f_i^{(j)}$ over j, and let

$$\bar{B} = \frac{1}{N} \sum_{i=1}^{N} (\bar{f_i} - O_i)^2$$

be the group's score, if the group mean is taken as the group forecast. Sanders [1963] claims that, from data he analyzes, the group score is better than the mean of the individual scores; that is,

$$\bar{B} < \frac{1}{K} \sum_{j=1}^{K} B^{(j)},$$

where K is the number of assessors in the group. In fact, $\bar{B}$ is better than each $B^{(j)}$ and $\bar{B}$ is more than 5% better than the best individual score. He takes this as evidence that groups make better forecasts than individuals. Consider the meaningfulness of these assertions.

8.4 Subjective Probability from Preferences among Lotteries: The SEU Hypothesis

8.4.1 *If Utility Is Known*[†]

Sometimes it is possible to estimate subjective probabilities from preferences among choices, acts, or lotteries. The procedure we shall describe is used frequently in experimental situations, going back to the work of Davidson, Suppes, and Siegel [1957]. The basic idea goes back to Ramsey

---

*Much of the literature of scoring rules has been concerned with the design of rules that encourage the assessor to be careful in his assessments and to report only his true beliefs, rather than to give answers that will improve his score.

[†]This subsection and the next are based on Luce and Suppes [1965, Section 3.3].

[1931]. Surveys of the approach can be found in Goodman [1973], Hampton *et al.* [1973], Huber [1974], Luce and Suppes [1965], and Staël von Holstein [1970a]. We shall make use of the version of the Expected Utility Hypothesis which uses subjective probabilities. This *Subjective EU Hypothesis*, or *SEU Hypothesis*, says that individuals make choices *as if* they were subjectively calculating both probabilities and utilities, and choosing that alternative with the larger (subjective) expected utility. We make use of the concepts and notation of Section 7.2.1.[†]

Suppose we wish to estimate the probability $p(A)$ of an event A. We shall first see how to do it if we know how to measure the utility of consequences. Suppose we compare the two choices ℓ and ℓ' described as follows:

ℓ : either event A occurs or it does not. If A occurs, the outcome is c, and if A does not occur, the outcome is c'.

ℓ': either event A occurs or it does not. If A occurs, the outcome is d, and if it does not occur, the outcome is d'.

If you are indifferent between ℓ and ℓ', then it follows by the SEU Hypothesis that $E(\ell) = E(\ell')$, that is, that

$$p(A)u(c) + p(A^c)u(c') = p(A)u(d) + p(A^c)u(d'),$$

where p gives the subjective probability we are trying to discover, and u is the known utility. Assuming that $u(c) - u(c') + u(d') - u(d) \neq 0$, and using $p(A^c) = 1 - p(A)$, we conclude that

$$p(A) = \frac{u(d') - u(c')}{u(c) - u(c') + u(d') - u(d)}. \tag{8.9}$$

Thus, if consequences c, c', d, and d' can be found so that choices ℓ and ℓ' are judged indifferent and $u(c) - u(c') + u(d') - u(d) \neq 0$, then subjective probabilities can be calculated from known utilities, using Eq. (8.9). Another method for finding subjective probabilities from known utilities is described in Exer. 1 below.

8.4.2 *If Utility Is Not Known*

There is an alternative, not quite so direct approach to determine subjective probabilities from preferences among choices, which does not require explicit knowledge of a utility function over consequences. This procedure is to derive the comparative probability relation $\succ$ of Section

---

[†]We shall not distinguish between the Expected Utility Hypothesis and the Expected Value Hypothesis, or between utility functions and value functions.

8.2 using the SEU Hypothesis and preferences among acts or choices, and then to use the methods of Section 8.5 to calculate subjective probability from $\succ$. Suppose A and A' are two disjoint events and B is the event $(A \cup A')^c$. Let c, c', and d be consequences such that c is preferred to c'. Note that we do not need to know how to calculate utilities over consequences to find such c and c'. However, we do know that if u is a utility function over K, the set of consequences, then $u(c) - u(c') > 0$. Let choices ℓ and ℓ' be defined as follows:

ℓ : either event A, event A', or event B occurs, and the consequences are, respectively, c, c', and d.

ℓ': either event A, event A', or event B occurs, and the consequences are, respectively, c', c, and d.

Then if R is strict preference among choices, we have by the SEU Hypothesis:

$$\ell R \ell' \Leftrightarrow p(A)u(c) + p(A')u(c') + p(B)u(d) >$$
$$p(A)u(c') + p(A')u(c) + p(B)u(d)$$
$$\Leftrightarrow p(A)u(c) + p(A')u(c') > p(A)u(c') + p(A')u(c)$$
$$\Leftrightarrow p(A)[u(c) - u(c')] > p(A')[u(c) - u(c')]$$
$$\Leftrightarrow p(A) > p(A') \qquad (\text{since } u(c) - u(c') > 0)$$
$$\Leftrightarrow A \succ A'.$$

Thus, we can derive $\succ$ among disjoint events from preferences on choices (lotteries). From $\succ$ among disjoint events we can easily derive it among nondisjoint events, if we are willing to assume the following monotonicity axiom: if $A \cap B = A \cap C = \varnothing$, then

$$B \succ C \Leftrightarrow A \cup B \succ A \cup C.$$

We shall encounter this axiom again in the next section, where we shall discuss it. To see the role it plays, consider events D and E. We can write

$$D = (D \cap E) \cup (D \cap E^c)$$

and

$$E = (D \cap E) \cup (D^c \cap E).$$

By the monotonicity axiom,

$$D \succ E \Leftrightarrow D \cap E^c \succ D^c \cap E,$$

and the events $D \cap E^c$ and $D^c \cap E$ are disjoint. Thus, we can determine $\succ$ between all pairs of events. Note that in our determination of $\succ$, we do not really need to know the utilities of consequences, but only that there are consequences c and c' such that $u(c) > u(c')$, that is, such that c is preferred to c'.

8.4.3 *The Savage Axioms for Expected Utility*

In Section 7.5, we observed that often one must make choices between complex acts or lotteries when the probabilities are not known. We have already seen that we can derive subjective probabilities from preferences among such acts, provided we invoke the SEU Hypothesis and we know how to measure utility. We have also seen that we can derive the qualitative probability ordering, even if utilities are not known. One very interesting approach to utility theory states conditions on preferences among acts, which are sufficient to allow the derivation of *both* a subjective probability measure and a utility function, which together satisfy the SEU Hypothesis. This approach, which builds on earlier work of de Finetti [1931, 1937], Ramsey [1931], and von Neumann and Morgenstern [1944], is due to Savage [1954]. We briefly discuss it. Good treatments of Savage's approach can be found in Luce and Raiffa [1957] and Fishburn [1970].

Suppose $(X, \mathcal{E})$ is the algebra of events where $\mathcal{E}$ consists of all subsets of X, and K is a set of consequences. Generalizing our discussion of Chapter 7, an *act* or *choice* will be thought of as a function f which assigns to each element of X a consequence of K. A *finite* act or choice f is an act with the property that there is a partition $A_1, A_2, \ldots, A_n$ of X, with each $A_i \in \mathcal{E}$, and there are elements $c_1, c_2, \ldots, c_n$ of K, so that on A_i, f always assigns the consequence c_i. These are the types of acts or choices or lotteries we studied in Chapter 7. We shall denote such a finite act as

$$f(A_1, A_2, \ldots, A_n; c_1, c_2, \ldots, c_n).$$

Let $\mathcal{Q}$ be the set of acts and let R be a binary relation on $\mathcal{Q}$ interpreted as strict preference. Savage states axioms on the system $(X, \mathcal{E}, K, \mathcal{Q}, R)$ sufficient to guarantee the existence of real-valued functions p on $\mathcal{E}$ and u on K satisfying the following conditions:

$$(X, \mathcal{E}, p) \text{ is a finitely additive probability space} \qquad (8.10)$$

and

$$fRf' \Leftrightarrow \sum_{i=1}^{n} p(A_i)u(c_i) > \sum_{j=1}^{m} p(B_j)u(d_j), \qquad (8.11)$$

for all finite acts

$$f = f(A_1, A_2, \ldots, A_n; c_1, c_2, \ldots, c_n),$$

$$f' = f(B_1, B_2, \ldots, B_m; d_1, d_2, \ldots, d_m)$$

in $\mathcal{Q}$.* If utility is bounded, we may define expected utility by

$$EU(f) = \sum_{x \in X} p(\{x\}) u[f(x)].$$

As Fishburn [1970, p. 194] points out, Savage's axioms actually imply that utility is bounded, and for all $f, f' \in \mathcal{Q}$,

$$fRf' \Leftrightarrow EU(f) > EU(f').$$

Suppose $\succ$ is a binary preference relation on K. If we identify a consequence c with the act $f(X; c)$, then it is reasonable to assume that

$$c \succ c' \Leftrightarrow f(X; c) Rf(X; c').$$

To use a notion from Chapter 7, we say that $(\mathcal{Q}, R)$ *extends* $(K, \succ)$. If $(\mathcal{Q}, R)$ extends $(K, \succ)$, it follows from (8.10) and (8.11) that u is an order-preserving utility function[†] for $(K, \succ)$. Presentation of Savage's axioms would take us too far afield, and so we simply refer the reader to one of the references above for such a presentation. See also Exers. 8 and 9.

A different approach than Savage's, with much the same goals, can be found in Luce and Krantz [1971] (see Krantz *et al.* [1971, Chapter 8]).

Sometimes a representation (8.10), (8.11) is sought with utility u additive with respect to an operation o. Axioms for this representation and a related one are studied in Roberts [1974] and in Luce [1972].

Exercises

1. If utilities are known, and if A is an event whose probability is unknown, you can estimate the subjective probability $p(A)$ as follows: Find consequences c and c' and a lottery ℓ with known probabilities, which is judged indifferent to the lottery ℓ' which gives c if A occurs and c' if A does not occur. Show that $p(A)$ can be calculated from the EU Hypothesis and $E(\ell)$.

---

*The representation (8.11) alone can be looked at as a special case of the polynomial conjoint measurement representation studied in Section 5.5.1.

[†]See Section 7.2.2 for the definition of order-preserving utility function.

2. Suppose $u(\$n) = n$. Suppose you wish to estimate your subjective probability that it will be fair weather tomorrow. Show how you can do this if you are indifferent between the following lottery ℓ and the lottery that gives you $1 if it is fair tomorrow and takes $1 from you if it is not:

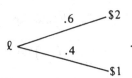

3. Suppose there are three candidates running for election, a Republican, a Democrat, and an Independent. Using dollar payoffs $100, $200, and $300, design lotteries to determine whether an individual thinks election of the Republican is more likely than election of the Democrat.

4. In the situation of Exer. 3, suppose $u(\$n) = n$. Discuss how to use the methods of Section 8.4.1 to determine the subjective probability of a Republican's being elected. ($\mathscr{E}$ can be taken to be all subsets of the set {Republican, Democrat, Independent}.)

5. Suppose $u(a) \neq u(b)$ and you are indifferent between the two acts $f[A^*, (A^*)^c; a, b]$ and $f[A^*, (A^*)^c; b, a]$, to use the notation of Section 8.4.3. Show from the SEU Hypothesis that you think A^* and $(A^*)^c$ are equally likely.

6. Show that under the SEU Hypothesis,

$$f\big[A^*, (A^*)^c; a, d\big] Rf\big[A^*, (A^*)^c; b, c\big] \Leftrightarrow u(a) - u(b) > u(c) - u(d).$$

(If u is unknown, we can define a quaternary relation D on the set of consequences K by

$$abDcd \Leftrightarrow f\big[A^*, (A^*)^c; a, d\big] Rf\big[A^*, (A^*)^c; b, c\big].$$

Then we can use the methods of Section 3.3 to find u. This idea goes back to Ramsey [1931].)

7. (Tversky [1967]) Let X be a set of events and C a set of monetary amounts. For every x in X and c in C, consider the act or choice or gamble that gives payoff c if x occurs and 0 if x does not occur. Let L be the set of all such gambles and R a binary relation of preference on L. Then L can be thought of as a Cartesian product $X \times C$. Show that there are functions $p: X \to Re$ and $u: C \to Re$ satisfying the SEU Hypothesis if and only if the pair $(L, R) = (X \times C, R)$ satisfies additive conjoint measurement (Section 5.4).

8. One of Savage's axioms is the following: Suppose $(\mathcal{Q}, R)$ extends $(K, \succ)$. Fix c and c' so that $c \succ c'$. Define $\succ^*$ on $\mathcal{E}$ by

$$A \succ^* A' \Leftrightarrow f(A, A^c; c, c') Rf[A', (A')^c; c, c'].$$

Let $A \succsim^* A'$ hold if and only if $\sim [A' \succ^* A]$. Then for all $A, A' \in \mathcal{E}$, either $A \succsim^* A'$ or $A' \succsim^* A$. Show that if $(\mathcal{Q}, R)$ extends $(K, \succ)$, then this axiom follows from the representation (8.10), (8.11). (The relation $\succ^*$ is the comparative probability relation.)

9. Another of Savage's axioms is the following: Suppose f and f' agree on A, g and g' agree on A, f and g agree on A^c, and f' and g' agree on A^c. Then

$$fRg \Leftrightarrow f'Rg'.$$

This says that preference between two acts should not depend on those outcomes that have identical consequences for the two acts. Show that this axiom follows from the representation (8.10), (8.11).

10. Consider the following acts:

$$f = f(C, D, E; a, b, b),$$
$$g = f(C', D, E'; a, b, b),$$
$$f' = f(C, D, E; a, a, b),$$
$$g' = f(C', D, E'; a, a, b).$$

Recall that C, D, E partition X and C', D, E' partition X. Show that according to the axiom in Exer. 9, with $A = (C \cap E') \cup (C' \cap E)$,

$$fRg \Leftrightarrow f'Rg'.$$

(However, as Krantz et al. [1971, p. 210] point out, if D adds more to C' than it does to C, we could have fRg and $g'Rf'$. They illustrate this point with the following game, adapted from Ellsberg [1961]. Consider three urns, one containing 200 white balls, one containing 200 black balls, and one containing 100 red balls. A coin is flipped, and, depending on its outcome, one of the first two urns is selected. Without informing the player, the balls from the selected urn are mixed with the balls from the third urn, the red balls. The player draws one ball from the mixture. Suppose a denotes a valuable prize, and b no prize. Consider the gambles with consequences based on the color of the ball drawn:

$$f = f(\text{Red, Black, White}; a, b, b)$$

and

$$g = f(\text{White, Black, Red}; a, b, b).$$

In each case, the (objective) probability of winning prize a is $1/3$. However, the player may prefer f to g. At least in f, he knows that after the toss of a coin, he has a chance of winning. In g, his action after the coin toss might have no effect, if the black urn had been chosen in the coin toss. Compare the gambles

$$f' = f(\text{Red, Black, White; } a, a, b)$$

and

$$g' = f(\text{White, Black, Red; } a, a, b).$$

Now, the (objective) probability of winning prize a is $2/3$ in each case. However, in f', the player's chances of winning with his own action, after the coin toss, can go as low as $1/3$ (if the white urn was chosen), while in g', his chances of winning by his own action are always $2/3$, regardless of the urn chosen as a result of the coin toss. Thus, he may prefer g' to f'. Both Ellsberg [1961] and Raiffa [1961] reported that sophisticated subjects (in informal questioning) had exactly the preferences fRg and $g'Rf'$.)

11. The Ellsberg example of Exer. 10 can be thought of as an argument against the additivity of subjective probability (and against the Monotonicity Axiom [Eq. (8.16)] we shall introduce in Section 8.5.1). For, let $A \cap B = A \cap C = \varnothing$, and consider

$$f = f(B, A, (A \cup B)^c; a, b, b)$$
$$g = f(C, A, (A \cup C)^c; a, b, b)$$
$$f' = f(B, A, (A \cup B)^c; a, a, b)$$
$$g' = f(C, A, (A \cup C)^c; a, a, b).$$

Then show that fRg can be interpreted as "B is subjectively more probable than C" and $g'Rf'$ can be interpreted as "$A \cup C$ is subjectively more probable than $A \cup B$."

8.5 The Existence of a Measure of Subjective Probability

8.5.1 *The de Finetti Axioms*

Let us return to the algebra of events $(X, \mathcal{E})$ and the binary relation of comparative probability $\succ$ on $\mathcal{E}$. The comparative probability relation $\succ$ was apparently first studied by Bernstein [1917]. Later references include Keynes [1921], de Finetti [1931, 1937], Koopman [1940a,b], Carnap [1950], Savage [1954], Suppes [1956], Kraft, *et al* [1959], Scott [1964], Luce [1967, 1968], Domotor [1969], Fishburn [1969, 1970, 1975], Fine [1971a, b, 1973, 1977], Kaplan [1971, 1974], Roberts [1973], Narens [1974],

Fine and Gill [1976], Suppes and Zanotti [1976] and Fine and Kaplan [1977].

We shall seek conditions on $(X, \mathcal{E}, \succ)$ sufficient to guarantee the existence of a subjective probability measure p, that is, a real-valued function p on $\mathcal{E}$ so that $(X, \mathcal{E}, p)$ is a finitely additive probability space and so that for all A, B in $\mathcal{E}$,

$$A \succ B \Leftrightarrow p(A) > p(B). \tag{8.8}$$

(Some authors have studied a weakening of condition (8.8) to the representation

$$A \succ B \Rightarrow p(A) > p(B).$$

For references, see for example Savage [1954], Fine [1973], Fishburn [1969, 1975] and the recent work of Cohen [1978].) Calculation of a subjective probability measure based on judgments of comparative probability is again subject to the difficulties we have discussed with the other methods: how questions are asked, availability of payoffs, and extraneous information can affect answers.

We shall use the notation

$$A \succsim B \Leftrightarrow \sim (B \succ A) \tag{8.12}$$

and

$$A \sim B \Leftrightarrow \sim (A \succ B) \,\&\, \sim (B \succ A) \tag{8.13}$$

$A \succsim B$ means that A is judged at least as probable as B, and $A \sim B$ means A and B are judged equally probable.

The following conditions on $(X, \mathcal{E}, \succ)$ are necessary for the existence of a subjective probability measure: for all A, B, and C in $\mathcal{E}$,

$$(\mathcal{E}, \succ) \text{ is a strict weak order,} \tag{8.14}$$

$$X \succ \varnothing \text{ and } A \succsim \varnothing, \tag{8.15}$$

$$\text{If } A \cap B = A \cap C = \varnothing, \text{ then } B \succ C \Leftrightarrow A \cup B \succ A \cup C. \tag{8.16}$$

Condition (8.16) is sometimes called a *Monotonicity Axiom*, and it is similar to the Monotonicity Axiom we encountered in studying extensive measurement in Section 3.2. That Condition (8.14) is necessary follows directly from Eq. (8.8). To see that Condition (8.15) is necessary, note that

$$1 = p(X) = p(X \cup \varnothing) = p(X) + p(\varnothing) = 1 + p(\varnothing),$$

so

$$p(\varnothing) = 0 < p(X),$$

so $X \succ \varnothing$. And, using Eqs. (8.5) and (8.8),

$$p(A) \geqq 0 \Rightarrow p(A) \geqq p(\varnothing) \Rightarrow A \succsim \varnothing.$$

Finally, to see that Condition (8.16) is necessary, note that if

$$A \cap B = A \cap C = \varnothing,$$

then

$$p(B) > p(C) \Leftrightarrow p(A) + p(B) > p(A) + p(C) \Leftrightarrow p(A \cup B) > p(A \cup C).$$

Conditions (8.14) through (8.16) were apparently first listed by de Finetti [1937], and they are called the *de Finetti axioms*.

Before going further, let us discuss whether the de Finetti axioms are reasonable ones for judgments of comparative probability. The first two axioms seem to be accepted in most studies. However, we can question Condition (8.14) in much the same way that we have questioned whether preference is a strict weak order. In particular, much as we did in Chapter 6, we can question whether or not the relation $A \sim B$, "seems equally probable," is transitive. This would follow if $(\mathcal{E}, \succ)$ were strict weak. The third axiom, (8.16), has been questioned by writers such as Edwards [1962], who argues that

$$p(A \cup B) = p(A) + p(B)$$

for $A \cap B = \varnothing$ is not necessarily satisfied by subjective probabilities. This is really an argument against the representation we are studying. See Exer. 11 of Section 8.4 for another argument against (8.16).

8.5.2 *Insufficiency of the de Finetti Axioms*

De Finetti asked whether his axioms were sufficient to guarantee the existence of a subjective probability measure. One example showing they are not was given by Savage [1954, p. 41]. Savage's example used an infinite collection of events $\mathcal{E}$. Later, Kraft, Pratt, and Seidenberg [1959] gave a counterexample with a finite set $\mathcal{E}$. We present their counterexample.

Let $X = \{a, b, c, d, e\}$, and let $\mathcal{E}$ be the collection of all subsets of X. Choose ϵ such that $0 < \epsilon < 1/3$, and define p on X by

$$p(a) = 4 - \epsilon, \quad p(b) = 1 - \epsilon, \quad p(c) = 3 - \epsilon, \quad p(d) = 2, \quad p(e) = 6.$$

Extend p to $\mathcal{E}$ by using additivity. Let $\succ$ be the strict weak order induced on $\mathcal{E}$ by p, using Eq. (8.8). Then $(X, \mathcal{E}, \succ)$ satisfies Conditions (8.14) through (8.16), the de Finetti axioms. [If the reader wants, he can "normalize" p by dividing by $16 - 3\epsilon$, which is $p(X)$.]

Note that $p(\{d, e\}) = 8$ and $p(\{a, b, c\}) = 8 - 3\epsilon$, so $p(\{d, e\}) > p(\{a, b, c\})$. We observe that there is no set $A \neq \{d, e\}$ or $\{a, b, c\}$ such that

$$\{d, e\} \succsim A \succsim \{a, b, c\}.$$

For if there is such an A, then $8 \geqq p(A) > 7$. The reader can easily convince himself that none of the other subsets of X has such a value. Now let $\succ'$ be obtained from $\succ$ by changing the order of $\{d, e\}$ and $\{a, b, c\}$. Since there is no element in between $\{d, e\}$ and $\{a, b, c\}$, no additional changes need be made. The new order $\succ'$ still satisfies the de Finetti axioms. Conditions (8.14) and (8.15) are trivial. Condition (8.16) follows, since there is no $A \neq \varnothing$ such that $A \cap \{d, e\} = A \cap \{a, b, c\} = \varnothing$. There is no subjective probability measure p' on $(X, \mathcal{E}, \succ')$. For, suppose there is. Note that

$$\{a\} \succ' \{b, c\},$$

$$\{c, d\} \succ' \{a, b\},$$

and

$$\{b, e\} \succ' \{a, c\}.$$

Thus

$$p'(a) > p'(b) + p'(c),$$
$$p'(c) + p'(d) > p'(a) + p'(b),$$

and

$$p'(b) + p'(e) > p'(a) + p'(c).$$

Adding these three inequalities and canceling $p'(a) + p'(b) + p'(c)$ gives

$$p'(d) + p'(e) > p'(a) + p'(b) + p'(c),$$

or

$$\{d, e\} \succ' \{a, b, c\},$$

which is a contradiction. Thus, the de Finetti axioms are not sufficient to imply the existence of a subjective probability measure, even if X is finite.

8.5.3 *Conditions Sufficient for a Subjective Probability Measure*

Let us return to the de Finetti axioms and ask for additional axioms sufficient to guarantee the existence of a subjective probability measure. As in Section 3.1.4, some additional restrictions on the binary relation are required. Indeed, suppose that $\mathcal{E}^*$ is the collection of equivalence classes in $\mathcal{E}$ under the equivalence relation $\sim$ defined by

$$A \sim B \Leftrightarrow \sim (A \succ B) \,\&\, \sim (B \succ A). \tag{8.13}$$

Suppose A^* denotes the equivalence class containing A and suppose that $A^* \succ^* B^*$ holds if and only if $A \succ B$. If $(\mathcal{E}, \succ)$ is strict weak, Theorem 1.4 implies that $\succ^*$ is well-defined. It follows from Theorem 3.4, Corollary 1, of Section 3.1.4 that if there is a measure of subjective probability, then $(\mathcal{E}^*, \succ^*)$ must have a countable order-dense subset. The de Finetti axioms plus this new assumption are still not sufficient to guarantee the existence of a measure of subjective probability. The Kraft–Pratt–Seidenberg example still applies.

There are a number of additional axioms which, when added to the de Finetti axioms and the assumption that there exists a countable order-dense subset of $(\mathcal{E}^*, \succ^*)$, are sufficient to guarantee the existence of a measure of subjective probability. Most of these amount to a rather strong assumption, which goes back to de Finetti [1937] and Koopman [1940a, b], and is also used by Savage [1954]. This extra assumption says that for all n, there is an *n-fold almost uniform partition* of $\mathcal{E}$, a collection of sets $X_1, X_2, \ldots, X_n$ in $\mathcal{E}$ that are disjoint and whose union is X (that is, a partition of X) and such that whenever $1 \leqq k < n$, the union of no k of these sets is more probable than the union of any $k + 1$ of them.

An *n-fold uniform partition* of $\mathcal{E}$ is a collection of n disjoint sets $X_1, X_2, \ldots, X_n$ in $\mathcal{E}$ whose union is X and such that for all i, j, $X_i \sim X_j$. If $\mathcal{E}$ has an n-fold uniform partition, then it has an n-fold almost uniform partition.

If for all n, $\mathcal{E}$ has an n-fold almost uniform partition, then it follows that $\mathcal{E}$ is infinite. Hence, assuming that $\mathcal{E}$ has such a partition rules out examples like that of Kraft–Pratt–Seidenberg. Even for infinite $\mathcal{E}$, this is a rather strong assumption.

The following theorem is due to Roberts [1973] and is based on one of Fine [1971a]. For other results, see Savage [1954], Suppes [1956], Kraft, Pratt, and Seidenberg [1959], and Luce [1967]; see the book by Fine [1973] for an extensive discussion of related results. See also Sections 8.5.4 and 8.5.5 and Exers. 14 through 17.

THEOREM 8.1. *Suppose X is a set, $\mathcal{E}$ is an algebra of subsets of X, and $\succ$ is a binary relation on $\mathcal{E}$. Then the following conditions are sufficient to* *guarantee the existence of a subjective probability measure on $(X, \mathcal{E}, \succ)$.*

(a) *The de Finetti axioms [Conditions (8.14) through (8.16)].*

(b) *$(\mathcal{E}^*, \succ^*)$ has a countable order-dense subset.*

(c) *For all n, there is an n-fold almost uniform partition of $\mathcal{E}$.*

The proof of this theorem is omitted. A major objection to the theorem is that it does not apply to finite situations. In the next two subsections and in Exers. 14 to 17, we discuss how to remedy that problem. We close this subsection with a uniqueness theorem.

THEOREM 8.2. *Suppose X is a set, $\mathcal{E}$ is an algebra of subsets of X, and $\succ$ is a binary relation on $\mathcal{E}$. Suppose p and p' are two measures of subjective probability on $\mathcal{E}$, and for all n, there is an n-fold almost uniform partition of $\mathcal{E}$. Then $p = p'$.*

Proof. The proof mimics the proof of the uniqueness theorem for extensive measurement (Theorem 3.7 of Section 3.2.3). (The basic idea is that without the assumption $p(X) = p'(X) = 1$, we can show $p' = \alpha p$, some $\alpha > 0$. Then this extra assumption gives us $\alpha = 1$.) Let us assume that $p \neq p'$. Then for some A, $p(A) \neq p'(A)$, so without loss of generality, $p'(A) < p(A)$. Hence, there are positive integers m and n such that

$$p'(A) < m/n < (m + 2)/n < p(A). \tag{8.17}$$

Now let $X_1, X_2, \ldots, X_n$ be an n-fold almost uniform partition of $\mathcal{E}$. Since the numbers $p'(X_1), p'(X_2), \ldots, p'(X_n)$ sum to 1, the average sum of m of these numbers is m/n. Hence, there are $X_{i_1}, X_{i_2}, \ldots, X_{i_m}$ such that

$$p'(X_{i_1}) + p'(X_{i_2}) + \cdots + p'(X_{i_m}) \geq m/n.$$

Similarly, there are $X_{j_1}, X_{j_2}, \ldots, X_{j_{m+2}}$ such that

$$p(X_{j_1}) + p(X_{j_2}) + \cdots + p(X_{j_{m+2}}) \leq (m + 2)/n.$$

By the almost uniformness of the partition,

$$p'(X_1 \cup X_2 \cup \cdots \cup X_{m+1}) \geq p'(X_{i_1} \cup X_{i_2} \cup \cdots \cup X_{i_m}) \geq m/n \tag{8.18}$$

and

$$p(X_1 \cup X_2 \cup \cdots \cup X_{m+1}) \leq p(X_{j_1} \cup X_{j_2} \cup \cdots \cup X_{j_{m+2}}) \leq (m + 2)/n. \tag{8.19}$$

Thus, by (8.17), (8.18), and (8.19),

$$p'(A) < p'(X_1 \cup X_2 \cup \cdots \cup X_{m+1}) \qquad (8.20)$$

and

$$p(X_1 \cup X_2 \cup \cdots \cup X_{m+1}) < p(\dot{A}). \qquad (8.21)$$

But now (8.20) implies that

$$X_1 \cup X_2 \cup \cdots \cup X_{m+1} \succ A$$

and (8.21) implies that

$$A \succ X_1 \cup X_2 \cup \cdots \cup X_{m+1},$$

which is impossible. ∎

It should be observed that the uniqueness of the measure of subjective probability is false without an extra assumption such as the one that there is an n-fold almost uniform partition. To see this, it suffices to consider the situation where $X = \{a, b\}$, $\mathcal{E} =$ all subsets of X, and $\{a, b\} \succ \{a\} \succ \{b\} \succ \varnothing$. Details are left to the reader. It would be very interesting to find, for a system $(X, \mathcal{E}, \succ)$ which has a measure of subjective probability, necessary and sufficient conditions for that measure to be unique.* It would also be interesting and practically useful (from the point of view of meaningfulness results) to systematically describe the types of admissible transformations of subjective probability measures which can arise.

8.5.4 Necessary and Sufficient Conditions for a Subjective Probability Measure in the Finite Case

In their paper, Kraft, Pratt, and Seidenberg [1959] present a set of conditions that are necessary and sufficient to guarantee the existence of a subjective probability measure on a structure $(X, \mathcal{E}, \succ)$, in the case that X is finite. In a later paper, Scott [1964] presents much the same conditions in a clearer format. These conditions are closely related to Scott's axioms for difference measurement, which we discussed in Section 3.3.2, and to Scott's axioms for conjoint measurement, which we discussed in Section 5.4.5. We shall present these conditions. For other related formulations, see Domotor [1969], Fishburn [1969], and Krantz et al. [1971, Section 9.2.2].

---

*The theorem of Suppes and Zanotti [1976], which we present in Section 8.5.5, gives conditions necessary and sufficient for the existence of a unique measure of subjective probability for the set we shall call $\mathcal{E}^+$. However, the conditions given in that theorem do not give rise to a unique subjective probability measure on $\mathcal{E}$. See Exer. 11.

To present the axioms, we need to use the *characteristic function* χ_A of a set $A \in \mathcal{E}$. χ_A is a function from X into $\{0, 1\}$ such that

$$\chi_A(a) = \begin{cases} 1 & \text{if} \quad a \in A, \\ 0 & \text{if} \quad a \notin A. \end{cases}$$

(The reader will easily verify that

$$\chi_{A \cup B} = \chi_A + \chi_B - \chi_{A \cap B}$$

and that

$$\chi_{A \cap B} = 0 \Leftrightarrow A \cap B = \varnothing.)$$

Suppose X is a set, $\mathcal{E}$ an algebra of subsets of X, and $\succ$ a binary relation on $\mathcal{E}$. Then the triple $(X, \mathcal{E}, \succ)$ is a *subjective probability structure in the sense of Scott* if the following axioms are satisfied:

AXIOM SP1. $X \succ \varnothing$ and $A \succsim \varnothing$ for all $A \in \mathcal{E}$.

AXIOM SP2. $A \succsim B$ or $B \succsim A$ for all $A, B \in \mathcal{E}$.

AXIOM SP3. Whenever $A_1, A_2, \ldots, A_r, B_1, B_2, \ldots, B_r \in \mathcal{E}$ and $A_i \succsim B_i$ for all $i < r$ and

$$\chi_{A_1} + \chi_{A_2} + \cdots + \chi_{A_r} = \chi_{B_1} + \chi_{B_2} + \cdots + \chi_{B_r}, \qquad (8.22)$$

then $B_r \succsim A_r$.

The reader should note that Axiom SP3 is really an infinite "schema" of axioms, one for each r. As in the case of difference and conjoint measurement, it does not reduce to a finite schema: elements A_i may be repeated. Axiom SP3 is unpleasant because it is stated in terms of characteristic functions rather than in terms of union or other primitive notions. However, Eq. (8.22) has a rather simple interpretation; it says that for each $a \in X$, a is in exactly as many A_i as B_i.

THEOREM 8.3 (Scott [1964]). *Suppose X is a finite set, $\mathcal{E}$ is an algebra of subsets of X, and $\succ$ is a binary relation on $\mathcal{E}$. Then there is a measure of subjective probability on $(X, \mathcal{E}, \succ)$ if and only if $(X, \mathcal{E}, \succ)$ is a subjective probability structure in the sense of Scott.*

We shall present a proof of the necessity of Scott's axioms. The proof of sufficiency, like those for Scott's Theorems 3.9 and 5.4, uses a clever variant of the separating hyperplane theorem. Axioms SP1 and SP2 are special cases of the de Finetti axioms, and so follow from the representation. Turning to Axiom SP3, let $X = \{x_1, x_2, \ldots, x_k\}$. Assume that $\{x_i\}$ is in $\mathcal{E}$ for each i. The following argument is easy to fix up if this assumption does not hold, by use of minimal sets in $\mathcal{E}$. We leave details of that to the reader. Define a linear functional L on Re^k by

$$L(0, \ldots, 0, 1, 0, \ldots, 0) = p(\{x_i\}),$$

where there is a 1 in the ith place and p is a measure of subjective probability. Identify χ_A with the vector that has a 1 in the ith place if and only if x_i is in A. Then, since p is finitely additive,

$$L(\chi_A) = \sum_{x_i \in A} p(\{x_i\}) = p(A).$$

Moreover,

$$L(\chi_{A_1} + \chi_{A_2} + \cdots + \chi_{A_r}) = L(\chi_{B_1} + \chi_{B_2} + \cdots + \chi_{B_r}) \Rightarrow$$

$$\sum_{i=1}^{r} L(\chi_{A_i}) = \sum_{i=1}^{r} L(\chi_{B_i}) \Rightarrow$$

$$\sum_{i=1}^{r} p(A_i) = \sum_{i=1}^{r} p(B_i). \tag{8.23}$$

Now $A_i \succsim B_i$ for $i < r$ implies

$$p(A_i) \geqq p(B_i), \qquad i = 1, 2, \ldots, r - 1.$$

Thus, (8.23) implies $p(A_r) \leqq p(B_r)$, or $B_r \succsim A_r$.

As we observed after the proof of Theorem 8.2, the measure of subjective probability may not be unique in the finite case. Sufficient conditions for uniqueness in the finite case are given by Suppes [1969].

Extending the methods used to prove Theorem 8.3, Scott has obtained some conditions that are necessary and sufficient for the existence of a measure of subjective probability, even if $\mathcal{E}$ is infinite. But these conditions are rather complicated and have not been published.

To the author's knowledge, most people are willing to accept Scott's first and second axioms, and no one has subjected Scott's third axiom to a test.

8.5.5 Necessary and Sufficient Conditions for a Subjective Probability Measure in the General Case

Suppes and Zanotti [1976] have recently given necessary and sufficient conditions for the existence of a subjective probability measure without the assumption of finiteness. Their basic idea is to use the axioms for extensive measurement. This is accomplished by passing to a more general structure than the algebra of subsets $\mathcal{E}$.

In particular, Suppes and Zanotti start with the correspondence between the elements of $\mathcal{E}$ and their characteristic functions, as does Scott. They define an *extended characteristic function of* $\mathcal{E}$ to be a finite sum of characteristic functions (repetitions allowed) and $\mathcal{E}^+$ to be the set of extended characteristic functions of $\mathcal{E}$.

Suppose p is a measure of subjective probability for $(X, \mathcal{E}, \succ)$. Then we may define a function E on $\mathcal{E}^+$ by defining $E(A^+)$ to be $\Sigma p(A_i)$ if $A^+ = \Sigma \chi_{A_i}$. Thus, in particular, if $A^+ = \chi_A$, $E(A^+) = p(A)$. We define $\succ^+$ on $\mathcal{E}^+$ by

$$A^+ \succ^+ B^+ \Leftrightarrow E(A^+) > E(B^+). \qquad (8.24)$$

Then $\succ^+$ restricted to $\mathcal{E}$—that is, characteristic functions of $\mathcal{E}$—is just $\succ$. We say $\succ^+$ *extends* $\succ$. The function E obviously satisfies

$$E(A^+ + B^+) = E(A^+) + E(B^+). \qquad (8.25)$$

Hence, the triple $(\mathcal{E}^+, \succ^+, +)$ satisfies the necessary and sufficient conditions for extensive measurement developed in Section 3.2.2; that is, it satisfies the axioms for an extensive structure. Using the notation $\succeq^+$ and $\sim^+$ defined from $\succ^+$ as $\succeq$ and $\sim$ were from $\succ$ in Eqs. (8.12) and (8.13), we may write these axioms as follows:

AXIOM SZ1. *For all* $A^+, B^+, C^+ \in \mathcal{E}^+$,

$$A^+ + (B^+ + C^+) \sim^+ (A^+ + B^+) + C^+.$$

AXIOM SZ2. $(\mathcal{E}^+, \succ^+)$ *is a strict weak order.*

AXIOM SZ3. *For all* $A^+, B^+, C^+ \in \mathcal{E}^+$,

$$A^+ \succ^+ B^+ \Leftrightarrow A^+ + C^+ \succ^+ B^+ + C^+ \Leftrightarrow C^+ + A^+ \succ^+ C^+ + B^+.$$

AXIOM SZ4. *For all* $A^+, B^+, C^+, D^+ \in \mathcal{E}^+$, *if* $A^+ \succ^+ B^+$, *then there is a positive integer* n *such that*

$$nA^+ + C^+ \succ^+ nB^+ + D^+.$$

In addition, it is clear that the following axioms also hold:

AXIOM SZ5. $\chi_X \succ^+ \chi_\emptyset$.

AXIOM SZ6. $A^+ \succeq^+ \chi_\emptyset$.

THEOREM 8.4 (Suppes and Zanotti [1976]). *Suppose* X *is a set,* $\mathcal{E}$ *is an algebra of subsets of* X, $\succ$ *is a binary relation on* $\mathcal{E}$, *and* $\mathcal{E}^+$ *is the set of* *extended characteristic functions of $\mathcal{E}$. Then there is a measure of subjective probability on $(X, \mathcal{E}, \succ)$ if and only if there is a binary relation $\succ^+$ on $\mathcal{E}^+$ extending $\succ$ and such that $(\mathcal{E}^+, \succ^+, +)$ satisfies Axioms SZ2, SZ3, SZ4, SZ5, and SZ6.*

Proof. We have already shown the necessity of these axioms. To show their sufficiency, we note that Axiom SZ1, with $=$ instead of $\sim^+$, follows from the definition of function addition. Then Axiom SZ1 follows, for $=$ implies $\sim^+$, since by Axiom SZ2, $\succ^+$ is strict weak. Now, by Theorem 3.6, Axioms SZ1 through SZ4 imply that there is a function E on $\mathcal{E}^+$ satisfying Eqs. (8.24) and (8.25). Note that $E(\chi_\varnothing) = 0$, since by Eq. (8.25),

$$E(\chi_\varnothing) + E(\chi_\varnothing) = E(\chi_\varnothing + \chi_\varnothing) = E(\chi_\varnothing).$$

By Axiom SZ5 and Eq. (8.24),

$$E(\chi_x) > E(\chi_\varnothing) = 0.$$

Let $p(A) = E(\chi_A)/E(\chi_x)$. We verify that p defines a measure of subjective probability. Since $\succ^+$ extends $\succ$, Eq. (8.24) implies that

$$A \succ B \Leftrightarrow p(A) > p(B).$$

Axiom SZ6 implies that

$$p(A) = E(\chi_A)/E(\chi_x) \geqq E(\chi_\varnothing)/E(\chi_x) = 0.$$

Also,

$$p(X) = E(\chi_x)/E(\chi_x) = 1.$$

Finally, if $A \cap B = \varnothing$, then

$$\begin{aligned}
p(A \cup B) &= E(\chi_{A \cup B})/E(\chi_x) \\
&= E(\chi_A + \chi_B)/E(\chi_x) \\
&= E(\chi_A)/E(\chi_x) + E(\chi_B)/E(\chi_x) \\
&= p(A) + p(B).
\end{aligned}$$

∎

One unpleasant aspect of the Suppes–Zanotti result is that it speaks in terms of the existence of an extension to a larger set $\mathcal{E}^+$, and so is not explicitly a representation theorem in terms of $(X, \mathcal{E}, \succ)$ alone. To the author's knowledge, it remains an open problem to state nice, necessary and sufficient conditions for the existence of a subjective probability measure which are stated in terms of $(X, \mathcal{E}, \succ)$ alone, and which do not require the assumption of finiteness.

Exercises

1. Show that if $(X, \mathcal{E}, \succ)$ has a measure of subjective probability and if

$$\{a, e, f\} \succ \{b, e\}$$

and

$$\{b, c, e\} \sim \{a, e, c, d\},$$

then

$$\{f\} \succ \{d\}.$$

2. Suppose $X = \{A, B, C, D, E\}$, $(X, \mathcal{E})$ is an algebra of subsets, $\succ$ is a binary relation on $\mathcal{E}$, and it is known that
    (α) $\{A\} \sim \{E\}$,
    (β) $\{B\} \sim \{C\}$,
    (γ) $\{B, C\} \sim \{D\}$,
    (δ) $\{A, E\} \sim \{B\}$.
  (a) Show that these hypotheses imply the following:
    (i) If there is a measure p of subjective probability, then

$$p(D) > p(A) + p(B).$$

    (ii) Moreover, the measure p is determined uniquely.
  (b) However, show that the conditions (α), (β), and (γ) do not determine p uniquely.

3. (a) Suppose $X = \{a, b\}$, $\mathcal{E}$ = all subsets of X, and $\succ$ is a strict weak order on $\mathcal{E}$ defined by

$$\{a, b\} \succ \{a\} \succ \{b\} \succ \varnothing.$$

Show that there is a continuum of subjective probability measures on $(X, \mathcal{E}, \succ)$.
  (b) Use one of Scott's axioms to show that, since $\{a\} \succ \{b\}$ and $\{a, b\} \succ \{a\}$, we must have $\{a, b\} \succ \{b\}$.

4. Raiffa [1968] discusses the proportion λ of medical doctors who are non-teetotalers and consume more scotch than bourbon. This exercise is based on some of his discussion of subjective estimates of the probability that λ is at least or at most a certain amount. Let X be the set of all outcomes $\lambda = r$ for $r \in [0, 1]$. Let $\mathcal{E}$ be the family of the following subsets of X:

    $\varnothing$
    A, which says that $\lambda \geq .6$
    B, which says that $\lambda < .6$

C, which says that $\lambda \geq .9$
D, which says that $\lambda < .9$
E, which says that $\lambda < .6$ or $\lambda \geq .9$
F, which says that $\lambda \geq .6$ and $\lambda < .9$
X, which says that $\lambda \geq 0$ and $\lambda \leq 1$.

(a) Observe that $\mathcal{E}$ is an algebra of subsets of X.

(b) The comparative probability relation $\succ$ on $\mathcal{E}$ is the strict weak order defined by

$$X \succ D \succ E \succ A \sim B \succ F \succ C \succ \varnothing.$$

Does there exist a measure of subjective probability on $(X, \mathcal{E}, \succ)$?

(c) If the answer to part (b) is yes, is this measure unique?

5. Show that under the de Finetti axioms, if $A \subsetneq B$, then it follows that $B \succeq A$, but not necessarily that $B \succ A$.

6. Prove the following results from the de Finetti axioms:

(a) If $A \succ \varnothing$ and $A \cap B = \varnothing$, then $A \cup B \succ B$.
(b) If $A \succeq B$, then $B^c \succeq A^c$.
(c) If $A \succeq B$, $C \succeq D$, and $A \cap C = \varnothing$, then $A \cup C \succeq B \cup D$.
(d) If $A \cup B \succeq C \cup D$ and $C \cap D = \varnothing$, then $A \succeq C$ or $B \succeq D$.
(e) If $B \succeq B^c$ and $C^c \succeq C$, then $B \succeq C$.
(f) If $A \sim \varnothing$ and $B \subsetneq A$, then $B \sim \varnothing$.
(g) If $A \sim \varnothing$ and $B \sim \varnothing$, then $A \cap B \sim \varnothing$ and $A \cup B \sim \varnothing$.
(h) If $A \sim \varnothing$, then $A^c \succ \varnothing$.

7. Suppose $(X, \mathcal{E}, \succ)$ is as in Exer. 3.

(a) Verify Scott's axioms directly.

(b) Identify the extended characteristic functions $\mathcal{E}^+$ and the binary relation $\succ^+$ on $\mathcal{E}^+$ defined in Section 8.5.5.

8. Let $X = \{1, 2, \ldots, n\}$ and let $\mathcal{E}$ be all subsets of X. Define $\succ$ on $\mathcal{E}$ by

$$A \succ B \Leftrightarrow |A| > |B|.$$

Show that $(X, \mathcal{E}, \succ)$ has a measure of subjective probability.

9. Consider the uniqueness of the measure of subjective probability in Exer. 8.

10. (a) If $(X, \mathcal{E}, \succ)$ is as in Exer. 3, show that there is no n-fold almost uniform partition.

(b) Is there such a partition for $(X, \mathcal{E}, \succ)$ as in Exer. 4?

(c) What about for $(X, \mathcal{E}, \succ)$ of Exer. 8?

11. Given $(X, \mathcal{E}, \succ)$, suppose $\succ^+$ extends $\succ$ to $\mathcal{E}^+$ in such a way that Axioms SZ2 through SZ6 are satisfied.

(a) Show that there is a unique function E on $\mathcal{E}^+$ satisfying Eqs. (8.24) and (8.25) and the condition $E(\chi_X) = 1$.

(b) Show, however, that this does not imply that there is a unique subjective probability measure on $(X, \mathcal{E}, \succ)$.

12. Prove transitivity of $\succeq$ directly from Scott's Axiom SP3 by observing that for all A, B, C,

$$\chi_A + \chi_B + \chi_C = \chi_B + \chi_C + \chi_A.$$

13. Show that the de Finetti axioms are independent.

14. Luce [1967] has given conditions on $(X, \mathcal{E}, \succ)$ sufficient for the existence of a measure of subjective probability, which do not require that $\mathcal{E}$ be infinite. (For a discussion, see Krantz et. al. [1971, Section 5.2.3].) Exercises 14 to 18 discuss Luce's conditions. These conditions are the three de Finetti axioms and two additional axioms. The first is the following.

We say that a sequence $A_1, \ldots, A_i, \ldots$ from $\mathcal{E}$ is a *standard sequence* relative to A in $\mathcal{E}$ if for $i = 1, 2, \ldots$, there exist B_i, C_i in $\mathcal{E}$ such that
 (1) $A_1 = B_1$ and $B_1 \sim A$,
 (2) $B_i \cap C_i = \varnothing$,
 (3) $B_i \sim A_i$,
 (4) $C_i \sim A_i$,
 (5) $A_{i+1} = B_i \cup C_i$.

ARCHIMEDIAN AXIOM: *If $A \succ \varnothing$, then every standard sequence relative to A is finite.*

Show that this axiom is a necessary condition for the existence of a measure of subjective probability and that it does indeed express an Archimedean condition.

15. Luce's second additional axiom is the following:

AXIOM A. *Suppose that $(X, \mathcal{E}, \succ)$ satisfies the de Finetti axioms. Suppose A, B, C, D are in $\mathcal{E}$, $A \cap B = \varnothing$, $A \succ C$, and $B \succeq D$. Then there exist C', D', E in $\mathcal{E}$ such that*
 (1) $E \sim A \cup D$,
 (2) $C' \cap D' = \varnothing$,
 (3) $E \supseteq C' \cup D'$,
 (4) $C' \sim C$ and $D' \sim D$.

(As Krantz et al, [1971] point out, Axiom A is difficult to explain without simply restating it in words. We leave this restatement to the reader.) Axiom A is satisfied by some finite structures $(X, \mathcal{E}, \succ)$. For example, let $X = \{a, b, c, d\}$, let $\mathcal{E}$ be all subsets of X, and let $P(a) = P(b) = P(c) = 0.2$, $P(d) = 0.4$. Show that Axiom A is satisfied. (Luce proves that the de Finetti axioms, the Archimedean Axiom, and Axiom A are sufficient to prove the existence of a measure of subjective probability.)

16. Show that Axiom A is not a necessary condition for the existence of a measure of subjective probability. (For example, any structure $(X, \mathcal{E}, \succ)$ whose equivalence classes under $\sim$ fail to form a single standard sequence violates Axiom A.)

17. Show that if $(X, \succ)$ is a strict simple order, then Luce's axioms imply that $\mathcal{E}$ is infinite.

18. Show that the structure $(X, \mathcal{E}, \succ)$ of Exer. 8 satisfies Luce's Axiom A.

References

Bar-Hillel, M., "On the Subjective Probability of Compound Events," *Organizational Behavior and Human Performance*, 9 (1973), 396–406.

Bartos, J. A., "The Assessment of Probability Distributions for Future Security Prices," Ph.D. Dissertation, Indiana University, Bloomington, Indiana, 1969.

Bernstein, S. N., "Axiomatic Foundations of Probability Theory," Collected Works IV, Academy of Science, USSR, 1917, pp. 10–25.

Betaque, N. E., and Gorry, A., "Automating Judgmental Decision-Making for a Serious Medical Problem," *Management Sci.*, **B17** (1971), 421–434.

Brier, G. W., "Verification of Forecasts Expressed in Terms of Probability," *Monthly Weather Rev.*, **78** (1950), 1–3.

Brier, G. W., and Allen, R. A., "Verification of Weather Forecasts," *Compendium of Meteorology*, American Meteorological Society, Boston, Massachusetts 1951, pp. 841–848.

Brown, T. A., "An Experiment in Probabilistic Forecasting," Rept. R-944-ARPA, the RAND Corporation, Santa Monica, California, July 1973.

Carnap, R., *Logical Foundations of Probability*, University of Chicago Press, Chicago, Illinois, 1950, 1962.

Cohen, M. A., "On the Characterization of Qualitative Probability Spaces," mimeographed, Department of Psychology and Social Relations, Harvard Univ., Cambridge, Massachusetts, 1978.

Davidson, D., Suppes, P., and Siegel, S., *Decision-Making: An Experimental Approach*, Stanford University Press, Stanford, California, 1957.

de Finetti, B., "Sul Significato Soggettivo della Probabilita, *Fund. Math.*, **31** (1931), 298–329.

de Finetti, B., "La Prevision: Ses Lois Logiques, Ses Sources Subjectives," *Ann. Inst. H. Poincaré*, **7** (1937), 1–68. Translated into English in H. E. Kyburg and H. E. Smokler (eds.), *Studies in Subjective Probability*, Wiley, New York, 1964, pp. 93–158.

de Finetti, B., "Does It Make Sense To Speak of 'Good Probability Appraisers'?" in I. J. Good (ed.), *The Scientist Speculates—An Anthology of Partly-Baked Ideas*, Heineman, London, 1962, pp. 357–364.

de Neufville, R., and Keeney, R. L., "Use of Decision Analysis in Airport Development for Mexico City," in A. W. Drake, R. L. Keeney, and P. M. Morse (eds.), *Analysis of Public Systems*, M.I.T. Press, Cambridge, Massachusetts, 1972.

Domotor, Z., "Probabilistic Relational Structures and Their Applications," Tech. Rept. 144, Institute for Mathematical Studies in the Social Sciences, Stanford University, Stanford, California, 1969.

Edwards, W., "Subjective Probabilities Inferred from Decisions," *Psychol. Rev.*, **69** (1962), 109–135.

Ellsberg, D., "Risk, Ambiguity, and the Savage Axioms," *Quart. J. Econ.*, **75** (1961), 643–669.

Feller, W., *An Introduction to Probability Theory and Its Applications*, Vol. I, Wiley, New York, 1962.

Fine, T. L., "A Note on the Existence of Quantitative Probability," *Ann. Math. Stat.*, **42** (1971a), 1182–1186.

Fine, T. L., "Rational Decision Making with Comparative Probability," *Proc. IEEE Conf. on Decision and Control Miami Beach*, Department of Electrical Engineering, University of Florida, Gainesville, 1971b, pp. 355–356.

Fine, T. L., *Theories of Probability*, Academic Press, New York, 1973.

Fine, T. L., "An Argument for Comparative Probability," in R. Butts and J. Hintikka (eds.), *Basic Problems in Methodology and Linguistics*, D. Reidel Publ. Co., Dordrecht, Holland, 1977, pp. 105–119.

Fine, T. L., and Gill, J. T., "The Enumeration of Comparative Probability Relations," *Ann. of Prob.*, **4** (1976), 667–673.

Fine, T. L., and Kaplan, M., "Joint Orders in Comparative Probability," *Ann. of Prob.*, **5** (1977), 161–179.

Fishburn, P. C., *Decision and Value Theory*, Wiley, New York, 1964.

Fishburn, P. C., "Weak Qualitative Probability on Finite Sets," *Ann. Math. Stat.*, **40** (1969), 2118–2126.

Fishburn, P. C., *Utility Theory for Decision Making*, Wiley, New York, 1970.

Fishburn, P. C., "Weak Comparative Probability on Infinite Sets," *Ann. of Prob.*, **3** (1975), 889–893.

Fryback, D. G., "Use of Radiologists' Subjective Probability Estimates in a Medical Decision Making Problem," Tech. Rept. MMPP 74 -14, Department of Psychology, University of Michigan, Ann Arbor, Michigan, 1974.

Goodman, B. C., "Direct Estimation Procedures for Eliciting Judgments about Uncertain Events," Technical Report, University of Michigan Engineering Psychology Laboratory, Ann Arbor, Michigan, November 1973.

Hampton, J. M., Moore, P. G., and Thomas, H., "Subjective Probability and Its Measurement," *J. Roy. Stat. Soc.*, Ser A **136** (1973), 21–42.

Hogarth, R. M., "Cognitive Processes and the Assessment of Subjective Probability Distributions," *J. Amer. Stat. Assoc.*, **70** (1975), 271–289.

Huber, G. P., "Methods for Quantifying Subjective Probabilities and Multi-Attribute Utilities," *Decision Sci.*, **5** (1974), 430–458.

Kahneman, D., and Tversky, A., "Subjective Probability: A Judgment of Representativeness," *Cognitive Psychol.*, **3** (1972), 430–454.

Kaplan, M. A., "Independence in Comparative Probability," M. S. Thesis, Department of Electrical Engineering, Cornell University, Ithaca, New York, 1971.

Kaplan, M. A., "Extensions and Limits of Comparative Probability Orders," Ph.D. Thesis, Department of Electrical Engineering, Cornell University, Ithaca, New York, 1974.

Keeney, R. L., "A Decision Analysis with Multiple Objectives: The Mexico City Airport," *Bell J. Econ. Management Sci.*, **4** (1973), 101–117.

Kemeny, J. G., *A Philospher Looks at Science*, Van Nostrand, Princeton, New Jersey, 1959.

Keynes, J. M., *A Treatise on Probability*, Macmillan, New York, 1921, 1929, 1962.

Kidd, J. B., "The Utilization of Subjective Probabilities in Production Planning," *Acta Psychol.*, **34** (1970), 338–347.

Kolmogorov, A. N., *Grundbegriffe der Wahrscheinlichkeitsrechnung*, Springer, Berlin, 1933. English Translation by N. Morrison, *Foundations of the Theory of Probability*, Chelsea, New York, 1956.

Koopman, B. O., "The Axioms and Algebra of Intuitive Probability," *Ann. Math.*, **41** (1940a), 269–292.

Koopman, B. O., "The Bases of Probability," *Bull. Amer. Math. Soc.*, **46** (1940b), 763–774. Reprinted in H. E. Kyburg, Jr., and H. E. Smokler (eds.), *Studies in Subjective Probability*, Wiley, New York, 1964, pp. 159–172.

Kraft, C. H., Pratt, J. W., and Seidenberg, A., "Intuitive Probability on Finite Sets," *Ann.*

Math. Stat., **30** (1959), 408–419.

Krantz, D. H., Luce, R. D., Suppes, P., and Tversky, A., *Foundations of Measurement*, Vol. I, Academic Press, New York, 1971.

Kyburg, H. E., Jr., and Smokler, H. E. (eds.), *Studies in Subjective Probability*, Wiley, New York, 1964.

Luce, R. D., "Sufficient Conditions for the Existence of a Finitely Additive Probability Measure," *Ann. Math. Stat.*, **38** (1967), 780–786.

Luce, R. D. "On the Numerical Representation of Quantitative Conditional Probability," *Ann. Math. Stat.*, **39** (1968), 481–491.

Luce, R. D., "Conditional Expected, Extensive Utility," *Theory and Decision*, **3** (1972), 101–106.

Luce, R. D., and Krantz, D. H., "Conditional Expected Utility," *Econometrica*, **39** (1971), 253–271.

Luce, R. D., and Raiffa, H., *Games and Decisions*, Wiley, New York, 1957.

Luce, R. D., and Suppes, P., "Preference, Utility, and Subjective Probability," in R. D. Luce, R. R. Bush, and E. Galanter (eds.), *Handbook of Mathematical Psychology*, Vol. III, Wiley, New York, 1965, pp. 249–410.

Murphy, A. H., and Winkler, R. L., "Credible Interval Temperature Forecasting: Some Experimental Results," *Monthly Weather Rev.*, **102** (1974), 784–794.

Nagel, E., "Principles of the Theory of Probability," *International Encyclopedia of Unified Science*, Vol. I, University of Chicago Press, Chicago, Illinois, 1939.

Narens, L., "Minimal Conditions for Additive Conjoint Measurement and Quantitative Probability," *J. Math. Psychol.*, **11** (1974), 404–430.

Peterson, C. R., Snapper, K. J., and Murphy, A. H., "Credible Interval Temperature Forecasts," *Bull. Amer. Meteorol. Soc.*, **53** (1972), 966–970.

Phillips, L. D., and Edwards, W., "Conservatism in a Simple Probability Inference Task," *J. Exp. Psychol.*, **72** (1966), 346–357.

Preston, M. G., and Baratta, P., "An Experimental Study of the Auction-Value of an Uncertain Outcome," *Amer. J. Psychol.*, **61** (1948), 183–193.

Raiffa, H., "Risk, Ambiguity, and the Savage Axioms: Comment," *Quart. J. Econ.*, **75** (1961), 690–694.

Raiffa, H., *Decision Analysis*, Addison-Wesley, Reading, Massachusetts, 1968.

Ramsey, F. P., "Truth and Probability," in F. P. Ramsey, *The Foundations of Mathematics and Other Logical Essays*, Harcourt, Brace, New York, 1931, pp. 156–198. [Reprinted in H. E. Kyburg, and H. E. Smokler (eds), *Studies in Subjective Probability*, Wiley, New York, 1964, pp. 61–92.]

Roberts, F. S., "A Note on Fine's Axioms for Subjective Probability," *Ann. Prob.*, **1** (1973), 484–487.

Roberts, F. S., "Laws of Exchange and Their Applications," *SIAM J. Appl. Math.*, **26** (1974), 260–284.

Sanders, F., "On Subjective Probability Forecasting," *J. Appl. Meteorol.*, **2** (1963), 191–201.

Sanders, F., "The Verification of Probability Forecasts," *J. Appl. Meteorol.*, **6** (1967), 756–761.

Sanders, F., "Skill in Forecasting Daily Temperature and Precipitation: Some Experimental Results," *Bull. Amer. Meteorol. Soc.*, **54** (1973), 1171–1179.

Savage, L. J., *The Foundations of Statistics*, Wiley, New York, 1954.

Savage, L. J., "Elicitation of Personal Probabilities and Expectations," *J. Amer. Stat. Assoc.*, **66** (1971), 783–801.

Scott, D., "Measurement Models and Linear Inequalities," *J. Math. Psychol.*, **1** (1964), 233–247.

Smil, V., "Energy and the Environment—A Delphi Forecast," *Long Range Planning*, **5** (1972), 27–32.

Smil, V., "Energy and the Environment: Scenarios for 1985 and 2000," *The Futurist*, **8** (1974a), 4–13.

Smil, V., *Energy and the Environment: A Long Range Forecasting Study*, University of Manitoba, Winnipeg, Manitoba, April 1974b.

Staël von Holstein, C.-A. S., *Assessment and Evaluation of Subjective Probability Distributions*, The Economic Research Institute at the Stockholm School of Economics, Stockholm, 1970a.

Staël von Holstein, C.-A. S., "Some Problems in the Practical Application of Bayesian Decision Theory," *Behavioral Approaches to Management*, The Graduate School of Economics and Business Administration, Gothenburg, 1970b.

Staël von Holstein, C.-A. S., "An Experiment in Probabilistic Weather Forecasting," *J. Appl. Meteorol.*, **10** (1971), 635–645.

Staël von Holstein, C.-A. S., "Probabilistic Forecasting: An Experiment Related to the Stock Market," *Organizational Behavior and Human Performance*, **8** (1972), 139–158.

Stone, M., "The Opinion Pool," *Ann. Math. Stat.*, **32** (1961), 1339–1342.

Suppes, P., "The Role of Subjective Probability and Utility in Decisionmaking," in J. Neyman (ed.), *Proceedings of the Third Berkeley Symposium of Mathematical Statistics and Probability*, Vol. 5, University of California Press, Berkeley, California, 1956.

Suppes, P., *Studies in the Methodology and Foundations of Science*, Reidel, Dordrecht, 1969.

Suppes, P., and Zanotti, M., "Necessary and Sufficient Conditions for Existence of a Unique Measure Strictly Agreeing with a Qualitative Probability Ordering," *J. Phil. Logic*, **5** (1976), 431–438.

Tversky, A., "Additivity, Utility, and Subjective Probability," *J. Math. Psychol.*, **4** (1967), 175–201.

von Neumann, J., and Morgenstern, O., *The Theory of Games and Economic Behavior*, Princeton University Press, Princeton, N.J., 1944, 1947, 1953.

Winkler, R. L., "Scoring Rules and the Evaluation of Probability Assessors," *J. Amer. Stat. Assoc.*, **64** (1969), 1073–1078.

Winkler, R. L., and Murphy, A. H., "Evaluation of Subjective Precipitation Probability Forecasts," *Proceedings of the First National Conference on Statistical Meteorology*, American Meteorological Society, Boston, Massachusetts, 1968a, pp. 148–157.

Winkler, R. L., and Murphy, A. H., " 'Good' Probability Assessors," *J. Appl. Meteorol.*, **7** (1968b), 751–758.

Author Index

Numbers set in *italics* denote pages on which complete literate citations are given.

Subject Index

411